Invitation to
DYNAMICAL SYSTEMS

Edward R. Scheinerman

Department of Mathematical Sciences
The Johns Hopkins University

DOVER PUBLICATIONS, INC.
Mineola, New York

Bibliographical Note

This Dover edition, first published in 2012, is an unabridged republication of the work originally published by Prentice-Hall, Inc., Upper Saddle River, N.J., in 1996. The two pages of color plates, which originally appeared between pp. 334 and 335, have been reproduced here on the inside front and back covers.

Library of Congress Cataloging-in-Publication Data

Scheinerman, Edward R.
    Invitation to dynamical systems / Edward R. Scheinerman. — Dover ed.
        p. cm.
    Originally published: Upper Saddle River, N.J. : Prentice-Hall, 1996.
    Includes bibliographical references and index.
    ISBN-13: 978-0-486-48594-2
    ISBN-10: 0-486-48594-3
    1. Differentiable dynamical systems. I. Title.

QA614.8.S34 2012
003'.85—dc23
                                                    2011027234

Manufactured in the United States by Courier Corporation
48594301
www.doverpublications.com

To Amy

Contents

3 Nonlinear Systems 1: Fixed Points   101

4 Nonlinear Systems 2: Periodicity and Chaos   153

Preface

Popular treatments of chaos, fractals, and dynamical systems let the public know there is a party but provide no map to the festivities. Advanced texts assume their readers are already part of the club. This *Invitation*, however, is meant to attract a wider audience; I hope to attract my guests to the beauty and excitement of dynamical systems in particular and of mathematics in general.

You are cordially invited to explore the world of dynamical systems.

For this reason the technical prerequisites for this book are modest. Students need to have studied two semesters of calculus and one semester of linear algebra. Although differential equations are used and discussed in this book, no previous course on differential equations is necessary. Thus this *Invitation* is open to a wide range of students from engineering, science, economics, computer science, mathematics, and the like. This book is designed for the sophomore-junior level student who wants to continue exploring mathematics beyond linear algebra but who is perhaps not ready for highly abstract material. As such, this book can serve as a bridge between (for example) calculus and topology.

Prerequisites: calculus and linear algebra, but no differential equations. This *Invitation* is designed for a wide spectrum of students.

My focus is on ideas, and not on theorem-proof-remark style mathematics. Rigorous proof is the jealously guarded crown jewel of mathematics. But nearly as important to mathematics is intuition and appreciation, and this is what I stress. For example, a technical definition of *chaos* is hard to motivate or to grasp until the student has encountered chaos in person. Not everyone wants to be a mathematician—are such people to be excluded from the party? Dynamical systems has much to offer the nonmathematician, and it is my goal to make these ideas accessible to a wide range of students. In addition, I sought to

Philosophy.

- present both the "classical" theory of linear systems and the "modern" theory of nonlinear and chaotic systems;

- to work with both continuous and discrete time systems, and to present these two approaches in a unified fashion;

- to integrate computing comfortably into the text; and

- to include a wide variety of topics, including bifurcation, symbolic dynamics, fractals, and complex systems.

Chapter overview

Here is a synopsis of the contents of the various chapters.

- The book begins with basic definitions and examples. Chapter 1 introduces the concepts of state vectors and divides the dynamical world into the discrete and the continuous. We then explore many instances of dynamical systems in the real world—our examples are drawn from physics, biology, economics, and numerical mathematics.

- Chapter 2 deals with linear systems. We begin with one-dimensional systems and, emboldened by the intuition we develop there, move on to higher dimensional systems. We restrict our attention to diagonalizable systems but explain how to extend the results in the nondiagonalizable case.

- In Chapter 3 we introduce nonlinear systems. This chapter deals with fixed points and their stability. We present two methods for assessing stability: linearization and Lyapunov functions.

- Chapter 4 continues the study of nonlinear systems. We explore the periodic and chaotic behaviors nonlinear systems can exhibit. We discuss how periodic points change as the system is changed (bifurcation) and how periodic points relate to one another (Sarkovskii's theorem). Symbolic methods are introduced to explain chaotic behavior.

- Chapter 5 deals with fractals. We develop the notions of contraction maps and of distance between compact sets. We explain how fractals are formed as the attractive fixed points of iterated function systems of affine functions. We show how to compute the (box-counting) dimension of fractals.

- Finally, Chapter 6 deals with complex dynamics, focusing on Julia sets and the Mandelbrot set.

As the chapters progress, the material becomes more challenging and more abstract. Sections that are marked with an asterisk may be skipped without any effect on the accessibility of the sequel. Likewise, starred exercises are either based on these optional sections or draw on material beyond the normal prerequisites of calculus and linear algebra.

Starred sections may be skipped.

Two appendices follow the main material.

- Appendix A is a bare-bones reminder of important background material from calculus, linear algebra, and complex numbers. It also gives a gentle introduction to differential equations.

- Appendix B deals with computing and is designed to help students use some popular computing environments in conjunction with the material in this book.

Every section of every chapter ends with a variety of problems. The problems cover a range of difficulties. Some are best solved with the aid of a computer. Problems marked with an asterisk use ideas from starred sections of the text or require background beyond the prerequisites of calculus and linear algebra.

Examplifications

Whereas Chapter 1 contains many examples and applications, the subsequent chapters concentrate on the mathematical aspects of dynamical systems. However, each of Chapters 2–6 ends with an "Examplifications" section designed to provide additional examples, applications, and amplification of the material in the main portion of the chapter. Some of these supplementary sections require basic ideas from probability.

Examplification = Examples + Applications + Amplification.

In Chapter 2 we show how to use linear system theory to study Markov chains. In Chapter 3 we reexamine Newton's method from a dynamical system perspective. Chapter 4's examplification deals with the question, How many times should one shuffle a deck of cards in order to be sure it is thoroughly mixed? In Chapter 5 we explore the relevance of fractal dimension to real-world problems. We explore how to use fractal dimension to estimate the surface area of a nonsmooth surface and the utility of fractal dimension in image analysis. Finally, in Chapter 6 we have two examplifications: a third visit to Newton's method (but with a complex-numbers point of view) and a revisit of fractals by considering complex-number bases.

Because there may not be time to cover all these supplementary sections in a typical semester course, they should be encouraged as outside reading.

Computing

This book could be used for a course which does not use the computer, but such an omission would be a shame. The computer is a fantastic exploration tool for dynamical systems. Although it is not difficult to write simple computer programs to perform many of the calculations, it is convenient to have a basic stock of programs for this purpose.

A collection of programs written in MATLAB, is available as a supplement for this book. Complete and mail the postcard which accompanies this book to receive a diskette containing the software. See §B.3 on page 366 for more information, including how to obtain the software via ftp. Included in the software package is documentation explaining how to use the various programs.

The software requires MATLAB to run. MATLAB can be used on various computing environments including Macintosh, Windows, and X-windows (Unix). MATLAB is a product of The MathWorks, Inc. For more information, the company can be reached at (508) 653-1415, or by electronic mail at info@mathworks.com. A less expensive student version of MATLAB (which is sufficient to run the programs offered with this book) is available from Prentice-Hall.

Extras for instructors

In addition to the software available to everyone who purchases this book, instructors may also request the following items from Prentice-Hall:

- a solutions book giving answers to the problems in this book, and

- a figures book, containing all the figures from the book, suitable for photocopying onto transparencies.

Planning a course

There is more material in this book than can comfortably be covered in one semester, especially for students with less than ideal preparation. Here are some suggestions and options for planning a course based on this text.

The examplification sections at the end of each chapter may be omitted, but this would be a shame, since some of the more fun material is found therein. At a minimum, direct students to these sections as supplemental reading. All sections marked with an asterisk can be safely omitted; these sections are more difficult and their material is not used in the sequel.

It is also possible to concentrate on just discrete or just continuous systems, but be warned that the two theories are developed together, and analogies are drawn between the two approaches.

Some further, chapter-by-chapter suggestions:

- A quick review of eigenvalues/vectors at the start of the course (in parallel with starting the main material) is advisable. Have students read Appendix A.

- Chapter 1: Section 1.1 is critical and needs careful development. Section 1.2 contains many examples of "real" dynamical systems. To present all of them in class would be too time consuming. I suggest that one or two be presented and the others assigned as outside reading. The applications in this section can be roughly grouped into the following categories:

    (1) physics (1.2.1, 1.2.2, 1.2.3),

    (2) economics (1.2.4, 1.2.5),

    (3) biology (1.2.7, 1.2.8), and

    (4) numerical methods (1.2.6, 1.2.9, 1.2.10, 1.2.11).

  The Newton's method example (1.2.9) ought to be familiar to students from their calculus class. Newton's method is revisited in two of the examplification sections.

- In Chapter 2, section 2.2.3 can safely be omitted.

- In Chapter 3, section 3.3 (Lyapunov functions) may be omitted. Lyapunov functions are used occasionally in the sequel (e.g., in section 4.1.2 to show that a certain system tends to cyclic behavior).

- In Chapter 4, section 4.1.3 can be omitted (although it is not especially challenging). Presentation of section 4.1 can be very terse, as this material is not used later in the text.

  The section on Sarkovski's Theorem (4.2.4) is perhaps the most challenging in the text and may be omitted. Instructors can mention the "period 3 implies all periods" result and move on.

  The symbolic methods in section 4.2.5 resurface in Chapter 5 in explaining how the randomized fractal drawing algorithms work.

- Chapter 5 is long, and some streamlining can be accomplished. Section 5.1.4 can be omitted, but we do use the concept of *compact set* later in the chapter.

  Section 5.3 can be compressed by omitting some proofs or just giving an intuitive discussion of the contraction mapping theorem, which forms the theoretical basis for the next section.

  Section 5.4 is the heart of this chapter.

  Section 5.5 can be omitted, but students might be disappointed. It's great fun to be able to draw fractals.

  The cover-by-balls definition of fractal dimension in section 5.6 is quite natural, but time can be saved by just using the grid-box counting formula.

- In Chapter 6, it is possible to omit sections 6.1 and 6.2 and proceed directly to the examplifications.

On the Internet

Readers with access to the Internet using the World Wide Web (e.g., using Mosaic) can visit the home page for this book at

```
http://www.mts.jhu.edu/~ers/invite.html
```

There, readers can find further information about this book including a list of errata, a gallery of pretty pictures, and access to the accompanying software (see §B.3, especially page 367).

Acknowledgments

During the course of writing this book, I have been fortunate to have had wonderful assistance and advice from students, friends, family, and colleagues.

Thanks go first to my department chair, John Wierman, who manages (amazingly) to be simultaneously my boss, colleague, and friend. Some years ago—despite my protests—he assigned me to teach our department's Dynamical Systems course. To my suprise, I had a wonderful time teaching this course and this book is a direct outgrowth.

Next, I'd like to thank all my students who helped me to develop this course and gave comments on early versions of the book. In particular, I would like to thank Robert Fasciano, Hayden Huang, Maria Maroulis, Scott Molitor, Karen Singer, and Christine Wu. Special thanks to Gregory Levin for his close reading of the manuscript and for his work on the solutions manual and accompanying software.

Several colleagues at Hopkins gave me valuable input and I would like to thank James Fill, Don Giddens, Alan Goldman, Charles Meneveau, Carey Priebe, Wilson J. Rugh, and James Wagner.

I also received helpful comments and contributions from colleagues at other universities. Many thanks to Steven Alpern (London School of Economics), Terry McKee (Wright State University), K. R. Sreenivasan (Yale University), and Daniel Ullman (George Washington University).

Prentice-Hall arranged for early versions of this manuscript to be reviewed by a number of mathematicians. Their comments were very useful and their contributions improved the manuscript. Thanks to: Florin Diacu (University of Victoria), John E. Franke (North Carolina State), Jimmie Lawson (Louisiana State University), Daniel Offin (Queens University), Joel Robbin (University of Wisconsin), Klaus Schmitt (University of Utah), Richard Swanson (Montana State University), Michael J. Ward (University of British Columbia), and Andrew Vogt (Georgetown University).

Thanks also to George Lobell and Barbara Mack at Prentice-Hall for all their hard work and assistance.

Thanks to Naomi Bulock and Cristina Palumbo of The MathWorks for setting up the software distribution.

Many thanks to my sister-in-law Suzanne Reyes for her help with the economics material.

Extra special thanks to my wife, Amy, and to our children, Rachel, Daniel, Naomi, and Jonah, for their love, support, and patience throughout this whole project.

And many thanks to you, the reader. I hope you enjoy this *Invitation* and would appreciate receiving your RSVP. Please send your comments and RSVP suggestions by e-mail to ers@jhu.edu or by conventional mail to me at the Department of Mathematical Sciences, The Johns Hopkins University, Baltimore, Maryland 21218, USA.

This book was developed from a sophomore-junior level course in Dynamical Systems at Johns Hopkins.

—ES, Baltimore
May 24, 1995

Chapter 1

Introduction

1.1 What is a dynamical system?

A dynamical system is a function with an attitude. A dynamical system is doing the same thing over and over again. A dynamical system is always knowing what you are going to do next.

Cryptic? I apologize. The difficulty is that virtually *anything* that evolves over time can be thought of as a dynamical system. So let us begin by describing *mathematical* dynamical systems and then see how many physical situations are nicely modeled by mathematical dynamical systems.

A dynamical system has two parts: a *state vector* which describes exactly the state of some real or hypothetical system, and a *function* (i.e., a rule) which tells us, given the current state, what the state of the system will be in the next instant of time.

1.1.1 State vectors

Physical systems can be described by numbers. This amazing fact accounts for the successful marriage between mathematics and the sciences. For example, a ball tossed straight up can be described using two numbers: its height h above the ground and its (upward) velocity v. Once we know these two numbers, h and v, the fate of the ball is completely determined. The pair of numbers (h, v) is a *vector* which completely describes the *state* of the ball and hence is called the *state vector* of the system. Typically, we write vectors as columns of numbers, so more properly, the state of this

system is $\begin{bmatrix} h \\ v \end{bmatrix}$.

The state vector is a numerical description of the current configuration of a system.

1

It may be possible to describe the state of a system by a single number. For example, consider a bank account opened with $100 at 6% interest compounded annually (see §1.2.4 on page 16 for more detail). The state of this system at any instant in time can be described by a single number: the balance in the account. In this case, the state vector has just one component.

On the other hand, some dynamical systems require a great many numbers to describe. For example, a dynamical system modeling global weather might have millions of variables accounting for temperature, pressure, wind speed, and so on at points all around the world. Although extremely complex, the state of the system is simply a list of numbers—a vector.

Whether simple or complicated, the state of the system is a vector; typically we denote vectors by bold, lowercase letters, such as **x**. (Exception: When the state can be described by a single number, we may write x instead of **x**.)

1.1.2   The next instant: discrete time

Given the current state, where will the system be next?

The second part of a dynamical system is a rule which tells us how the system changes over time. In other words, if we are given the current state of the system, the rule tells us the state of the system in the next instant.

In the case of the bank account described above, the next instant will be one year later, since interest is paid only annually; time is *discrete*. That is to say, time is a sequence of separate chunks each following the next like beads on a string. For the bank account, it is easy to write down the rule which takes us from the state of the system at one instant to the state of the system in the next instant, namely,

$$\mathbf{x}(k + 1) = 1.06\mathbf{x}(k). \tag{1.1}$$

We write $\mathbf{x}(k)$ to denote the state of the system at discrete time k.

Some comments are in order. First, we have said that the state of the system is a vector[1] **x**. Since the state changes over time, we need a notation for what the state is at any specific time. The state of the system at time k is denoted by $\mathbf{x}(k)$. Second, we use the letter k to denote discrete time. In this example (since interest is only paid once a year) time is always a whole number. Third, equation (1.1) does not give a complete description of the dynamical system since it does not tell us the opening balance of the account. A complete description of the system is

$$\mathbf{x}(k + 1) = 1.06\mathbf{x}(k), \text{ and}$$

---

[1]In this case, our vector has only one component: the bank balance. In this example we are still using a boldface **x** to indicate that the state vector typically has several entries. However, since this system has only one state variable, we may write x in place of **x**.

$$\mathbf{x}(0) = 100.$$

It is customary to begin time at 0, and to denote the initial state of the system by $\mathbf{x}_0$. In this example $\mathbf{x}_0 = \mathbf{x}(0) = 100$.

The state of the bank account in all future years can now be computed. We see that $\mathbf{x}(1) = 1.06\mathbf{x}(0) = 1.06 \times 100 = 106$, and then $\mathbf{x}(2) = 1.06\mathbf{x}(1) = 1.06 \times 106 = 112.36$. Indeed, we see that

$$\mathbf{x}(k) = (1.06)^k \times 100,$$

or more generally,

$$\mathbf{x}(k) = 1.06^k \mathbf{x}_0. \tag{1.2}$$

Now it isn't hard for us to see directly that $1.06^k \mathbf{x}_0$ is a general formula for $\mathbf{x}(k)$. However, we can verify that equation (1.2) is correct by checking two things: (1) that it satisfies the initial condition $\mathbf{x}(0) = \mathbf{x}_0$, and (2) that it satisfies equation (1.1). Now (1) is easy to verify, since

$$\mathbf{x}(0) = (1.06)^0 \times \mathbf{x}_0 = \mathbf{x}_0.$$

Further, (2) is also easy to check, since

$$\mathbf{x}(k+1) = 1.06^{k+1}\mathbf{x}_0 = (1.06) \times (1.06)^k \mathbf{x}_0 = 1.06\mathbf{x}(k).$$

A larger context

Let us put this example into a broader context which is applicable to all discrete time dynamical systems. We have a state vector $\mathbf{x} \in \mathbf{R}^n$ and a function $f: \mathbf{R}^n \to \mathbf{R}^n$ for which

The general form of a discrete time dynamical system.

$$\mathbf{x}(k+1) = f(\mathbf{x}(k)).$$

In our simple example, $n = 1$ (the bank account is described by a single number: the balance) and the function $f: \mathbf{R} \to \mathbf{R}$ is simply $f(\mathbf{x}) = 1.06\mathbf{x}$. Later, we consider more complicated functions f. Once we are given that $\mathbf{x}(0) = \mathbf{x}_0$ and that $\mathbf{x}(k+1) = f(\mathbf{x}(k))$, we can, in principle, compute all values of $\mathbf{x}(k)$, as follows:

$$\mathbf{x}(1) = f(\mathbf{x}(0)) = f(\mathbf{x}_0)$$

$$\mathbf{x}(2) = f(\mathbf{x}(1)) = f(f(\mathbf{x}_0))$$

$$\mathbf{x}(3) = f(\mathbf{x}(2)) = f(f(f(\mathbf{x}_0)))$$

$$\mathbf{x}(4) = f(\mathbf{x}(3)) = f(f(f(f(\mathbf{x}_0))))$$

$$\vdots$$

$$\mathbf{x}(k) = f(\mathbf{x}(k-1)) = f(f(\ldots(f(\mathbf{x}_0))\ldots))$$

We write $f^k(\mathbf{x})$ to denote the result computed by k applications of the function f to the value $\mathbf{x}$.

where in the last line we have f applied k times to $\mathbf{x}_0$. We need a notation for repeated application of a function. Let us write $f^2(\mathbf{x})$ to mean $f(f(\mathbf{x}))$, write $f^3(\mathbf{x}) = f(f(f(\mathbf{x})))$, and in general, write

$$f^k(\mathbf{x}) = \underbrace{f(f(f(\ldots f(\mathbf{x}))\ldots))}_{k \text{ times}}.$$

WARNING: In this book, the notation $f^k(\mathbf{x})$ does not mean $(f(\mathbf{x}))^k$ (the number $f(\mathbf{x})$ raised to the kth power), nor does it mean the kth derivative of f.

1.1.3   The next instant: continuous time

Bank accounts which change only annually or computer chips which change only during clock cycles are examples of systems for which time is best viewed as progressing in discrete packets. Many systems, however, are better described with time progressing smoothly. Consider our earlier example of a ball thrown straight up. Its instantaneous status is given by its state vector $\mathbf{x} = \begin{bmatrix} h \\ v \end{bmatrix}$. However, it doesn't make sense to ask what its state will be in the "next" instant of time—there is no "next" instant since time advances continuously.

Continuous time is denoted by t.

We reflect this different perspective on time by using the letter t (rather than k) to denote time. Typically t is a nonnegative real number and we start time at $t = 0$.

Since we cannot write down a rule for the "next" instant of time, we instead describe how the system is changing at any given instant. First, if our ball has (upward) velocity v, then we know that $dh/dt = v$; this is the definition of velocity. Second, gravity pulls down on the ball and we have $dv/dt = -g$ where g is a positive constant.[2] The change in the system can thus be described by

$$h'(t) = v(t) \tag{1.3}$$

$$v'(t) = -g, \tag{1.4}$$

which can be rewritten in matrix notation:

$$\begin{bmatrix} h'(t) \\ v'(t) \end{bmatrix} = \begin{bmatrix} 0 & 1 \\ 0 & 0 \end{bmatrix} \begin{bmatrix} h(t) \\ v(t) \end{bmatrix} + \begin{bmatrix} 0 \\ -g \end{bmatrix}.$$

---

[2]Near the surface of the earth, g is approximately 9.8 m/s^2.

Since $\mathbf{x}(t) = \begin{bmatrix} h(t) \\ v(t) \end{bmatrix}$, this can all be succinctly written as

$$\mathbf{x}' = f(\mathbf{x}), \tag{1.5}$$

where $f(\mathbf{x}) = A\mathbf{x} + \mathbf{b}$, A is the 2×2 matrix $\begin{bmatrix} 0 & 1 \\ 0 & 0 \end{bmatrix}$, and $\mathbf{b}$ is the constant vector $\begin{bmatrix} 0 \\ -g \end{bmatrix}$.

Indeed, equation (1.5) is the form for *all* continuous time dynamical systems. A continuous time dynamical systems has a state vector $\mathbf{x}(t) \in \mathbf{R}^n$ and we are given a function $f : \mathbf{R}^n \to \mathbf{R}^n$ which specifies how quickly each component of $\mathbf{x}(t)$ is changing, i.e., $\mathbf{x}'(t) = f(\mathbf{x}(t))$, or more succinctly, $\mathbf{x}' = f(\mathbf{x})$.

The general form for a continuous time dynamical system.

Returning to the example at hand, suppose the ball starts at height h_0 and with upward velocity v_0, i.e., $\mathbf{x}_0 = \begin{bmatrix} h_0 \\ v_0 \end{bmatrix}$. We claim that the equations

$$h(t) = h_0 + v_0 t - \frac{1}{2} g t^2, \text{ and}$$

$$v(t) = v_0 - gt$$

describe the motion of the ball. We could derive these answers from what we already know[3], but it is simple to verify directly the following two facts: (1) when $t = 0$ the formulas give h_0 and v_0, and (2) these formulas satisfy the differential equations (1.3) and (1.4).

For (1) we observe that $h(0) = h_0 + v_0 0 - \frac{1}{2} 0^2 = h_0$ and, $v(0) = v_0 - g0 = v_0$. For (2) we see that

$$h'(t) = \frac{d}{dt} \left[h_0 + v_0 t - \frac{1}{2} g t^2 \right] = v_0 - gt = v(t),$$

verifying equation (1.3) and that

$$v'(t) = \frac{d}{dt} [v_0 - gt] = -g,$$

verifying equation (1.4).

---

[3]We could derive these answers by integrating equation (1.4) and then (1.3).

1.1.4   Summary

A dynamical system is specified by a state vector $\mathbf{x} \in \mathbf{R}^n$, (a list of numbers which may change as time progresses) and a function $f: \mathbf{R}^n \to \mathbf{R}^n$ which describes how the system evolves over time.

There are two kinds of dynamical systems: discrete time and continuous time.

For a discrete time dynamical system, we denote time by k, and the system is specified by the equations

$$\mathbf{x}(0) = \mathbf{x}_0, \text{ and}$$

$$\mathbf{x}(k+1) = f(\mathbf{x}(k)).$$

It thus follows that $\mathbf{x}(k) = f^k(\mathbf{x}_0)$, where f^k denotes a k-fold application of f to x_0.

For a continuous time dynamical system, we denote time by t, and the following equations specify the system:

$$\mathbf{x}(0) = \mathbf{x}_0, \text{ and}$$

$$\mathbf{x}' = f(\mathbf{x}).$$

Problems for §1.1

◆1.   Suppose you throw a ball up, but not straight up. How would you model the state of this system (the flying ball)? In other words, what numbers would you need to know in order to completely describe the state of the system? For example, the height of the ball is one of the state variables you would need to know. Find a complete description. Neglect air resistance and assume gravity is constant.

[Hint: Two numbers suffice to describe a ball thrown straight up: the height and the velocity. To model a ball thrown up, but not straight up, requires more numbers. What numerical information about the state of the ball do you require?]

◆2.   For each of the following functions f find $f^2(x)$ and $f^3(x)$.

   (a) $f(x) = 2x$.
   (b) $f(x) = 3x - 2$.
   (c) $f(x) = x^2 - 3$.
   (d) $f(x) = \sqrt{x + 1}$.
   (e) $f(x) = 2^x$.

◆3.   For each of the functions in the previous problem, compute $f^7(0)$. If you have difficulty, explain why.

◆4. Consider the discrete time system

$$x(k+1) = 3x(k); \quad x(0) = 2.$$

Compute $x(1), x(2), x(3),$ and $x(4)$.

Now give a formula for $x(k)$.

◆5. Consider the discrete time system

$$x(k+1) = ax(k), \quad x(0) = b$$

where a and b are constants. Find a formula for $x(k)$.

◆6. Consider the continuous time dynamical system

$$x' = 3x, \quad x(0) = 2.$$

Show that for this system $x(t) = 2e^{3t}$.

[To do this you should check that the formula $x(t) = 2e^{3t}$ satisfies (1) the equation $x' = 3x$ and (2) the equation $x(0) = 2$. For (1) you need to check that the derivative of $x(t)$ is exactly $3x(t)$. For (2) you should check that substituting 0 for t in the formula gives the result 2.]

◆7. Based on your experience with the previous problem, find a formula for $x(t)$ for the system

$$x' = ax; \quad x(0) = b,$$

where a and b are constants. Check that your answer is correct. Does your formula work in the special cases $a = 0$ or $b = 0$?

◆8. Killing time. Throughout this book we assume that the "rule" which describes how the system is changing does *not* depend on time. How can we model a system whose dynamics change over time? For example, we might have the system with state vector **x** for which

$$x_1' = 3x_1 + (2 - t)x_2$$
$$x_2' = x_1x_2 - t.$$

Thus the rate at which x_1 and x_2 change depends on the time t.

Create a new system which is equivalent to the above system for which the rule doesn't depend on t.

[Hint: Add an extra state variable which acts just like time.]

◆9. Killing time again. Use your idea from the previous problem to eliminate the dependence on time in the following discrete time system.

$$x_1(k+1) = 2x_1(k) + kx_2(k)$$
$$x_2(k+1) = x_1(k) - k - 3x_2(k)$$

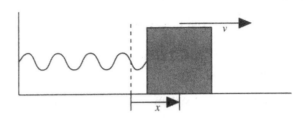

Figure 1.1. A mass on a frictionless surface attached to a wall by a spring.

♦10. The Collatz $3x + 1$ problem. Pick a positive integer. If it is even, divide
       it by two. Otherwise (if it's odd) multiply it by three and add one. Now
       repeat this procedure on your answer. In other words, consider the function

$$f(x) = \begin{cases} x/2 & \text{if } x \text{ is even,} \\ 3x + 1 & \text{if } x \text{ is odd.} \end{cases}$$

If we begin with $x = 10$ and we iterate f we get

$$10 \mapsto 5 \mapsto 16 \mapsto 8 \mapsto 4 \mapsto 2 \mapsto 1 \mapsto 4 \mapsto \cdots$$

Notice that from this point on we get an endless stream of 4,2,1,4,2,1,. . . .

Write a computer program to compute f and iterate f for various starting
values. Do the iterates always fall into the pattern 4,2,1,4,2,1,. . . regardless
of the starting value? No one knows!

1.2 Examples

In the previous section we introduced the concept of a dynamical system.
Here we look at several examples—some continuous and some discrete.

1.2.1 Mass and spring

Our first example of a continuous time dynamical system consists of a mass
sliding on a frictionless surface and attached to a wall by an ideal spring;
see Figure 1.1. The state of this system is determined by two numbers:
x, the distance the block is from its neutral position, and v, its velocity
to the right. When $x = 0$ we assume that the spring is neither extended
nor compressed and exerts no force on the block. As the block is moved
to the right ($x > 0$) of this neutral position, the spring pulls it to the left.

The spring exerts a force
proportional to the distance it
is compressed or stretched.
This is known as Hooke's law.

Conversely, if the block is to the left of the neutral position ($x < 0$), the spring is compressed and pushes the block to the right. Assuming we have an ideal spring, the force F on the block when it is at position x is $-kx$, where k is a positive constant. The minus sign reflects the fact that the direction of the force is opposite the direction of the displacement.

From basic physics, we recall that $F = ma$, where m is the mass of the block, and acceleration, a, is the rate of change of velocity (i.e., $a = dv/dt$). Substituting $F = -kx$, we have

$$v' = -\frac{k}{m}x. \tag{1.6}$$

By definition, velocity is the rate of change of position, that is,

$$x' = v. \tag{1.7}$$

We can simplify matters further by taking $k = m = 1$. Finally, we combine equations (1.6) and (1.7) to give

$$\begin{bmatrix} x' \\ v' \end{bmatrix} = \begin{bmatrix} 0 & 1 \\ -1 & 0 \end{bmatrix} \begin{bmatrix} x \\ v \end{bmatrix}, \tag{1.8}$$

or equivalently,

$$\mathbf{y}' = A\mathbf{y}, \tag{1.9}$$

where $\mathbf{y} = \begin{bmatrix} x \\ v \end{bmatrix}$ is the state vector and $A = \begin{bmatrix} 0 & 1 \\ -1 & 0 \end{bmatrix}$. Let us assume that the block starts in state $\mathbf{y}_0 = \begin{bmatrix} x_0 \\ v_0 \end{bmatrix} = \begin{bmatrix} 1 \\ 0 \end{bmatrix}$, i.e., the block is not moving but is moved one unit to the right. Then we claim that

$$\mathbf{y}(t) = \begin{bmatrix} \cos t \\ -\sin t \end{bmatrix} \tag{1.10}$$

describes the motion of the block at future times. Later (in Chapter 2) we show how to derive this. For now, let us simply verify that this is correct. There are two things to check: (1) that $\mathbf{y}(0) = \begin{bmatrix} 1 \\ 0 \end{bmatrix}$ and (2) that $\mathbf{y}$ satisfies equation (1.8), or equivalently, equation (1.9). To verify (1) we simply substitute $t = 0$ into equation (1.10) and we see that

$$\mathbf{y}(0) = \begin{bmatrix} \cos 0 \\ -\sin 0 \end{bmatrix} = \begin{bmatrix} 1 \\ 0 \end{bmatrix} = \mathbf{y}_0,$$

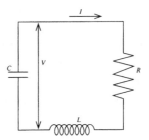

Figure 1.2. An electrical circuit consisting of a resistor, a capacitor, and an inductor (coil).

as required. For (2), we take derivatives as follows:

$$\mathbf{y}'(t) = \begin{bmatrix} \cos t \\ -\sin t \end{bmatrix}' = \begin{bmatrix} -\sin t \\ -\cos t \end{bmatrix} = \begin{bmatrix} 0 & 1 \\ -1 & 0 \end{bmatrix} \begin{bmatrix} \cos t \\ -\sin t \end{bmatrix} = A\mathbf{y}(t),$$

as required.

Since the position is $x(t) = \cos t$, we see that the block bounces back and forth forever. This, of course, is not physically realistic. Friction, no matter how slight, eventually will slow the block to a stop.

1.2.2   RLC circuits

Consider the electrical circuit in Figure 1.2. The capacitance of the capacitor C, the resistance of the resistor R, and the inductance of the coil L are constants; they are part of the circuit design. The current in the circuit I and the voltage drop V across the resistor and the coil vary with time.[4]

These can be measured by inserting an ammeter anywhere in the circuit and attaching a voltmeter across the capacitor (see the figure). Once the initial current and voltage are known, we can predict the behavior of the system. Here's how.

The charge on the capacitor is $Q = -CV$. The current is the rate of change in the charge, i.e., $I = Q'$. The voltage drop across the resistor is RI and the voltage drop across the coil is LI', so in all we have $V = LI' + RI$.

---

[4]We choose V to be positive when the upper plate of the capacitor is positively charged with respect to the bottom plate.

We can solve the three equations

$$Q = -CV,$$

$$I = Q', \quad \text{and}$$

$$V = LI' + RI$$

for V' and I'. We get

$$V' = -Q'/C = -\frac{1}{C}I$$

$$I' = \frac{1}{L}V - \frac{R}{L}I,$$

which can be rewritten in matrix notation as

$$\begin{bmatrix} V \\ I \end{bmatrix}' = \begin{bmatrix} 0 & -1/C \\ 1/L & -R/L \end{bmatrix} \begin{bmatrix} V \\ I \end{bmatrix}. \qquad (1.11)$$

Let's consider a special case of this system. If the circuit has no resistance ($R = 0$) and if we choose $L = C = 1$, then the system becomes

$$\begin{bmatrix} V \\ I \end{bmatrix}' = \begin{bmatrix} 0 & -1 \\ 1 & 0 \end{bmatrix} \begin{bmatrix} V \\ I \end{bmatrix},$$

which is nearly the same as equation (1.8) on page 9 for the mass-and-spring system. Indeed, if $V(0) = 1$ and $I(0) = 0$, you should check that

A resistance-free RLC circuit oscillates in just the same way as the frictionless mass and spring.

$$V(t) = \cos t$$

$$I(t) = \sin t$$

describes the state of the system for all future times t. The resistance-free RLC circuit and the frictionless mass-and-spring systems behave in (essentially) identical fashions.

In reality, of course, there are no friction-free surfaces or resistance-free circuits. In Chapter 2 (see pages 68-71) we revisit these examples and analyze the effect of friction/resistance on these systems.

1.2.3 Pendulum

Consider an ideal pendulum as shown in Figure 1.3. The bob has mass m and is attached by a rigid pole of length L to a fixed pivot. The state of this dynamical system can be described by two numbers: θ, the angle the

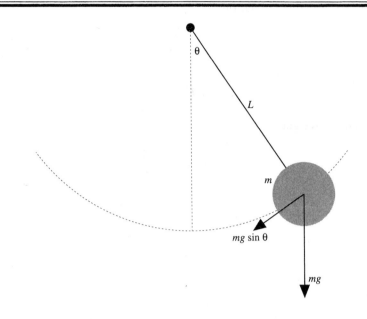

Figure 1.3. A simple pendulum.

pendulum makes with the vertical, and ω, the rate of rotation (measured, say, in radians per second). By definition, $\omega = d\theta/dt$.

Gravity pulls the bob straight down with force mg. This force can be resolved into two components: one parallel to the pole and one perpendicular. The force parallel to the pole does not affect how the pendulum moves. The component perpendicular to the pole has magnitude $mg \sin \theta$; see Figure 1.3.

Now we want to apply Newton's law, $F = ma$. We know that the force is $mg \sin \theta$. We need to relate a to the state variable θ. Since distance s along the arc of the pendulum is $L\theta$, and $a = s''$, we have $a = (L\theta)'' = L\omega'$. Thus $\omega' = a/L = (ma)/(mL) = -(mg \sin \theta)/(mL) = -(g/L) \sin \theta$. We can summarize what we know as follows:

$$\theta'(t) = \omega(t), \text{ and} \qquad (1.12)$$

$$\omega'(t) = -\frac{g}{L} \sin \theta(t). \qquad (1.13)$$

(The minus sign in equation (1.13) reflects the fact that when $\theta > 0$, the force tends to send the pendulum back to the vertical.) Let $\mathbf{x} = \begin{bmatrix} \theta \\ \omega \end{bmatrix}$ be the state vector; then equations (1.12) and (1.13) can be expressed

$$\mathbf{x}' = f(\mathbf{x}),$$

where $f: \mathbf{R}^2 \to \mathbf{R}^2$ is defined by

$$f\begin{bmatrix} x \\ y \end{bmatrix} = \begin{bmatrix} y \\ -\frac{g}{L}\sin x \end{bmatrix}. \tag{1.14}$$

Although we were able to present an exact description of the motion of the mass and spring of §1.2.1 (see equation (1.10)), we cannot give an exact formula for the dynamical system of equations (1.12) and (1.13).

This is a more complicated system because of the sine function. An exact solution is too hard.

   We can still gain a feel for the action of the pendulum, however, by two methods: (1) linear approximation and (2) numerical methods.

Linear approximation

The function f in equation (1.14) is nonlinear; it contains the sine function. However, if the angular displacement of the pendulum is very small, then $\sin\theta \approx \theta$; this is an instance of the limit

If θ is small, we can approximate $\sin\theta$ by θ.

$$\lim_{\theta\to 0} \frac{\sin\theta}{\theta} = 1.$$

Replacing $\sin\theta$ by θ in equation (1.13) we can rewrite our system as

$$\begin{bmatrix} \theta \\ \omega \end{bmatrix}' = \begin{bmatrix} 0 & 1 \\ -g/L & 0 \end{bmatrix}\begin{bmatrix} \theta \\ \omega \end{bmatrix} \tag{1.15}$$

If we take $L = g$ (e.g., assume the pole is 9.8 meters long), then equation (1.15) is exactly the same as equation (1.8); hence if we begin the pendulum with a slight displacement, we would expect the angle to vary with time sinusoidally. In other words, the pendulum will swing back and forth—*amazing!*

Numerical methods

Some systems of differential equations (such as equation (1.8)) can be solved exactly by analytic means, others (such as equation (1.14)) cannot. A computer, however, may be useful in such cases. Euler's method (see §1.2.10 on page 28) is one technique for working *numerically* with differential equations. Although Euler's method is easy to explain, it is not very accurate. Other methods, while more accurate are harder to analyze. Nonetheless, these more sophisticated methods are readily available

When an exact formula cannot be found, numerical methods may help.

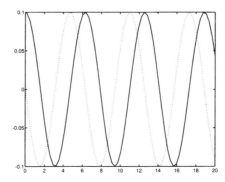

Figure 1.4. The motion of a pendulum with $\theta_0 = 0.1$ and $\omega_0 = 0$. The solid curve is the angle θ and the dotted curve is the rate of rotation ω. The horizontal axis is time, t.

in various mathematical computer environments such as MATLAB, *Maple*, *Mathematica*, and *Mathcad*.

There are various drawbacks to numerical methods (see §4.1.4 on page 163 where we discuss how they may be totally useless), including the fact they do not give us a formula from which we can make conclusions. However, we can still get a good idea of how a system behaves by using numerical methods.

In §B.1.2 on page 357 we show how to use packages such as MATLAB to find approximate (numerical) solutions to differential equations. With these methods, we can examine the pendulum system. To simplify matters, we take $g = L = 9.8$, so our system from equation (1.14) becomes

$$\left[\begin{array}{c} \theta \\ \omega \end{array} \right]' = \left[\begin{array}{c} \omega \\ -\sin\theta \end{array} \right]. \tag{1.16}$$

Let us start the pendulum system at $\mathbf{x}_0 = \left[\begin{array}{c} \theta(0) \\ \omega(0) \end{array} \right] = \left[\begin{array}{c} 0.1 \\ 0 \end{array} \right]$; physically, we move the weight a small distance away from the straight-down position. The result is illustrated in Figure 1.4. Notice that the curves look identical to sine and cosine waves, as we might expect from our discussion on linear approximations.

Next, let us try a large initial displacement. When $\theta = \pi$, the bob is straight up; we begin with $\theta = 3$ (nearly vertical). The resulting plot is shown in Figure 1.5. Although periodic, the curves do not look at all like sine waves.

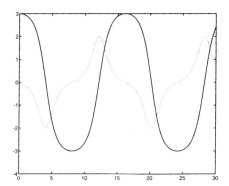

Figure 1.5. The motion of a pendulum with $\theta_0 = 3$ and $\omega_0 = 0$. The solid curve is θ and the dotted curve is ω.

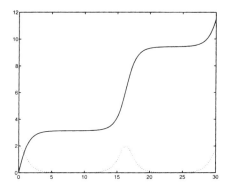

Figure 1.6. The motion of a pendulum with $\theta_0 = 0$ and $\omega_0 = 2$. The solid curve is θ and the dotted curve is ω.

Finally, let us begin with the bob hanging straight down ($\theta = 0$) but give it a hefty initial spin ($\omega = 2$). The result is Figure 1.6 The surprise is that the plot of θ appears to go up and up and is not periodic! What is going on? What we see is that the pendulum is continually rotating in the same direction (notice that ω is always positive) and so the pendulum is winding around and around. It is interesting to notice that the mass spends most of its time near the vertical position, where it is moving the most slowly.

1.2.4   Your bank account

Earlier we discussed bank accounts as examples of discrete time dynamical systems. Let us revisit this example. We discussed a deposit of $100 in the bank ($\mathbf{x}(0) = 100$) and we supposed that each year the bank pays 6% interest, i.e., $\mathbf{x}(k + 1) = 1.06\mathbf{x}(k)$.

Monthly compounding.
Let us try to make this example more realistic. Some banks pay interest monthly. If the annual interest rate is r, then the account increases by a factor of $(1+r/12)$ each month. Let us also suppose that we make a deposit each month in the amount b. Our system becomes

$$\mathbf{x}(k + 1) = (1 + r/12)\mathbf{x}(k) + b. \qquad (1.17)$$

Notice that equation (1.17) has the form $\mathbf{x}(k+1) = f(\mathbf{x}(k))$, where $f(x) = ax + b$—a linear equation. Such linear systems are discussed at length in Chapter 2.

Continuous compounding.
Now, many banks *post* interest monthly, but, in fact *pay* interest continuously. The instant the account earns another penny, interest on that penny starts to accumulate. If say, our account has $\mathbf{x}$ dollars and is paying 6% interest, then at this instant it is increasing in value at a rate of $0.06\mathbf{x}$ dollars per year. In symbols, $d\mathbf{x}/dt = 0.06\mathbf{x}$. Imagine we continuously deposit money into our account at a rate of b dollars per year, then we can view our bank deposit as a continuous time dynamical system for which

$$\frac{d\mathbf{x}}{dt} = r\mathbf{x} + b. \qquad (1.18)$$

Notice, again, that equation (1.18) is of the form $\mathbf{x}' = f(\mathbf{x})$, where f is a linear[5]function: $f(\mathbf{x}) = r\mathbf{x} + b$. In Chapter 2 we show how to solve this kind of system exactly. For now we can take advantage of the fact that this differential equation is readily handled by computer algebra systems such as *Mathematica*. Here we show the input and output to *Mathematica* to solve $\mathbf{x}' = r\mathbf{x} + b$ with $\mathbf{x}(0) = \mathbf{x}_0$:

Using the computer to get an exact formula. See §B.1.1.

```
DSolve[ {x'[t] == r x[t] + b, x[0] == x0}, x[t], t]


                      r t      r t
            -b + b E      + E      r x0
   {{x[t] -> -----------------------}}
                      r
```

---

[5]More properly, f is an *affine* function.

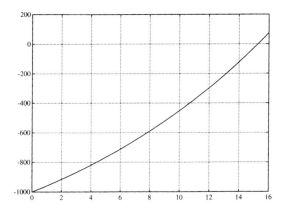

Figure 1.7. Indebtedness as a function of time on a $1000 loan at 6% interest, paid back at a rate of $100 per year.

Rewriting *Mathematica*'s answer we see that

$$\mathbf{x}(t) = e^{rt}\mathbf{x}_0 + \frac{b}{r}\left(e^{rt} - 1\right). \qquad (1.19)$$

This formula is especially interesting when $\mathbf{x}_0$ is *negative*. What does a negative bank balance mean? It might indicate that we are overdrawn (uh oh), or it might represent a loan we are paying off (such as a car loan or a mortgage). Given that the loan is at interest rate r and we are paying b dollars per year (typically as $b/12$ dollars per month), the expression for $\mathbf{x}(t)$ in equation (1.19) tells us our indebtedness at any given point in the loan.

A negative balance: a loan to repay.

For example, suppose we borrow $1000 at 6% interest and pay back at a rate of $100 per year (paid continuously over the course of the year). Figure 1.7 shows our indebtedness over time. We see that it takes just over 15 years to pay back the loan (for a total of over $1500 in payments).

1.2.5 Economic growth

Let us switch from economics on the small scale (a bank account) to economics on the grand scale: a nation's economy. Here we are concerned with the extent to which a nation invests in capital (the machinery and equipment it uses to produce goods and services).

A simplified version of a model of economic growth due to Solow.

We begin by listing the various economic quantities which are relevant to capitalization.

- K: the total amount the nation has invested in capital.

- d: the rate at which the capital depreciates. Thus K is decreasing at a rate dK. We assume that d is a constant.

- N: the population of the nation.

- ρ: the rate of growth of the population. We assume that ρ is constant. Thus $N' = \rho N$.

- Y: the output (total of goods and services) produced by the nation. The level of output depends on the total capital (equipment) K and total labor (population) N. In order to double the amount produced, *both* the amount of labor and amount of capital would need to be doubled. A reasonable formula for Y in terms of K and N is

$$Y = A\sqrt{KN}, \tag{1.20}$$

where A is a constant.

- k: the per capita capitalization, i.e., $k = K/N$.

- y: the per capita output, i.e., $y = Y/N$; this is a measure of worker productivity.

- s: the savings rate. Since savings are equivalent to investment (money deposited into bank accounts is loaned to firms to buy capital), K is increasing at a rate of sY.

We now organize these quantities into a dynamical system. This system has a single state variable, k, the per-capita capitalization.

We know that $k = K/N$, so we compute the derivative of k by the derivative of quotients rule:

$$k' = \frac{NK' - KN'}{N^2} = \frac{K'}{N} - \frac{KN'}{N^2}. \tag{1.21}$$

Now, $K' = sY - dK$ (capital increases thanks to savings but decreases due to depreciation). Also, $N' = \rho N$ (population increases at a fixed rate ρ). Substituting these into equation (1.21) we get

$$k' = \frac{sY - dK}{N} - \frac{K\rho N}{N^2}.$$

We substitute $Y = A\sqrt{KN}$ to get

$$k' = \frac{sA\sqrt{KN} - dK}{N} - \frac{K}{N},$$

and since $k = K/N$, we arrive at

$$k' = sA\sqrt{k} - (d + \rho)k. \tag{1.22}$$

We can use *Mathematica* to solve equation (1.22). It gives

```
  2  2                                    2
 A  s              2 A s C[1]           C[1]
-------- +  ---------------------- +  ----------
    2         ((d + p) t)/2            (d + p) t
(d + p)     E                (d + p)   E
```

(Here we use p to stand for ρ. The term C[1] stands for a constant; the value of this constant can be determined given the value of $k(0)$.)

We can now analyze what happens as $t \to \infty$. The second and third terms have denominators which go to infinity as $t \to \infty$, so these terms vanish. Thus as $t \to \infty$, we see that $k(t) \to [As/(d + \rho)]^2$. Our model predicts that per capita capitalization approaches a constant level.

To learn that $k(t)$ tends to a limit as $t \to \infty$ we relied on *Mathematica* to find an explicit formula for $k(t)$. However, this is not necessary. We explore (in Chapter 3) how to reach the same conclusion without solving any differential equations.

What if we can't solve?

1.2.6 Pushing buttons on your calculator

Do you ever just play with your calculator? One fun thing to do is to enter any number, and start pressing the $\boxed{\sqrt{}}$ button. What happens? After pressing the button many times, the display always reads 1.0000. Well, not always. If you put put in a negative number, you get an error. And if you start with 0, then you always have 0. But if you start with any positive number, you eventually reach 1. Try it!

Iterating a function is the same as repeatedly pushing the same button on a calculator.

Try playing with your cosine button. Set your calculator to Radians, enter any number, and keep pressing the $\boxed{\cos x}$ button. What happens? Try it!

It's not hard to explain why iterating the square-root key leads to 1. Let's recast this example as a dynamical system. The state of the system is simply the number on the display, x. The rule to get to the next state is simply $f: x \mapsto \sqrt{x}$, or in our usual notation,

$$x(k+1) = \sqrt{x(k)}, \qquad (1.23)$$

or, equivalently, $x(k+1) = x(k)^{1/2}$. Iterating, we have

$$x(0) = x_0$$
$$x(1) = [x(0)]^{1/2} = (x_0)^{1/2}$$
$$x(2) = [x(1)]^{1/2} = (x_0)^{1/4}$$
$$x(3) = [x(2)]^{1/2} = (x_0)^{1/8}$$
$$\vdots$$
$$x(k) = (x_0)^{1/2^k}.$$

Since $1/2^k \to 0$ as $k \to \infty$, we see that, provided $x_0 > 0$, $x(k) = (x_0)^{1/2^k} \to 1$ as $k \to \infty$.

The example of repeatedly pressing $\boxed{\cos x}$ is a bit harder to explain directly, but we look at it carefully in Chapter 3. Formally, we are looking at the dynamical system

$$x(k+1) = \cos x(k).$$

Let us plot a graph of what happens when we iterate $\cos x$ starting with, say, $x = 0$. Figure 1.8 is a plot of the values produced by successive iterates of $\cos x$. The horizontal axis counts the number of iterations.

Spread sheet programs are ideal for performing computations for discrete time dynamical systems.

Incidentally, the easiest computer software to use to produce this plot is spread sheet software, most commonly used for financial matters! Indeed, Figure 1.8 was created using Microsoft *Excel*, although other spreadsheet programs would work nicely as well; see Figure 1.9. We enter the values of the vector $\mathbf{x}(0)$ in the first row of the spread sheet. In the next row, we enter formulas to compute each component of $\mathbf{x}(1)$ from the entries in the previous row. Now comes the fun part. We use $\mathbf{x}(1)$ to find $\mathbf{x}(2)$ using exactly the same computations as those which brought us from $\mathbf{x}(0)$ to $\mathbf{x}(1)$. Thus we simply copy the formulas in the second row to the third row, and then to the fourth, fifth, etc. This can be done easily using "copy and paste" features of spread sheet software and does not require retyping.

In Figure 1.9 we entered the value "0" into cell B2. Then we entered the formula "=cos(B2)" into cell B3 and then copied it into subsequent cells in column B. Notice that each entry in column B is the cosine of the

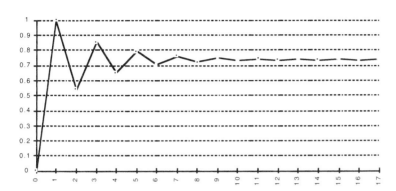

Figure 1.8. Iterating $\cos x$ starting with $x = 0$.

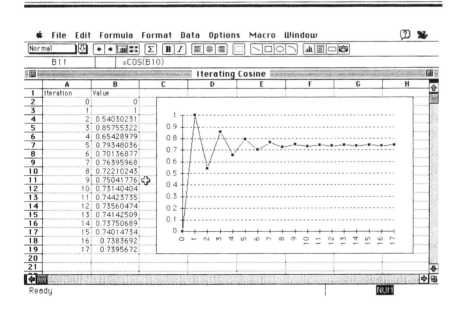

Figure 1.9. Computing iterates of the cosine function using a spreadsheet program.

number above it. Now, if we want to compute cosine iterates starting with a different x_0 (say $x_0 = 2$) we simply retype the entry in cell B2 and the entire spread sheet recomputes (and even updates the graph).

1.2.7 Microbes

A universe in a jar.

A jar is filled with a nutritive solution and some bacteria. As time progresses, the bacteria reproduce (by dividing) and die. Let b (for birth) be the rate at which the microbes reproduce and p (for perish) be the rate at which they die. Then, net, the population is growing at the rate $b - p$. This means that if there are x bacteria in the jar, then the rate at which the number of bacteria is increasing is $(b - p)x$, that is, $dx/dt = rx$, where $r = b - p$. If we begin with x_0 bacteria at time $t = 0$, then (see problems 6 and 7 on page 7)

$$x(t) = e^{rt}x_0. \tag{1.24}$$

In the short run, this makes sense; the formula says that there are x_0 bacteria at time 0 and then the number grows at an exponential rate. However, as time goes on, equation (1.24) implies that the number of bacteria will be exceedingly large (larger than the number of atoms in the universe if we take it literally). Thus the simple model $dx/dt = rx$ is not realistic in the long haul.

As the number of bacteria reproduce, they tend to crowd each other, produce toxic waste products, etc. It makes sense to postulate a death rate that increases with the population.

Again, let us assume a constant rate of reproduction b, so that if there are x bacteria, they are increasing in number at a rate bx. Now instead of a constant death rate, let us suppose that the death rate is px, and so if there are x bacteria, they are decreasing in number at a rate px^2. Combining these, we have the dynamical system

$$\frac{dx}{dt} = bx - px^2. \tag{1.25}$$

Now this is a differential equation for which an analytic solution is known (and we present that solution in a moment). Often, however, it is difficult or impossible to find analytic solutions to differential equations. We try not to rely on finding analytic solutions. Instead, let us see what we can learn directly from equation (1.25).

Let us consider the question, Is there a self-sustaining population in this model? We are looking for a number $\tilde{x}$ for which $b\tilde{x} - p\tilde{x}^2 = 0$; at this special level, the net reproduction/death rates are exactly in balance and

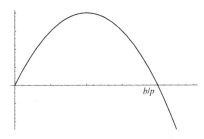

Figure 1.10. A graph of the right-hand side of equation (1.25).

(since this $\tilde{x}$ makes $dx/dt = 0$) the population is neither increasing nor decreasing.

By setting the right-hand-side of equation (1.25) equal to zero we get

$$\frac{dx}{dt} = bx - px^2 = 0 \quad \Rightarrow \quad x = 0 \quad \text{or} \quad x = \frac{b}{p}.$$

We see there are two self-sustaining population levels: $\tilde{x} = 0$ and $\tilde{x} = b/p$. These two values, of course, correspond to the two roots of the quadratic equation $bx - px^2 = 0$. This is the equation of a parabola, and its graph is given in Figure 1.10.

First, let's consider $\tilde{x} = 0$. Clearly this is self-sustaining! There are no bacteria, so none can be born and none can die. Forever there will be no bacteria in the jar. Of course, with the slightest contamination ($x > 0$, but smaller than b/p) we see that $dx/dt = bx - px^2 > 0$ (look at the graph in Figure 1.10). Thus the number of bacteria will start to increase as soon as the jar has been contaminated. The equilibrium value of $\tilde{x} = 0$ is *unstable*; slight perturbations away from this equilibrium will destroy the equilibrium.

An example of what we call an unstable fixed point.

On the other hand, consider $\tilde{x} = b/p$. At this population level, bacteria are being born at a rate $b\tilde{x} = b(b/p) = b^2/p$ and are dying at a rate $p\tilde{x}^2 = p(b/p)^2 = b^2/p$, so birth and death rates are exactly in balance. But let us consider what happens in case the population x is slightly above or slightly below $\tilde{x} = b/p$. If x is slightly above b/p, we see that dx/dt is negative (look at the graph in Figure 1.10); hence, the number of bacteria will drop back toward b/p. Conversely, if x is slightly below b/p, we see that dx/dt is positive, so the population will tend to increase back toward b/p. We see that b/p is a *stable* equilibrium. Small perturbations away

An example of what we call a stable fixed point.

from $\tilde{x} = b/p$ will self-correct back to b/p.

Now, as promised, we present an analytic solution to equation (1.25), using the computer algebra package *Mathematica*:

```
DSolve[{x'[t]  ==  b x[t] - p x[t]^2, x[0]==x0},x[t],t]
```

```
                    b t
              b E
{{x[t] ->  ----------------}}
              b t      b
        -p + E      p + --
                        x0
```

We can rewrite this answer in conventional notation as

$$x(t) = \frac{x_0 b e^{bt}}{(b - px_0) + px_0 e^{bt}}. \tag{1.26}$$

Examine equation (1.26) and observe that if x_0 is any positive number, then $x(t) \to b/p$ as $t \to \infty$. This confirms what we previously discussed: The system gravitates toward the stable fixed point b/p.

1.2.8 Predator and prey

A classical model of an ecological system developed by Lotka and Volterra.

In the previous section we considered a simple model of a biological system involving only one species. Now we consider a more complex model involving two species. The first (the prey) we imagine is some herbivore (say, rabbits) whose population at time t is $r(t)$. The second (the predator) feeds on the prey; let's say they are wolves and their population at time t is $w(t)$.

Left on their own the rabbits will reproduce, well, like rabbits: $dr/dt = ar$ for some positive constant a. The wolves, on the other hand, will starve without rabbits to eat and their population will decline: $dw/dt = -bw$ for some $b > 0$.

However, when brought into the same environment, the wolves will eat the rabbits with the expected effects on each population: more wolves, fewer rabbits. Suppose there are w wolves and r rabbits. What is the likelihood that a wolf will catch a rabbit? The more wolves or the more rabbits there are, the more likely that a wolf will meet a rabbit. For this reason, we assume there is loss to the rabbit population proportional to rw and a gain to the wolf population, also proportional to rw. We write these

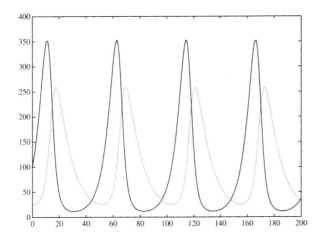

Figure 1.11. Variation in predator and prey populations over time. The solid curve is the prey (rabbit) and the dotted curve is the predator (wolf) population.

changes in the population as follows:

$$\frac{dr}{dt} = ar - grw \qquad (1.27)$$

$$\frac{dw}{dt} = -bw + hrw, \qquad (1.28)$$

where a, b, g, h are positive constants. We can write this pair of equations in the form $\mathbf{x}' = f(\mathbf{x})$, where $\mathbf{x} = \begin{bmatrix} r \\ w \end{bmatrix}$ and $f\begin{bmatrix} r \\ w \end{bmatrix} = \begin{bmatrix} ar - grw \\ -bw + hrw \end{bmatrix}$.

We can numerically approximate the solution to the system of differential equations (1.27) and (1.28), for example, using the ode45 routine of MATLAB. (See §B.1.2 on page 357.) For example, let

$$a = 0.2, \ b = 0.1, \ g = 0.002, \ h = 0.001, \ r_0 = 100, \ \text{and} \ w_0 = 25.$$

Looking at the results in Figure 1.11; we see that the rabbit and wolf populations fluctuate over time. You should notice that the population behavior is periodic—roughly every 50 time units the pattern repeats. When there are few wolves, the rabbit population soars; then, as food (i.e., rabbits) becomes more plentiful, the wolf population rises. But as the wolf population climbs, the wolves overhunt the rabbits, and the rabbit population falls.

While real populations of predators and prey have been observed to oscillate, the pattern is rarely this clean. This predator-prey model is too simple to capture the intricacies of an ecosystem.

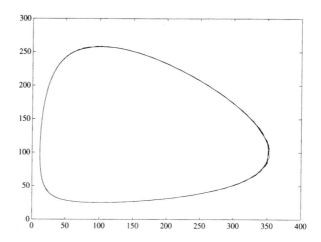

Figure 1.12. Phase diagram for predator-prey model. Horizontal axis is the number of prey (rabbits), and the vertical axis is the number of predators (wolves).

This causes food to become scarce for the wolves, and their numbers fall in turn. Finally, the number of wolves is low enough for the rabbit population to begin to recover, and the cycle begins again.

A phase diagram for the predator-prey system.

To fully appreciate the cyclic nature of this process, we can plot the rabbit and wolf population sizes on a single graph with the x-axis denoting the number of rabbits and the y-axis the number of wolves; see Figure 1.12. Each point on the curve represents a state of the system; the curve is called a *phase diagram*. Just as each point on the curve represents a snapshot of the system, the curve in its entirety represents the full story of how the system progresses. The state of the system is a point which travels counterclockwise around the curve. Trace your finge. counterclockwise around the curve and interpret what each population is doing at each point.

1.2.9 Newton's Method

Using dynamical systems to solve equations.

How much is $\sqrt[5]{9}$? For what number x does $\cos x = x$? Find a root of the equation $x^5 + x - 1 = 0$. Find a number x so that $x^x = 2$.

Each of these problems requires a numerical answer. How can we find it? Newton's method is a clever numerical procedure for solving equations of the form $g(x) = 0$.

Here is how it works. We begin with a guess, x_0, for a root to the

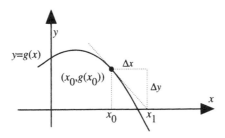

Figure 1.13. Newton's method for finding the root of an equation $g(x) = 0$.

equation $g(x) = 0$. If we are incredibly lucky, $g(x_0) = 0$; otherwise, we use the following procedure to find (we hope) a better guess x_1. This method is illustrated in Figure 1.13. We know x_0, so we compute $y_0 = g(x_0)$. We would like to find the magic number $\tilde{x}$ so that $g(\tilde{x}) = 0$. *If the curve were a straight line*, then, since we know the slope of the curve at the point $(x_0, g(x_0))$ is $g'(x_0)$, we could find exactly where the curve $y = g(x)$ crossed the x-axis. Regrettably, the curve is not straight, but perhaps it is not too far off. We pretend, for the moment, the curve is straight and we seek the point (x_1, y_1) where $y_1 = g(x_1) = 0$. Again, if the curve were straight, we'd have

$$g'(x_0) = \frac{\Delta y}{\Delta x} = \frac{y_1 - y_0}{x_1 - x_0} = \frac{0 - g(x_0)}{x_1 - x_0}.$$

Solving this equation for x_1 we get

$$x_1 = x_0 - \frac{g(x_0)}{g'(x_0)}.$$

The hope is that although x_1 is not really going to be a root of the equation $g(x) = 0$, it is closer to being a root than x_0 was. (Of course, for this method to have any chance of working, we need g to be differentiable and $g'(x_0) \neq 0$.)

Now, if x_1 is a better guess than x_0, how can we gain even more accuracy? Simple! By using exactly the same procedure, this time starting with x_1. You should smell a discrete time dynamical system. Here it is:

$$x(k + 1) = x(k) - \frac{g(x(k))}{g'(x(k))}. \tag{1.29}$$

In other words, if we let $f(x) = x - g(x)/g'(x)$, then we iterate f starting with x_0 with the hope that $f^k(x)$ converges to a root of the equation $g(x) = 0$.

In §3.4 (page 146) we show that if x_0 is a reasonable guess, then this procedure converges quickly to a root of $g(x)$. For now, let's do an example.

Suppose we wish to compute $\sqrt[5]{9}$. In other words, we want a root of the equation $x^5 - 9 = 0$, i.e., let $g(x) = x^5 - 9$. What is a reasonable first guess? Well $g(1) = -9 < 0$, and $g(2) = 32 > 0$, so there must be a root between 1 and 2. Let's start with $x(0) = 1.5$. Our next guess is

$$x(1) = x(0) - \frac{g(x(0))}{g'(x(0))} = x(0) - \frac{x(0)^5 - 9}{5x(0)^4} = 1.5 - \frac{1.5^5 - 9}{5 \times 1.5^4} = 1.55555\ldots$$

Repeating this procedure, we compile the following results:

$$x(0) = 1.5$$
$$x(1) = 1.55555555\ldots$$
$$x(2) = 1.55186323\ldots$$
$$x(3) = 1.55184557\ldots$$

Amazingly, $x(3)$ is the *correct* value of $\sqrt[5]{9}$ to the number of digits shown! Further iterations of Newton's method changes only less significant digits.

Although Newton's method does not converge this rapidly for all problems, it is still a very quick and powerful method.

1.2.10   Euler's method

Using the discrete to approximate the continuous.

Consider the differential equation

$$\frac{dy}{dx} = x + y. \tag{1.30}$$

In other words, we seek a function $f(x)$ (also called y) for which $f'(x)$ equals $x + f(x)$. Courses on differential equations give a variety of tools for finding such functions. Computer algebra systems such as *Mathematica* or *Maple* can actually solve this equation analytically. Here is how *Maple* solves it:

```
dsolve( diff(y(x),x) = y(x) + x, y(x));
```

```
y(x) = - x - 1 + exp(x) _C1
```

In common notation, the solution is $y = ae^x - x - 1$, where a is any constant. To see that this is correct, just observe that

$$y' = (ae^x - x - 1)' = ae^x - 1 = (ae^x - x - 1) + x = y + x.$$

If we are also told that $y(0) = 1$, then, since $y = ae^x - x - 1$, we can solve for a:

$$1 = y(0) = ae^0 - 0 - 1 \quad \Rightarrow \quad a = 2.$$

Thus if the problem is

$$y' = x + y; \qquad y(0) = 1, \tag{1.31}$$

then the answer is

$$y = 2e^x - x - 1. \tag{1.32}$$

If, ultimately, we just want to know the value of $y(1)$, we simply plug in 1 for x and get $y(1) = 2e - 2 \approx 3.4366$.

The differential equation (1.31) is easy to solve either by a computer or by the human who has had a course in differential equations. It is not hard, however, to write a differential equation which can stump the best human or computer differential equation solver. In such cases, we often rely on numerical methods (see §B.1.2).

What if I don't know how to solve differential equations?

Euler's method is a simple method for finding numerical solutions to differential equations. It is simple but, regrettably, not very accurate.

Here is how Euler's method works. We are given a differential equation of the form

$$\frac{dy}{dx} = f(x, y) \qquad \text{with} \qquad y(0) = y_0 \tag{1.33}$$

and we want to know the value of, say, $y(1)$. [The initial condition we give need not be at $x = 0$, and the value we seek need not be at $x = 1$; these choices were made to simplify the exposition.]

We divide the interval between 0 and 1 into n equal-size pieces, where n is a large number (the larger n is, the more accurate the answer, but it requires more computations). We are given that $y(0) = y_0$, and we use this to estimate $y(1/n)$. *If the function y were a straight line, then*

$$y(1/n) = y(0) + y(1/n) - y(0)$$
$$= y(0) + \frac{y(1/n) - y(0)}{(1/n) - 0} \cdot \frac{1}{n}$$
$$= y(0) + (1/n)\frac{\Delta y}{\Delta x}$$
$$= y(0) + (1/n)y'(0)$$
$$= y(0) + (1/n)f(0, y_0).$$

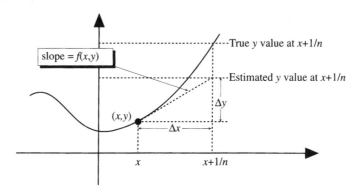

Figure 1.14. Taking a single step of the Euler method for numerically approximating the solution to a differential equation.

Of course, $\Delta y / \Delta x$ is only approximately $dy/dx = f'(0)$; however, we know how to compute $f'(0)$ but not $\Delta y / \Delta x$.

How do we find $y(2/n)$? By the same method:

$$y(2/n) = y(1/n) + (1/n) f(1/n, y(1/n)).$$

Because we don't really know $y(1/n)$, we use the approximation from before. In this manner we compute $y(3/n), y(4/n), \ldots, y(n/n) = y(1)$.

We can express this as a discrete time dynamical system. Let $\mathbf{z}$ be our state variable with

$$\mathbf{z}(k) = \left[\begin{array}{c} z_1(k) \\ z_2(k) \end{array} \right] = \left[\begin{array}{c} k/n \\ y(k/n) \end{array} \right].$$

Then the system is

$$\mathbf{z}(k+1) = \left[\begin{array}{c} z_1(k+1) \\ z_2(k+1) \end{array} \right] = \left[\begin{array}{c} z_1(k) + 1/n \\ z_2(k) + (1/n) f(z_1(k), z_2(k)) \end{array} \right]. \quad (1.34)$$

Let's look at our example, equation (1.31), which we repeat here:

$$y' = x + y; \qquad y(0) = 1.$$

We want to compute $y(1)$. [We know that $y = 2e^x - x - 1$ and so $y(1) = 2e - 2 \approx 3.4366$, but we'll ignore that for the moment.] Instead, we'll use Euler's method with

$$f(x, y) = x + y, \quad y(0) = 1, \quad \text{and} \quad n = 10.$$

Thus from $y(0) = 1$, we have

$$y(0.1) = y(0) + (1/n)f(0, 1) = 1 + 0.1 \times (0 + 1) = 1.1$$

$$y(0.2) = y(0.1) + (1/n)f(0.1, 1.1) = 1.1 + 0.1 \times (0.1 + 1.1) = 1.22$$

$$y(0.3) = y(0.2) + (1/n)f(0.2, 1.22) = 1.22 + 0.1 \times (0.2 + 1.22) = 1.362$$

$$\vdots$$

Continuing in this fashion, we obtain the following list of values:

x	y
0.0	1.0000
0.1	1.1000
0.2	1.2200
0.3	1.3620
0.4	1.5282
0.5	1.7210
0.6	1.9431
0.7	2.1974
0.8	2.4872
0.9	2.8159
1.0	3.1875

Thus Euler's method (with step size 0.1) computes $y(1) \approx 3.1875$, when, in fact, $y(1) = 3.4366$—a relative error of over 7%, which is pretty bad.

Euler's method is not very accurate. Better methods are available. See §B.1.2.

Figure 1.15 shows the curve (actually only 11 points joined by line segments) we found and the actual solution (shown as a dotted curve).

If we decrease the step size to 0.01, then Euler's method predicts $y(1) \approx 3.4096$, which has relative error under 1%. With 1000 steps of size 0.001, we arrive at $y(1) \approx 3.4338$, which is pretty good, but we had to do *a lot* of computation. By contrast, sophisticated routines for computing numerical solutions to differential equations (such as MATLAB's ode45) attain greater accuracy with much less computation.

1.2.11 "Random" number generation

Random number generation is a feature of many computer programming languages and environments. MATLAB, *Mathematica*, adn the like all have ways to provide users with random values.

Totally deterministic random numbers!

How does a computer make a random number? Interestingly, the numbers are not random at all! They are produced by a deterministic procedure which we will recognize as a discrete time dynamical system.

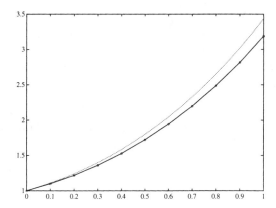

Figure 1.15. Ten steps of Euler's method (solid) and the true solution (dotted) to differential equation (1.31).

The Unix operating system provides the C language computer programmer with a function called `lrand48` for producing random integers in the set $\{0, 1, 2, \ldots, 2^{48} - 1\}$. The manual page for `lrand48` explains how these "random" values are produced.

If the last produced random number was x, then the next random number will be

$$(ax + b) \bmod m,$$

where a, b, and m are positive integers. The value of m is 2^{48}, the value of a is 11 (eleven), and the value of b is given in base-16 as 5DEECE66D and in base-8 as 273673163155.

Thus if $x(0) = x_0$ is the initial "random" number, we have that

$$x(k + 1) = ax(k) + b \bmod m.$$

[Note: $a \bmod b$ means the remainder in the division problem $a \div b$. For example, 10 mod 4 = 2, 30 mod 8 = 6 and 12 mod 3 = 0.]

Problems for §1.2

♦1. Explain the comments at the beginning of this chapter: "A dynamical system is a function with an attitude. A dynamical system is doing the same thing over and over again. A dynamical system is always knowing what you are going to do next."

◆2.   Play with your scientific calculator. Pick a button (such as $\boxed{\sin x}$ or $\boxed{e^x}$)
       and see what happens if you press it repeatedly. Try combinations, such as
       $\boxed{\sin x}$ $\boxed{\cos x}$ $\boxed{\sin x}$ $\boxed{\cos x}$ ....

◆3.   Weird springs. Suppose you construct a mass-and-spring system (as in
       §1.2.1), but your spring behaves rather strangely. Instead of exerting a force
       proportional to displacement ($F = -kx$), it exerts a force proportional to
       the *cube* of its displacement ($F = -kx^3$). What is the new dynamical
       system?

◆4.   Weird springs continued. Use numerical differential equation software to
       show how the system from the previous problem behaves. Plot graphs.

◆5.   Checking/savings. Suppose a person has three accounts: checking, savings
       and retirement. Each month, the checking account is credited with a pay
       check. Each month the person pays rent, utilities, and other expenses from
       the checking account and makes a deposit into savings and into retirement
       (assume all these amounts are the same from month to month). The checking
       account has a monthly fee and earns no interest. The savings and retirement
       accounts earn interest. Furthermore, the person has a car loan (paid from
       checking).

       Create a dynamical system to model this situation.

◆6.   Discuss the effects of changing the various parameters (population growth,
       depreciation, savings rate) on steady-state per capita capitalization in the
       economic growth model of §1.2.5 on page 17.

◆7.   Cobb-Douglas output functions. In the economic growth model (§1.2.5)
       we postulated that output Y depends on capital K and labor N. We set
       $Y = A\sqrt{KN}$, where A is a constant.

       A more general form is $Y = AK^a N^{1-a}$ where a is a constant between 0
       and 1.

       Under this new output function, does output still double if we double labor
       and capital?

       What is the new dynamical system in this case? You should derive a new
       version of equation (1.22).

◆8.   In the microbe model (§1.2.7 on page 22) we assumed that the birth rate b
       did not change with the population x, but the death rate was proportional
       to x. Devise a new model in which the birth rate decreases as population
       increases (say, because of less favorable environmental conditions).

       Does your model feature a stable fixed point?

◆9.   Ecosystem. Create a dynamical system to model the following ecosystem.
       There are four types of species: (1) scavengers, (2) herbivores, (3) car-
       nivores, and (4) top-level carnivores. The herbivores eat plants (which
       you may assume are always in abundant supply), the carnivores eat only

the herbivores, and the top-level carnivores eat both the herbivore and the carnivores. The scavengers eat dead carnivores (both kinds).

◆10. Use numerical differential equation software to study how the system you created in the previous problem behaves. Plot graphs. Do you see cyclic behavior? Tweak your parameters and see what you can learn.

◆11. Use Newton's method to solve the following equations:

  (a) $\sin x = \cos x$.

  (b) $e^x = \tan x$.

  (c) $x^3 - 5 = 0$.

  (d) $x^5 - x + 4 = 0$.

◆12. Use Euler's method to estimate $y(1)$ for each of the following differential equations:

  (a) $y' - y = 2$, $y(0) = \frac{1}{2}$.

  (b) $y' + e^t = y$, $y(0) = 1$.

  (c) $y' + y^2 = 3$, $y(0) = -2$.

◆13. **Reduction of Order.** The famous Fibonacci sequence is

$$1, 1, 2, 3, 5, 8, 13, \cdots.$$

Each term is the sum of the previous two terms, i.e.,

$$x(k+2) = x(k+1) + x(k)$$

where $x(k)$ is the kth Fibonacci number. The above relation is not in the form we require for a (discrete time) dynamical system. Indeed, knowing just the single number $x(k)$ does not tell us what the next "state" $x(k+1)$ will be. To convert this recurrence into a dynamical system, let the state vector be $\begin{bmatrix} x(k) \\ y(k) \end{bmatrix}$, where $y(k) = x(k+1)$. With this new notation, we have

$$x(k+1) = y(k)$$
$$y(k+1) = x(k) + y(k)$$

and observe that we now have a proper dynamical system. [What should the starting vector be?]

Use the above idea (adding extra state variables) to convert each of the following into proper dynamical systems.

  (a) $x(k+2) = x(k)x(k+1)$.

  (b) $x(k+2) = y(k)x(k+1)$ and $y(k+2) = x(k) + y(k) + y(k+1)$.

  (c) $x'' = 3x' - 2x + 2$.

1.3   What we want; what we can get

We have introduced the ideas of discrete and continuous time dynamical systems and we hope it is clear that the notion of a dynamical system can be useful in modeling many different kinds of phenomena. Once we have created a model, we would like to use it to make predictions. Given a dynamical system either of the discrete form $\mathbf{x}(k + 1) = f(\mathbf{x}(k))$ or of the continuous sort $\mathbf{x}' = f(\mathbf{x})$, and an initial value $\mathbf{x}_0$, we would very much like to know the value of $\mathbf{x}(k)$ [or, $\mathbf{x}(t)$] for all values of k [or t]. In some rare instances, this is possible. For example, if f is a linear function we develop methods in Chapter 2 for computing all future states $\mathbf{x}$.

Unfortunately, it is all too common that the dynamical system in which we are interested does not yield an analytic solution. What then? One option is numerical methods. However, we can also determine the qualitative nature of the solution. For example, in the predator-prey model, we saw that our system goes into a periodic behavior. In the microbes example (§1.2.7 on page 22), we didn't need the analytic solution to find a *stable fixed point* to which the system is led. In later chapters we explore the notion of fixed points and of periodic behavior of these systems. And we encounter behaviors besides periodicity: A system might blow up (its state vector goes to infinity with time) or it might become *chaotic*, and we'll see what that means!

Chapter 2

Linear Systems

In Chapter 1 we introduced discrete $[\mathbf{x}(k + 1) = f(\mathbf{x}(k))]$ and continuous $[\mathbf{x}' = f(\mathbf{x})]$ time dynamical systems. The function $f: \mathbf{R}^n \to \mathbf{R}^n$ might be quite simple or terribly complicated.

In this chapter we study dynamical systems in which the function f is particularly nice: we assume f is linear. When f is a function of one variable (i.e., $f: \mathbf{R} \to \mathbf{R}$), we mean that $f(x) = ax + b$, where a and b are constants. When f is a function of several variables (i.e., $f: \mathbf{R}^n \to \mathbf{R}^n$), we mean that $f(\mathbf{x}) = A\mathbf{x} + \mathbf{b}$, where A is an $n \times n$ matrix and $\mathbf{b}$ is a fixed n-vector.

We call functions of the form $f(x) = ax + b$ *linear* because their graphs are lines. This is the usual language in the engineering world. Mathematicians call these functions *affine*.

To gain intuition, we begin with the case where f is a function of just one variable ($f(x) = ax + b$) and then proceed to the multivariable case.

2.1 One dimension

We begin by considering both discrete and continuous time dynamical systems in which $f(x) = ax + b$:

$$x(k + 1) = ax(k) + b \qquad \text{and} \qquad x' = ax + b.$$

2.1.1 Discrete time

First, let us consider discrete time systems. The system has the form

$$x(k + 1) = ax(k) + b; \qquad x(0) = x_0.$$

We discuss this case first analytically (i.e., by equations) and then geometrically (with graphs).

Analysis

Suppose first that $b = 0$, i.e., $x(k + 1) = ax(k)$. It is very clear that for any k we have simply that $x(k) = a^k x_0$. (See problem 5 on page 7.)

If $|a| < 1$, then $a^k \to 0$ as $k \to \infty$ and so $x(k) \to 0$.

If $|a| > 1$, then a^k explodes as $k \to \infty$. Thus unless $x_0 = 0$, we have $|x(k)| \to \infty$.

Finally, suppose $|a| = 1$. If $a = 1$ then we have just that $x(0) = x(1) = x(2) = x(3) = \cdots$, i.e., $x(k) = x_0$ for all eternity. If $a = -1$, then $x(0) = -x(1) = x(2) = -x(3) = \cdots$, that is, $x(k)$ alternates between x_0 and $-x_0$ forever.

Now we consider the full case of $x(k + 1) = ax(k) + b$. We begin by working out the first few values:

$$x(0) = x_0,$$

$$x(1) = ax(0) + b = ax_0 + b,$$

$$x(2) = ax(1) + b = a(ax_0 + b) + b = a^2 x_0 + ab + b,$$

$$x(3) = ax(2) + b = a(a^2 x_0 + ab + b) + b = a^3 x_0 + a^2 b + ab + b,$$

$$x(4) = ax(0) + b = a(a^3 x_0 + a^2 b + ab + b) + b =$$

$$= a^4 x_0 + a^3 b + a^2 b + ab + b.$$

Do you see the pattern? We have

$$x(k) = a^k x_0 + (a^{k-1} + a^{k-2} + \cdots + a + 1)b. \qquad (2.1)$$

We can simplify equation (2.1) by noticing that

$$a^{k-1} + a^{k-2} + \cdots + a + 1$$

is a geometric series which equals

$$\frac{a^k - 1}{a - 1}$$

provided $a \neq 1$. If $a = 1$, the series $a^{k-1} + a^{k-2} + \cdots + a + 1$ simply equals k. We summarize this as

$$x(k) = \begin{cases} a^k x_0 + \left(\frac{a^k - 1}{a - 1}\right) b & \text{when } a \neq 1, \text{ and} \\ x_0 + kb & \text{when } a = 1. \end{cases}$$

Let's examine this answer closely for the cases $|a| < 1$, $|a| > 1$, and $|a| = 1$.

If $|a| < 1$, then $a^k \to 0$ as $k \to \infty$, and so

$$x(k) \to \frac{b}{1-a}$$

regardless of the value of x_0. Notice that this special number $\tilde{x} = b/(1-a)$ has the property that $f(\tilde{x}) = \tilde{x}$; here are the calculations:

$$f(\tilde{x}) = a\tilde{x} + b$$

$$= a \left[\frac{b}{1-a} \right] + b$$

$$= \frac{ab + (1-a)b}{1-a}$$

$$= \frac{b}{1-a}$$

$$= \tilde{x}.$$

That is, $\tilde{x}$ is a *fixed point* of the function f, meaning that $f(\tilde{x}) = \tilde{x}$. Further, we call $\tilde{x}$ an *attractive* or *stable* fixed point of the dynamical system because the system is attracted to this point.

Next, suppose that $|a| > 1$. In this case $a^k \to \infty$ as k gets large. To see how this affects $x(k)$ we collect the a^k terms:

$$x(k) = a^k x_0 + \left(\frac{a^k - 1}{a - 1} \right) b = a^k \left(x_0 - \frac{b}{1-a} \right) + \frac{b}{1-a}.$$

Now, if $x_0 \neq b/(1-a)$ (our $\tilde{x}$ from before!) then $|x(k)| \to \infty$ as $k \to \infty$. However, if $x_0 = \tilde{x}$, then $x(k) = \tilde{x}$ for all time.

Finally, we consider $|a| = 1$. First, if $a = 1$, then $x(k) = x_0 + kb$. So if $b \neq 0$, then $|x(k)| \to \infty$; otherwise ($b = 0$) $x(k)$ is stuck at x_0 regardless of the value of x_0. Second, if $a = -1$, then observe that

$$x(0) = x_0$$

$$x(1) = -x_0 + b$$

$$x(2) = -(-x_0 + b) + b = x_0$$

$$x(3) = -x_0 + b$$

$$x(4) = x_0$$

$$\vdots$$

Margin notes:

When $|a| < 1$, $x(k)$ tends to a limit.

When $|a| < 1$, $x(k)$ usually tends to infinity.

When $|a| = 1$, many behaviors are possible.

Discrete: $x(k+1) = ax(k) + b$ with $x(0) = x_0$			
Conditions on			Behavior of
a	b	x_0	$x(k)$ as $k \to \infty$
$|a| < 1$	—	—	$\to \frac{b}{1-a}$
$|a| > 1$	—	$\neq \frac{b}{1-a}$	blows up
		$= \frac{b}{1-a}$	fixed at $\frac{b}{1-a}$
$a = 1$	$b \neq 0$	—	blows up
	$b = 0$	—	fixed at x_0
$a = -1$	—	$\neq b/2$	oscillates: $x_0, b - x_0, \ldots$
	—	$= b/2$	fixed at $b/2$

Table 2.1. The possible behaviors of one-dimensional discrete time linear systems.

Thus we see that $x(k)$ oscillates between two values, x_0 and $b - x_0$. But if $x_0 = b - x_0$, i.e., $x_0 = b/2 = b/(1 - (-1)) = \tilde{x}$, then $x(k)$ is stuck at the fixed point $\tilde{x}$.

Table 2.1 summarizes the behavior of the system

$$x(k+1) = ax(k) + b; \quad x(0) = x_0$$

given various conditions on a, b, and x_0.

Geometry

We can understand the behavior of linear systems by plotting graphs; this is known as *graphical analysis*.

Let us revisit systems of the form $x(k+1) = ax(k) + b$ from a geometric point of view. Graphs will make clear why $|a| < 1$ causes the iterations to converge to $\tilde{x}$, while $|a| > 1$ causes the iterates to explode.

Figures 2.1 through 2.6 illustrate most of the cases in Table 2.1. In each case we have plotted the function $y = f(x) = ax + b$, varying the value of a. On these graphs we have drawn the line $y = x$ which enables us to draw pictures for iterating f. Examine one of the figures (say Figure 2.1) as we explain. Choose a starting point $x(0)$ on the x-axis. Draw a line straight up to the line $y = f(x)$. The y-coordinate of this point is the next number to which you wish to apply f, so you want to find the point $y = f(x(0))$ on the x-axis. To do this, draw a line horizontally from $(x(0), f(x(0)))$ to the line $y = x$; you have found the point $(f(x(0)), f(x(0))) = (x(1), x(1))$. If you drop a line down to the x-axis, this point is $x(1)$.

Now the cycle repeats. Given $x(1)$ on the x-axis, draw a line up to the graph of $y = f(x)$—this is the point $(x(1), f(x(1))) = (x(1), x(2))$. Now

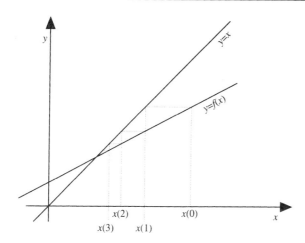

Figure 2.1. Iterating $f(x) = ax + b$ with $0 < a < 1$.

draw a line horizontally to the line $y = x$; this locates the point $(x(2), x(2))$. Drop a line from this point down to the x-axis, and voilà!—you are ready to repeat.

Let's look at the figures in detail. You should study each carefully and redraw them for yourself to be sure you understand what is going on.

Figure 2.1 illustrates what happens when we iterate $y = f(x) = ax + b$ with $0 < a < 1$. We chose $x(0)$ larger than $\tilde{x} = b/(1 - a)$. With each iteration it is easy to see that the values $x(0)$, $x(1)$, $x(2)$, etc. get smaller and step inexorably toward $\tilde{x}$. If we had chosen $x(0) < \tilde{x}$, we would see the successive iterates increasing, marching steadily to $\tilde{x}$; you should draw in this part of the diagram.

Notice that the lines $y = x$ and $y = ax + b$ intersect at the point $(\tilde{x}, \tilde{x})$.

Next, consider Figure 2.2. In this diagram we have plotted the graph of $y = f(x) = ax + b$ with $-1 < a < 0$. The line slopes downward, but not very steeply (at an angle less than $45°$ to the horizontal). We start with $x(0)$ a good bit to the right of $\tilde{x}$. Observe that $x(1)$ is to the *left* of $\tilde{x}$, but not nearly as far. Successive iterations take us to alternate sides of $\tilde{x}$, but getting closer and closer—and ultimately converging—to $\tilde{x}$.

Now look at Figure 2.3. In this case we have $a > 1$, so the line $y = f(x) = ax + b$ is sloped steeply upward. We start $x(0)$ just slightly greater than $\tilde{x}$. Observe that $x(1)$ is now to the right of $x(0)$, and then $x(2)$ is farther right, etc. It should be clear that the successive iterates are marching off to $+\infty$. If we had taken $x(0)$ just to the left of $\tilde{x}$ (i.e.,

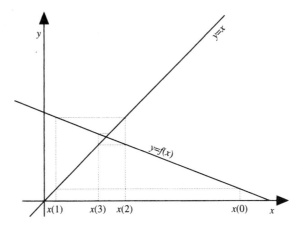

Figure 2.2. Iterating $f(x) = ax + b$ with $-1 < a < 0$.

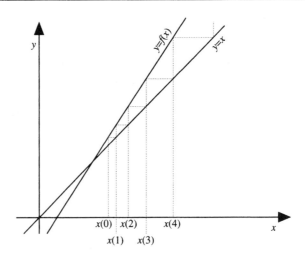

Figure 2.3. Iterating $f(x) = ax + b$ with $a > 1$.

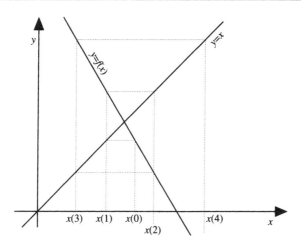

Figure 2.4. Iterating $f(x) = ax + b$ with $a < -1$.

$x(0) < \tilde{x}$), we would see the iterates moving faster and faster to the left and heading toward $-\infty$. Draw this portion for yourself.

In Figure 2.4 we have $a < -1$; hence the line $y = f(x) = ax + b$ is sloped steeply downward. We begin with $x(0)$ just to the right of $\tilde{x}$. Observe that $x(1) < \tilde{x}$ but at a greater distance from $\tilde{x}$ than $x(0)$. Next, $x(2)$ is to the right of $\tilde{x}$, $x(3)$ is to the left, etc., with each at increasing distance from $\tilde{x}$ and diverging to ∞.

Figure 2.5 illustrates what happens when we iterate $f(x) = 1x + b$ with $b \neq 0$. (In the diagram we have chosen $b > 0$; you should sketch a graph with $b < 0$ to compare.) We see that each iteration moves the point $x(k)$ a step (of length b) to the right and heads to $+\infty$.

Finally, Figure 2.6 considers the case $f(x) = -1x + b$. The starting value $x(0)$ is taken to be to the left of $\tilde{x}$. Next, $x(1)$ is to the right of $\tilde{x}$ and then $x(2)$ is back at exactly the same location as $x(0)$. In this manner, $x(0), x(2), x(4)$, etc. all have the same value (denoted by "x(even)" in the figure), and, likewise, $x(1) = x(3) = x(5) = \cdots$ (denoted by "x(odd)").

The various figures illustrate the possible behaviors of discrete time linear systems in one variable. As you examine each case, be sure to compare the geometric and the analytic approaches.

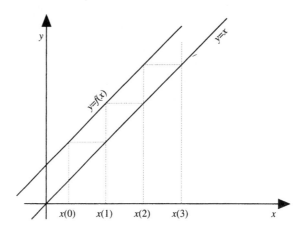

Figure 2.5. Iterating $f(x) = ax + b$ with $a = 1$ and $b \neq 0$.

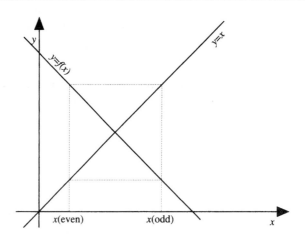

Figure 2.6. Iterating $f(x) = ax + b$ with $a = -1$.

2.1.2 Continuous time

Now we turn to *continuous time* linear systems in one variable, that is, systems of the form

$$x' = ax + b; \qquad x(0) = x_0.$$

As in the discrete time case, let's begin with the special case of $b = 0$, i.e., $x' = ax$. (See problem 7 on page 7.) We write this as:

$$\frac{x'}{x} = a. \tag{2.2}$$

Recall that x (as well as x') is a function of t. We integrate both sides of equation (2.2):

$$\int \frac{x'}{x} \, dt = \int a \, dt$$

to get

$$\log x = at + C \tag{2.3}$$

where C is a constant of integration.

We write $\log x$ to stand for base e logarithm. Some write $\ln x$ for this.

We undo the logarithm in equation (2.3) by exponentiating both sides to give

$$x = e^{at+C} = e^{at} e^C. \tag{2.4}$$

Finally, we need to figure out what e^C should be. We know that $x(0) = x_0$. Substituting $t = 0$ into equation (2.4), we have $x(0) = e^0 e^C$, and therefore the solution to the system $x' = ax$ with $x(0) = x_0$ is

$$x = e^{at} x_0. \tag{2.5}$$

This analytic solution enables us to find the long-term behavior of $x' = ax$. When $a < 0$, the term $e^{at} \to 0$ as $t \to \infty$, and therefore $x(t) \to 0$ *regardless* of the initial value x_0. When $a > 0$, the term $e^{at} \to \infty$ and so, unless $x_0 = 0$, we see that $|x(t)| \to \infty$. Finally, if $a = 0$, we see that $x(t)$ is stuck at x_0 for all time.

The behavior of the continuous system depends on the sign of a.

Now we are ready to discuss the full case $x' = ax + b$ (with nonzero b allowed). We will observe that the general nature of the solution is quite similar to the $b = 0$ case. Our method is to reduce the problem we don't know how to solve ($x' = ax + b$) to one we do know how to solve ($x' = ax$) by a sneaky substitution. (The substitution requires $a \neq 0$; we handle the case $a = 0$ separately below.) Let

$$u = x + \frac{b}{a}.$$

Thus $u' = x'$ and $x = u - b/a$. Observe that $u(0) = x_0 + b/a$. We replace the x's in $x' = ax + b$ to get

$$u' = a\left(u - \frac{b}{a}\right) + b = au.$$

Since $u' = au$, we have $u = e^{at}u(0)$. Now, recasting this in terms of x we arrive at

$$x + \frac{b}{a} = e^{at}\left(x_0 + \frac{b}{a}\right),$$

which rearranges to

$$x = e^{at}\left(x_0 + \frac{b}{a}\right) - \frac{b}{a}. \tag{2.6}$$

Let's see what this solution predicts about the behavior of the system $x' = ax + b$.

When a is negative, the system tends to a stable fixed point.

If $a < 0$, then $e^{at} \to 0$ as $t \to \infty$. Thus $x(t) \to -b/a$ regardless of the value of x_0. Indeed, we call $\tilde{x} = -b/a$ a *stable fixed point* of the system. It is called *fixed* because if the system is in state $\tilde{x}$, then it will be there for all time. It is *stable* because the system gravitates toward that value.

When a is positive, the system explodes.

Now let's consider $a > 0$. In this case $e^{at} \to \infty$ as $t \to \infty$. Examining equation (2.6) we notice that if $x_0 \neq \tilde{x} = -b/a$, then $x(t)$ explodes as $t \to \infty$. However, if $x_0 = -b/a$, then the system is stuck at $-b/a$. The value $\tilde{x} = -b/a$ is an *unstable fixed point* of the system. It is *unstable* because even if we start very near (but not at) $\tilde{x}$, the system moves far away from $\tilde{x}$.

Finally, we consider $a = 0$; equation (2.6) does not apply because we assumed $a \neq 0$ to solve the differential equation $x' = ax + b$. But the system $x' = b$ is easy to solve. Integrating both sides we see that $x = bt + C$, where C is a constant. Since $x(0)$ needs to equal x_0, we note that $C = x_0$. Thus we have $x = bt + x_0$. If $b = 0$, we note (as we discussed before) that the system is stuck at x_0. Otherwise ($b \neq 0$) we see that $x(t)$ blows up regardless of the value of x_0.

Table 2.2 summarizes these results for one-dimensional continuous time linear systems.

Geometry

Graphical analysis of continuous time systems.

Next we consider the system $x' = ax + b$ from a geometric viewpoint. We consider the three cases ($a < 0$, $a > 0$, and $a = 0$) in turn.

Figure 2.7 shows a graph of the function $y = ax + b$ with $a < 0$. The

Continuous: $x'(t) = ax(t) + b$ with $x(0) = x_0$			
Conditions on			Behavior of
a	b	x_0	$x(t)$ as $t \to \infty$
$a < 0$	—	—	$\to -\frac{b}{a}$
$a > 0$	—	$\neq -\frac{b}{a}$	blows up
	—	$= -\frac{b}{a}$	fixed at $-\frac{b}{a}$
$a = 0$	$b \neq 0$	—	blows up
	$b = 0$	—	fixed at x_0

Table 2.2. The possible behaviors of one-dimensional continuous time linear systems.

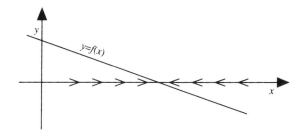

Figure 2.7. The dynamical system $x' = f(x) = ax + b$ with $a < 0$.

line $y = f(x)$ crosses the x-axis at $\tilde{x} = -b/a$. To the left of $\tilde{x}$ observe that $f(x) > 0$, and since $x' = f(x)$, that means that x is *increasing* in this region. Thus if $x < \tilde{x}$, then, as time progresses, x heads toward $\tilde{x}$; this is indicated by the arrows on the x-axis. Conversely, if $x > \tilde{x}$, then $f(x) < 0$, and since $x' = f(x)$, x is *decreasing* in this region; hence, as time progresses, x heads toward $\tilde{x}$. Thus we see that regardless of where we begin (i.e., for any x_0) the system gravitates toward $\tilde{x}$.

Now examine Figure 2.8 to see the opposite situation. In this example we consider the system $x' = ax + b$ with $a > 0$. When $x > \tilde{x}$, we observe that $f(x) > 0$ and so x is increasing toward $+\infty$. However, if $x < \tilde{x}$, then $f(x) < 0$ and therefore x is decreasing toward $-\infty$. The arrows on the x-axis illustrate this. In the special case that x is exactly $\tilde{x}$ we have

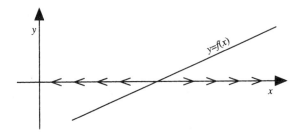

Figure 2.8. The dynamical system $x' = f(x) = ax + b$ with $a > 0$.

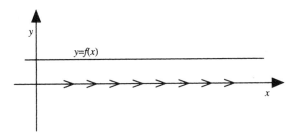

Figure 2.9. The dynamical system $x' = f(x) = ax + b$ with $a = 0$ and $b > 0$.

$f(x) = f(\tilde{x}) = f(-b/a) = 0$, and so x is unchanging.

Finally, consider Figure 2.9. In this case $f(x) = b$ (so $a = 0$), and we have drawn the case for $b > 0$. At every value of x we observe that x will be increasing (at rate b). Thus $x \to +\infty$ as time progresses. When $b < 0$ (please draw or at least imagine what the graph would look like), we see $x \to -\infty$. And, if b is exactly 0, then the system never goes anywhere, since $dx/dt = 0$.

2.1.3 Summary

We have covered all possible behaviors of one-dimensional linear systems:

$$
\begin{aligned}
x(k+1) &= ax(k) + b \quad \text{discrete time, and} \\
x'(t) &= ax(t) + b \quad \text{continuous time.}
\end{aligned}
$$

In the discrete case, when $|a| < 1$, the system gravitates toward the value $\tilde{x} = b/(1-a)$. When $|a| > 1$, the system is repelled from this same value, $\tilde{x}$. The case $|a| = 1$ is marginal, and the behavior depends on b.

In the continuous case, when $a < 0$, the system goes to $\tilde{x} = -b/a$. When $a > 0$, the system diverges away from $\tilde{x}$. The case $a = 0$ is marginal, and the behavior depends on b.

Problems for §2.1

◆1.   Find an exact formula for $x(k)$, where $x(k+1) = ax(k)+b$, $x(0) = x_0 = c$, and a, b, and c have the following values:

  (a) $a = 1, b = 1, c = 1$.
  (b) $a = 1, b = 0, c = 2$.
  (c) $a = \frac{3}{2}, b = -1, c = 0$.
  (d) $a = -1, b = 1, c = 4$.
  (e) $a = -\frac{1}{2}, b = 1, c = \frac{3}{2}$.

◆2.   For each of the discrete time systems in the previous problem, determine whether or not $|x(k)| \to \infty$. Determine if the system has a fixed point and whether or not the system is approaching that fixed point.

◆3.   Find the exact value of $x(t)$ where $x' = ax + b$, $x(0) = x_0 = c$, and a, b, and c have the following values:

  (a) $a = 1, b = 0, c = 1$.
  (b) $a = 0, b = 1, c = 0$.
  (c) $a = 0, b = 0, c = 1$.
  (d) $a = -1, b = 1, c = 2$.
  (e) $a = 2, b = 3, c = 0$.

◆4.   For each of the continuous time systems in the previous problem, determine whether or not $|x(t)| \to \infty$. Determine if the system has a fixed point and whether or not the system is approaching that fixed point.

◆5.   Verify that equation (2.6) on page 46 is the solution to the system $x' = ax+b$ with $x(0) = x_0$ by direct substitution.

◆6.   Suppose $f(x) = ax + b$ and $g(x) = cx + d$. Let $h(x) = f[g(x)]$ (i.e., $h = f \circ g$). Suppose we iterate h, i.e., we compute $h^k(x)$. Find conditions on a, b, c, and d under which $|h^k(x)|$ either stays bounded or else, goes to infinity. Determine the fixed point(s) of h in terms of a, b, c, and d.

◆7.   A population of fish reproduces and dies at rates proportional to the total population. If $x(t)$ is the number of fish at time t, then we have $x' = cx$, where c is a constant (which we will assume is positive).

  Now suppose the fish are harvested at an absolute rate, say r, hence $x' = cx - r$. Find an analytic solution to this dynamical system and give a qualitative discussion of the long-term state of the population.

2.2 Two (and more) dimensions

In this section we consider discrete and continuous time linear systems in several variables. The systems, of course, have the form $\mathbf{x}(k+1) = f(\mathbf{x}(k))$ (discrete) or $\mathbf{x}' = f(\mathbf{x})$ (continuous). The state vector $\mathbf{x}$ is no longer a single number but rather is a vector with n components (i.e., $\mathbf{x} \in \mathbf{R}^n$). The function f (which has the form $f(x) = ax + b$ above) is of the form $f(\mathbf{x}) = A\mathbf{x} + \mathbf{b}$, where A is an $n \times n$ matrix and $\mathbf{b} \in \mathbf{R}^n$ is a (constant) vector.

As the old jokes go, I have some good news and some bad news.

First, the good news. The behavior of one-dimensional systems (§2.1) provides *excellent* motivation for the material we present here. In many respects, multidimensional systems behave just like their one-dimensional counterparts. For example, discrete time systems gravitate toward a unique fixed point when—well I'd like to say $|A| < 1$, but that doesn't make sense, since what do we mean by the "absolute value of a matrix"? Instead, we just have to check that the absolute value of something else[1] is less than 1. And, in continuous time we saw that $x' = ax$ has the simple solution $x = e^{at} x_0$; the same is true in several variables. The system $\mathbf{x}' = A\mathbf{x}$ has solution $\mathbf{x} = e^{At} \mathbf{x}_0$; naturally, we're going to have to make sense out of "e to a matrix".

But now, the bad news. The entrance of matrices into the story requires us to dust off our linear algebra. You'll especially need to review the ideas of diagonalization and eigenvalues/vectors. For your convenience, the major results we need from linear algebra are recounted with minimal discussion in Appendix A. It makes sense at this point to scan Appendix A and to review your linear algebra text.

Diagonalization makes our work easier.

One of the key steps in our treatment of linear systems is diagonalization (see §A.1.4). One difficulty with this approach is that *not all matrices diagonalize!*

> In §2.2.1 and §2.2.2 we assume that the matrix A diagonalizes.

In §2.2.3 we discuss how the theory carries over to nondiagonalizable matrices.

---

[1]OK. If you like, I'll spill the beans here. The condition is that all the eigenvalues of A have absolute value less than 1.

2.2.1 Discrete time

In this section we consider linear discrete time dynamical systems, i.e., systems of the form

$$\mathbf{x}(k+1) = A\mathbf{x}(k) + \mathbf{b}; \qquad \mathbf{x}(0) = \mathbf{x}_0.$$

Analysis

As in the one-dimensional case $[x(k+1) = ax(k) + b]$, we begin by dropping the $+\mathbf{b}$ term and concentrating on the system

$$\mathbf{x}(k+1) = A\mathbf{x}(k), \qquad (2.7)$$

where $\mathbf{x} \in \mathbf{R}^n$ and A is an $n \times n$ matrix.

Now it is simple to calculate that

$$\mathbf{x}(1) = A\mathbf{x}(0) = A\mathbf{x}_0$$
$$\mathbf{x}(2) = A\mathbf{x}(1) = A^2\mathbf{x}_0$$
$$\mathbf{x}(3) = A\mathbf{x}(2) = A^3\mathbf{x}_0$$
$$\vdots$$

Thus we see that $\mathbf{x}(k) = A^k\mathbf{x}_0$.

This is all well and good, and the equation $\mathbf{x}(k) = A^k\mathbf{x}_0$ gives us an exact formula for $\mathbf{x}(k)$ but it does not yet tell us the general behavior of the system. Recall that in one dimension we considered the cases $|a| < 1$ and $|a| > 1$ to judge the long-term behavior. We wish to do the same analysis here.

What condition on A is analogous to $|a| < 1$? One guess is that all the entries of A should have absolute value less than 1. For example, if $A = \begin{bmatrix} 0.1 & 0.2 \\ 0.3 & 0.4 \end{bmatrix}$ and $\mathbf{x}_0 = \begin{bmatrix} 1 \\ 1 \end{bmatrix}$, then computer calculations (try it!) show that the entries in $\mathbf{x}(1000) = A^{1000}\mathbf{x}_0$ are on the order of 10^{-269}; it's a safe bet that $\mathbf{x}(k) \to \mathbf{0}$. However, if $A = \begin{bmatrix} 0.2 & 0.4 \\ 0.6 & 0.8 \end{bmatrix}$ and $\mathbf{x}_0 = \begin{bmatrix} 1 \\ 1 \end{bmatrix}$ we are distressed to learn that the entries in $\mathbf{x}(1000) = A^{1000}\mathbf{x}_0$ are on the order of 10^{31}, and it's a safe bet that the entries are going to infinity. Thus our reasonable guess that the entries of A need to have absolute value less than 1 is wrong. We need a better guess: look at the eigenvalues...

Generalizing the condition $|a| < 1$.

We assume that A diagonalizes[2] (be sure to read your linear algebra text book—see also §A.1.4). We assume that A has n linearly independent eigenvectors $\mathbf{v}_1, \ldots, \mathbf{v}_n$ with associated eigenvalues $\lambda_1, \ldots, \lambda_n$. Let Λ be the diagonal matrix with diagonal entries $\lambda_1, \ldots, \lambda_n$, and let S be the $n \times n$ matrix whose i^{th} column is $\mathbf{v}_i$. Thus we may write $A = S\Lambda S^{-1}$. Notice that

$$A^k = (S\Lambda S^{-1})(S\Lambda S^{-1})(S\Lambda S^{-1}) \cdots (S\Lambda S^{-1}).$$

Since matrix multiplication is associative, we rewrite this expression as

$$A^k = S\Lambda(S^{-1}S)\Lambda(S^{-1}S)\Lambda(S^{-1}S) \cdots (S^{-1}S)\Lambda S^{-1}.$$

Notice that the $(S^{-1}S)$ terms evaluate to I and therefore disappear, leaving

$$A^k = S\Lambda^k S^{-1}.$$

Now, Λ is a diagonal matrix whose diagonal entries are A's eigenvalues: $\lambda_1, \ldots, \lambda_n$. Raising a diagonal matrix to a power is easy:

$$\Lambda^k = \begin{bmatrix} \lambda_1 & 0 & 0 & \cdots & 0 \\ 0 & \lambda_2 & 0 & \cdots & 0 \\ 0 & 0 & \lambda_3 & \cdots & 0 \\ \vdots & \vdots & \vdots & \ddots & \vdots \\ 0 & 0 & 0 & \cdots & \lambda_n \end{bmatrix}^k = \begin{bmatrix} \lambda_1^k & 0 & 0 & \cdots & 0 \\ 0 & \lambda_2^k & 0 & \cdots & 0 \\ 0 & 0 & \lambda_3^k & \cdots & 0 \\ \vdots & \vdots & \vdots & \ddots & \vdots \\ 0 & 0 & 0 & \cdots & \lambda_n^k \end{bmatrix}.$$

What happens when we raise a matrix to a large power? The answer depends on its eigenvalues.

Thus to understand the behavior of A^k we need to understand the behavior of the λ_j^k's. If λ_j is a real number, then $\lambda_j^k \to 0$ if $|\lambda_j| < 1$, and λ_j^k explodes if $|\lambda_j| > 1$. However, the eigenvalues of A might be complex numbers, and then we need to know how λ_j^k behaves for complex λ_j.

To do this, we find it easiest to write λ_j in its polar form: $\lambda_j = r_j e^{i\theta_j}$ (see §A.2, especially equation (A.5) on page 349). Thus

$$\left| \lambda_j^k \right| = \left| r_j^k e^{ik\theta_j} \right| \to \begin{cases} 0 & \text{if } r < 1, \text{ and} \\ \infty & \text{if } r > 1. \end{cases}$$

What are the implications for the system $\mathbf{x}(k+1) = A\mathbf{x}(k)$? If all the eigenvalues of A have absolute value less than 1, then A^k tends to the zero matrix as $k \to \infty$. Thus $\mathbf{x}(k) = A^k \mathbf{x}_0 \to \mathbf{0}$.

On the other hand, if some eigenvalue of A has absolute value greater than 1, entries in A^k are diverging to ∞. Let's examine how this affects the values $\mathbf{x}(k)$.

---

[2]Not all matrices diagonalize. We assume A does to make our life easier at this point. The theory we are about to develop also works for matrices which do not diagonalize, but the analysis is more difficult. An approach is presented in §2.2.3.

We are assuming that the eigenvectors $\mathbf{v}_1, \ldots, \mathbf{v}_n$ are linearly independent. Any family of n linearly independent vectors in $\mathbf{R}^n$ forms a basis, hence every vector ($\mathbf{x}_0$ in particular) can be written uniquely as a linear combination of the $\mathbf{v}_i$'s. Thus we may write

$$\mathbf{x}(0) = \mathbf{x}_0 = c_1\mathbf{v}_1 + c_2\mathbf{v}_2 + \cdots + c_n\mathbf{v}_n,$$

where the c_i's are scalars (numbers). Multiplying both sides by A, we get

$$\mathbf{x}(1) = A\mathbf{x}(0) = c_1 A\mathbf{v}_1 + c_2 A\mathbf{v}_2 + \cdots + c_n A\mathbf{v}_n.$$

Since each $\mathbf{v}_i$ is an eigenvector of A, we have $A\mathbf{v}_i = \lambda_i\mathbf{v}_i$. We can therefore rewrite the preceding equation as:

$$\mathbf{x}(1) = c_1\lambda_1\mathbf{v}_1 + c_2\lambda_2\mathbf{v}_2 + \cdots + c_n\lambda_n\mathbf{v}_n.$$

That was fun; let's do it again! If we multiply both sides of the previous equation by A, we get

$$\mathbf{x}(2) = A\mathbf{x}(1) = c_1\lambda_1 A\mathbf{v}_1 + c_2\lambda_2 A\mathbf{v}_2 + \cdots + c_n\lambda_n A\mathbf{v}_n$$
$$= c_1\lambda_1^2\mathbf{v}_1 + c_2\lambda_2^2\mathbf{v}_2 + \cdots + c_n\lambda_n^2\mathbf{v}_n.$$

When we iterate this process we get

$$\mathbf{x}(k) = c_1\lambda_1^k\mathbf{v}_1 + c_2\lambda_2^k\mathbf{v}_2 + \cdots + c_n\lambda_n^k\mathbf{v}_n. \tag{2.8}$$

Now, if $|\lambda_i| > 1$, then $|\lambda_i^k| \to \infty$ as $k \to \infty$. Hence, unless $c_i = 0$, we have $|\mathbf{x}(k)| \to \infty$.

In summary, if some of the eigenvalues of A have absolute value greater than 1, then for *typical* $\mathbf{x}_0$ we have $|\mathbf{x}(k)| \to \infty$. For some very special $\mathbf{x}_0$ (those with $c_i = 0$ if $|\lambda_i| > 1$), $\mathbf{x}(k)$ doesn't explode.

One case remains.[3] What happens if all eigenvalues λ_i have absolute value less than *or equal* to 1, and some have absolute value exactly 1? In this case, we again write $\mathbf{x}_0$ as a linear combination of the eigenvectors:

$$\mathbf{x}(0) = \mathbf{x}_0 = c_1\mathbf{v}_1 + c_2\mathbf{v}_2 + \cdots + c_n\mathbf{v}_n$$

---

[3]The Mishna states: If a fledgling bird is found within fifty cubits of a dovecote, it belongs to the owner of the dovecote. If it is found outside the limit of fifty cubits, it belongs to the person who finds it.

Rabbi Jeremiah asked: If one foot of the fledgling is within fifty cubits, and one foot is outside it, what is the law?

It was for this question that Rabbi Jeremiah was thrown out of the House of Study.

Baba Batra 23b

from which it follows (equation (2.8)) that

$$\mathbf{x}(k) = c_1 \lambda_1^k \mathbf{v}_1 + c_2 \lambda_2^k \mathbf{v}_2 + \cdots + c_n \lambda_n^k \mathbf{v}_n.$$

The terms involving λ_i's with absolute value less than 1 disappear, but the other components neither vanish nor explode. A complex number with absolute value 1 has the form $z = e^{i\theta}$ and therefore $z^k = e^{ik\theta}$, which also has absolute value 1. Thus $\mathbf{x}(k)$ typically neither vanishes nor explodes but dances about at a modest distance from $\mathbf{0}$.

We have considered the special system $\mathbf{x}(k + 1) = A\mathbf{x}(k)$, where A is assumed to be diagonalizable. If the eigenvalues of A all have absolute value less than 1, then $\mathbf{x}(k) \rightarrow \mathbf{0}$ as $k \rightarrow \infty$. If some eigenvalue has absolute value bigger than 1, then typically $|\mathbf{x}(k)| \rightarrow \infty$. Finally, if some eigenvalues have absolute value equal to 1, and the rest have absolute value less than 1, then typically $\mathbf{x}(k)$ neither explodes nor vanishes.

When $\mathbf{b} \neq \mathbf{0}$.

The situation for the more general case $\mathbf{x}(k + 1) = A\mathbf{x}(k) + \mathbf{b}$ is quite similar. First, recall the one-dimensional case: the system either gravitated to, or was repelled from, the fixed point $\tilde{x} = b/(1 - a)$. We will see the same behavior here. The question is, What serves the role of $b/(1 - a)$? Take a guess! (I'll tell you in a few minutes, but take a guess anyway.)

As in the one-dimensional case, let us compute the iterates $\mathbf{x}(0)$, $\mathbf{x}(1)$, $\mathbf{x}(2)$, etc. to gain a feel for the general case:

$$\mathbf{x}(0) = \mathbf{x}_0,$$

$$\mathbf{x}(1) = A\mathbf{x}(0) + \mathbf{b} = A\mathbf{x}_0 + \mathbf{b},$$

$$\mathbf{x}(2) = A\mathbf{x}(1) + \mathbf{b} = A^2\mathbf{x}_0 + A\mathbf{b} + \mathbf{b},$$

$$\mathbf{x}(3) = A\mathbf{x}(2) + \mathbf{b} = A^3\mathbf{x}_0 + A^2\mathbf{b} + A\mathbf{b} + \mathbf{b},$$

$$\mathbf{x}(4) = A\mathbf{x}(3) + \mathbf{b} = A^4\mathbf{x}_0 + A^3\mathbf{b} + A^2\mathbf{b} + A\mathbf{b} + \mathbf{b}.$$

Is the pattern clear? We have

$$\mathbf{x}(k) = A^k\mathbf{x}_0 + \left(A^{k-1} + A^{k-2} + \cdots + A + I\right)\mathbf{b}.$$

To simplify this, observe that

$$\left(A^{k-1} + A^{k-2} + \cdots + A + I\right)(I - A) = I - A^k,$$

and so, *provided $I - A$ is invertible*, we have

$$\mathbf{x}(k) = A^k\mathbf{x}_0 + (I - A^k)(I - A)^{-1}\mathbf{b}. \tag{2.9}$$

Discrete: $\mathbf{x}(k+1) = A\mathbf{x}(k) + \mathbf{b}$ with $\mathbf{x}(0) = \mathbf{x}_0$	
Conditions on eigenvalues of A	Behavior of $\mathbf{x}(k)$
All have $|\lambda| < 1$	converges to $\tilde{\mathbf{x}} = (I - A)^{-1}\mathbf{b}$
Some have $|\lambda| > 1$	typically, $|\mathbf{x}(k)| \to \infty$
All have $|\lambda| \le 1$; and some have $|\lambda| = 1$	stays near, but does not approach $\tilde{\mathbf{x}}$, or $|\mathbf{x}(k)| \to \infty$

Table 2.3. The possible behaviors of multidimensional discrete time linear systems.

The formula, of course, is valid provided $I - A$ is invertible, which is equivalent to saying that 1 is not an eigenvalue of A.

Now, what happens as $k \to \infty$? If the absolute values of A's eigenvalues are all less than 1 (hence $I - A$ is invertible), then A^k tends to the zero matrix, hence $\mathbf{x}(k) \to \tilde{\mathbf{x}} = (I - A)^{-1}\mathbf{b}$. [Aha! The value $\tilde{\mathbf{x}} = (I - A)^{-1}\mathbf{b}$ is precisely analogous to the one-dimensional $\tilde{x} = b/(1 - a) = (1 - a)^{-1}b$.] *The case: all $|\lambda| < 1$.*

Alternatively, if some eigenvalues have absolute value bigger than 1, then A^k blows up, and for most $\mathbf{x}_0$ we have $|\mathbf{x}(k)| \to \infty$. (There are exceptional $\mathbf{x}_0$'s, of course. For example, if 1 is not an eigenvalue of A and if $\mathbf{x}_0 = \tilde{\mathbf{x}} = (I - A)^{-1}\mathbf{b}$, then $\mathbf{x}(k) = \tilde{\mathbf{x}}$ for all k. See problems 4 and 5 on page 88.) *The case: some $|\lambda| > 1$.*

Finally, if some eigenvalues have absolute value equal to 1 and the other eigenvalues have absolute value less than 1, we see a range of behaviors (see problems 13–15 on page 89). The system might stay near $\tilde{\mathbf{x}}$, or it might blow up.

Table 2.3 summarizes the behavior of discrete time, multivariate linear systems.

Geometry

We noted that there are three basic cases for discrete time linear systems: (a) all eigenvalues have absolute value less than 1, (b) some eigenvalues have absolute value exceeding 1, and (c) all eigenvalues have absolute value at most 1 and some have absolute value equal to 1.

The eigenvalues, since they are complex numbers, can be represented as points in the plane (the number $a + bi$ is placed at the point (a, b); see §A.2). Figure 2.10 illustrates the three cases geometrically.

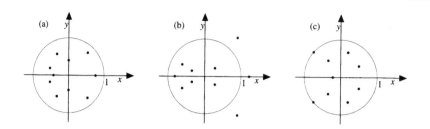

Figure 2.10. The behavior of discrete linear systems. Three cases are shown: (a) all $|\lambda| < 1$, (b) some $|\lambda| > 1$, and (c) all $|\lambda| \leq 1$ and some $|\lambda| = 1$.

A special case has an interesting geometric interpretation. Let us say that A has a unique eigenvalue of maximum absolute value; since eigenvalues come in conjugate pairs, such an eigenvalue is necessarily real. Let us also assume that this eigenvalue is positive[4] and, furthermore, is greater than 1. Let λ_1 be this eigenvalue. In symbols, our assumptions are

$$1 < \lambda_1 > |\lambda_j| \quad \text{for all } j \neq 1.$$

Let $\mathbf{v}_1, \mathbf{v}_2, \ldots, \mathbf{v}_n$ be the n linearly independent eigenvectors associated with $\lambda_1, \lambda_2, \ldots, \lambda_n$. Then we can write

$$\mathbf{x}_0 = \mathbf{x}(0) = c_1 \mathbf{v}_1 + c_2 \mathbf{v}_2 + \cdots + c_n \mathbf{v}_n,$$

from which it follows that

$$\mathbf{x}(k) = c_1 \lambda_1^k \mathbf{v}_1 + c_2 \lambda_2^k \mathbf{v}_2 + \cdots + c_n \lambda_n^k \mathbf{v}_n,$$

from which we factor out λ_1^k to get

$$\mathbf{x}(k) = \lambda_1^k \left(c_1 \mathbf{v}_1 + c_2 \frac{\lambda_2^k}{\lambda_1^k} \mathbf{v}_2 + \cdots + c_n \frac{\lambda_n^k}{\lambda_1^k} \mathbf{v}_n \right).$$

Since $|\lambda_1| > |\lambda_j|$ for all $j > 1$, the ratios λ_j^k/λ_1^k all go to 0. Thus if $c_1 \neq 0$, and for k large we have

$$\mathbf{x}(k) \approx c_1 \lambda_1^k \mathbf{v}_1.$$

Geometrically, this means that the state vector is heading to infinity, in essentially a straight line, in the direction of the eigenvector $\mathbf{v}_1$.

---

[4]If all entries in a matrix are positive, then it can be proven that the eigenvalue of maximum absolute value is unique, real, and positive.

2.2.2   Continuous time

We now consider multidimensional continuous time linear systems:

$$\mathbf{x}'(t) = A\mathbf{x}(t) + \mathbf{b}.$$

Analysis

We begin our study of multidimensional continuous time linear systems
with our usual simplification, namely, that $\mathbf{b} = \mathbf{0}$. The system becomes

$$\mathbf{x}' = A\mathbf{x}.$$

Now, instead of having a single differential equation in one unknown quantity, we have n differential equations in n variables! For example,

$$\begin{bmatrix} x_1 \\ x_2 \end{bmatrix}' = \begin{bmatrix} 2 & 1 \\ 1 & 2 \end{bmatrix} \begin{bmatrix} x_1 \\ x_2 \end{bmatrix},$$

which can be written out as

$$\frac{dx_1}{dt} = 2x_1 + x_2,$$

$$\frac{dx_2}{dt} = x_1 + 2x_2.$$

These equations are difficult to solve because each is dependent on the
other. It makes us long for the simplicity of the equation $x' = ax$, whose
solution is just $x = e^{at}x_0$. Wouldn't it be wonderful if more complicated
systems had such an easy solution?

Good news! The system $\mathbf{x}' = A\mathbf{x}$ has nearly as simple a solution! Here
it is: $\mathbf{x} = e^{At}\mathbf{x}_0$. The question is, What do we mean by "e raised to a
matrix??"

Understanding "e raised to a matrix".

Here is what we need to do:

- First, we have to understand what e^{At} means.

- Second, we need to see that $e^{At}\mathbf{x}_0$ is indeed a solution to the system
  of equations.

- Finally, we need to understand how to convert the simple solution
  $e^{At}\mathbf{x}_0$ into standard functions of single variables.

Let us begin, then, with understanding what "e to a matrix" means by first considering what e^x means. We can define e^x by its power series:

$$e^x = 1 + x + \frac{x^2}{2!} + \frac{x^3}{3!} + \frac{x^4}{4!} + \cdots .$$

Now, "all" we have to do is drop a matrix in for the x. Let A be a square matrix. It is clear what, say, $\frac{x^3}{3!}$ should become: simply $\frac{1}{3!}A^3$. But what about the "1" at the beginning of the series? The best choice is simply I, an identity matrix of the same size as A. Thus we define e^A (alternate notation: $\exp A$) by

$$e^A = \exp A = I + A + \frac{1}{2!}A^2 + \frac{1}{3!}A^3 + \frac{1}{4!}A^4 + \cdots .$$

Let's do some simple examples: First, let $A = \begin{bmatrix} 0 & 0 \\ 0 & 0 \end{bmatrix}$. Then

$$\exp \begin{bmatrix} 0 & 0 \\ 0 & 0 \end{bmatrix} = \begin{bmatrix} 1 & 0 \\ 0 & 1 \end{bmatrix} + \begin{bmatrix} 0 & 0 \\ 0 & 0 \end{bmatrix} + \frac{1}{2!}\begin{bmatrix} 0 & 0 \\ 0 & 0 \end{bmatrix}^2 + \frac{1}{3!}\begin{bmatrix} 0 & 0 \\ 0 & 0 \end{bmatrix}^3 + \cdots$$

$$= \begin{bmatrix} 1 & 0 \\ 0 & 1 \end{bmatrix} .$$

Thus $\exp \begin{bmatrix} 0 & 0 \\ 0 & 0 \end{bmatrix} = \begin{bmatrix} 1 & 0 \\ 0 & 1 \end{bmatrix}$. In general, e to a zero matrix gives the identity matrix.

For another example let's compute e^I.

$$e^I = I + I + \frac{1}{2!}I^2 + \frac{1}{3!}I^3 + \frac{1}{4!}I^4 + \cdots .$$

Thus the off-diagonal entries in e^I are all 0, while the diagonal entries are all $1 + 1 + 1/2! + 1/3! + 1/4! + \cdots = e$. Thus $e^I = eI$.

e to a matrix is not the same as exponentiating each entry!

These two examples illustrate that "e to a matrix" is *not* made by simply raising e to the entries of the matrix: $\exp \begin{bmatrix} a & b \\ c & d \end{bmatrix} \neq \begin{bmatrix} e^a & e^b \\ e^c & e^d \end{bmatrix}$.

Let's do one more example. Let $A = \begin{bmatrix} 2 & 0 \\ 0 & -3 \end{bmatrix}$ and let us compute e^{At}, where t is a scalar. Then

$$e^{At} = \exp(At) = I + (At) + \frac{1}{2!}(At)^2 + \frac{1}{3!}(At)^3 + \cdots .$$

Let's work out each of these terms:

$$At = t \begin{bmatrix} 2 & 0 \\ 0 & -3 \end{bmatrix} = \begin{bmatrix} 2t & 0 \\ 0 & -3t \end{bmatrix},$$

$$(At)^2 = t^2 A^2 = t^2 \begin{bmatrix} 4 & 0 \\ 0 & 9 \end{bmatrix} = \begin{bmatrix} 4t^2 & 0 \\ 0 & 9t^2 \end{bmatrix},$$

$$(At)^3 = t^3 A^3 = t^3 \begin{bmatrix} 8 & 0 \\ 0 & -27 \end{bmatrix} = \begin{bmatrix} 8t^3 & 0 \\ 0 & -27t^3 \end{bmatrix},$$

$$\vdots$$

$$(At)^j = t^j A^j = t^j \begin{bmatrix} 2^j & 0 \\ 0 & (-3)^j \end{bmatrix} = \begin{bmatrix} (2t)^j & 0 \\ 0 & (-3t)^j \end{bmatrix}.$$

Thus e^{At} equals

$$e^{At} = \exp \begin{bmatrix} 2t & 0 \\ 0 & -3t \end{bmatrix}$$

$$= \begin{bmatrix} 1 & 0 \\ 0 & 1 \end{bmatrix} + \begin{bmatrix} 2t & 0 \\ 0 & -3t \end{bmatrix} + \frac{1}{2!} \begin{bmatrix} 4t^2 & 0 \\ 0 & 9t^2 \end{bmatrix} + \frac{1}{3!} \begin{bmatrix} 8t^3 & 0 \\ 0 & -27t^3 \end{bmatrix} + \cdots$$

$$= \begin{bmatrix} 1 + 2t + \frac{1}{2!}(2t)^2 + \cdots & 0 \\ 0 & 1 + (-3t) + \frac{1}{2!}(-3t)^2 + \cdots \end{bmatrix}$$

$$= \begin{bmatrix} e^{2t} & 0 \\ 0 & e^{-3t} \end{bmatrix}.$$

The only thing special about the matrix $A = \begin{bmatrix} 2 & 0 \\ 0 & -3 \end{bmatrix}$ is that it's a diagonal matrix. Indeed, the preceding examples can be generalized. For any diagonal matrix Λ:

$$\exp(\Lambda t) = \exp \left(t \begin{bmatrix} \lambda_1 & 0 & \cdots & 0 \\ 0 & \lambda_2 & \cdots & 0 \\ \vdots & \vdots & \ddots & \vdots \\ 0 & 0 & \cdots & \lambda_n \end{bmatrix} \right) = \begin{bmatrix} e^{\lambda_1 t} & 0 & \cdots & 0 \\ 0 & e^{\lambda_2 t} & \cdots & 0 \\ \vdots & \vdots & \ddots & \vdots \\ 0 & 0 & \cdots & e^{\lambda_n t} \end{bmatrix}$$

In every example so far A has been a diagonal matrix. We return to consider how to compute e^A for nondiagonal matrices after we justify the usefulness of matrix exponentials in solving $\mathbf{x}' = A\mathbf{x}$. The operation e^A

Computers can calculate
matrix exponentials.

is actually built into computer packages such as MATLAB (as the `expm`
function) and *Mathematica* (as the `MatrixExp` function). Here is a bit of
a MATLAB session:

```
>>a=[2 0 ; 0 -3]

a =
      2      0
      0     -3

>>expm(a)

ans =
      7.3891              0
            0       0.0498
```

Use a calculator to check that $e^2 \approx 7.3891$ and $e^{-3} \approx 0.0498$, as we would
expect.

   Mathematica gives us the exact (i.e., symbolic) result:

```
MatrixExp[{{2t,0},{0,-3t}}] // MatrixForm

  2 t
E           0


            -3 t
0           E
```

Does $e^{At}\mathbf{x}_0$ solve our
problem?

   Our next step is to understand why $e^{At}\mathbf{x}_0$ is a solution to the dynamical
system $\mathbf{x}' = A\mathbf{x}$ with $\mathbf{x}(0) = \mathbf{x}_0$.

   First, we do a quick sanity check: e^{At} is an $n \times n$ matrix (which depends
on t) and $\mathbf{x}_0$ is a fixed n-vector. Thus $e^{At}\mathbf{x}_0$ is an n-vector which varies
over time. Also, if we plug in 0 for t we have $\mathbf{x}(0) = e^{A \cdot 0}\mathbf{x}_0 = I\mathbf{x}_0 = \mathbf{x}_0$
(recall that e raised to a zero matrix is the identity); therefore, the formula
$\mathbf{x}(t) = e^{At}\mathbf{x}_0$ works for $t = 0$.

   Now comes the more complicated part. We need to see that $\mathbf{x} = e^{At}\mathbf{x}_0$
satisfies the equation $\mathbf{x}' = A\mathbf{x}$. The left side of this differential equation is
$\mathbf{x}'$, so we take derivatives of the entries in $\mathbf{x}$ and work to show that

$$\frac{d}{dt}\left(e^{At}\right)\mathbf{x}_0 = Ae^{At}\mathbf{x}_0.$$

We compute as follows:

$$\mathbf{x}' = \frac{d}{dt}\left(e^{At}\mathbf{x}_0\right)$$

$$= \frac{d}{dt}\left[\left(I + (At) + \frac{1}{2!}(At)^2 + \frac{1}{3!}(At)^3 + \frac{1}{4!}(At)^4 + \cdots\right)\mathbf{x}_0\right]$$

$$= \frac{d}{dt}\left(I\mathbf{x}_0 + (At)\mathbf{x}_0 + \frac{1}{2!}(At)^2\mathbf{x}_0 + \frac{1}{3!}(At)^3\mathbf{x}_0 + \frac{1}{4!}(At)^4\mathbf{x}_0 + \cdots\right).$$

To continue this calculation, we compute[5] the derivative $\frac{d}{dt}$ of individual terms which look like $\frac{1}{3!}(At)^3\mathbf{x}_0$. Observe that this term can be rewritten as $t^3\frac{1}{3!}(A^3\mathbf{x}_0)$. Now, $\frac{1}{3!}A^3\mathbf{x}_0$ is just a vector, and t^3 is just a scalar multiplying that vector. We continue our computation:

$$\mathbf{x}' = \frac{d}{dt}\left(I\mathbf{x}_0 + (At)\mathbf{x}_0 + \frac{1}{2!}(At)^2\mathbf{x}_0 + \frac{1}{3!}(At)^3\mathbf{x}_0 + \frac{1}{4!}(At)^4\mathbf{x}_0 + \cdots\right)$$

$$= \frac{d}{dt}\left[I\mathbf{x}_0 + t\,(A\mathbf{x}_0) + t^2\left(\frac{1}{2!}A^2\mathbf{x}_0\right) + t^3\left(\frac{1}{3!}A^3\mathbf{x}_0\right) + t^4\left(\frac{1}{4!}A^4\mathbf{x}_0\right) + \cdots\right]$$

$$= \mathbf{0} + (A\mathbf{x}_0) + 2t\left(\frac{1}{2!}A^2\mathbf{x}_0\right) + 3t^2\left(\frac{1}{3!}A^3\mathbf{x}_0\right) + 4t^3\left(\frac{1}{4!}A^4\mathbf{x}_0\right) + \cdots$$

$$= A\,(\mathbf{x}_0) + At\left(\frac{1}{1!}A^1\mathbf{x}_0\right) + At^2\left(\frac{1}{2!}A^2\mathbf{x}_0\right) + At^3\left(\frac{1}{3!}A^3\mathbf{x}_0\right) + \cdots.$$

(In the last step, we took a term such as $3t^2\left(\frac{1}{3!}A^3\mathbf{x}_0\right)$, canceled the 3 in front with the 3! in the denominator, leaving 2! downstairs, and pulled a factor of A out to get $At^2\left(\frac{1}{2!}A^2\mathbf{x}_0\right)$.) Observe that all the terms have a factor of A, which we now collect, continuing the computation:

$$\mathbf{x}' = A\left[I\mathbf{x}_0 + t\left(\frac{1}{1!}A^1\mathbf{x}_0\right) + t^2\left(\frac{1}{2!}A^2\mathbf{x}_0\right) + t^3\left(\frac{1}{3!}A^3\mathbf{x}_0\right) + \cdots\right]$$

$$= A\left[I + t\,(A) + t^2\left(\frac{1}{2!}A^2\right) + t^3\left(\frac{1}{3!}A^3\right) + \cdots\right]\mathbf{x}_0$$

$$= A\left[I + (At) + \frac{1}{2!}(At)^2 + \frac{1}{3!}(At)^3 + \cdots\right]\mathbf{x}_0$$

$$= A\left(e^{At}\mathbf{x}_0\right)$$

$$= A\mathbf{x}.$$

---

[5]The term-by-term differentiation of a power series is not always valid. The validity in this case can be proved but requires ideas from real analysis which are beyond the scope of this text.

Whew! That was a long trip, but we have seen that if we let $\mathbf{x} = e^{At}\mathbf{x}_0$, then we have first, that $\mathbf{x}(0) = \mathbf{x}_0$ and (with much more effort) that $\mathbf{x}' = A\mathbf{x}$. This shows that $\mathbf{x} = e^{At}\mathbf{x}_0$ is the desired solution.

Great! Now we know that a problem such as

$$\begin{bmatrix} x_1 \\ x_2 \end{bmatrix}' = \begin{bmatrix} 2 & 1 \\ 1 & 2 \end{bmatrix}\begin{bmatrix} x_1 \\ x_2 \end{bmatrix}; \qquad \begin{bmatrix} x_1(0) \\ x_2(0) \end{bmatrix} = \begin{bmatrix} 1 \\ 0 \end{bmatrix}$$

Recall that the notation $\exp A$ means the same thing as e^A.

has, as its solution,

$$\begin{bmatrix} x_1(t) \\ x_2(t) \end{bmatrix} = \exp\left(\begin{bmatrix} 2 & 1 \\ 1 & 2 \end{bmatrix}t\right)\begin{bmatrix} 1 \\ 0 \end{bmatrix}.$$

Somehow, this is not very satisfying. True, we can use MATLAB, or the like, to compute specific values of $x_1(t)$ or $x_2(t)$ for any t. For example, to get the values at $t = 2$ we would simply type

```
>>a = [2 1 ; 1 2]

a =
        2      1
        1      2

>>expm(2*a) * [1;0]

ans =
     205.4089
     198.0199
```

The numbers look rather big, and it would be safe to bet that $|\mathbf{x}(t)| \to \infty$ as $t \to \infty$, but the $e^{At}\mathbf{x}_0$ form of the solution doesn't tell us (yet) what is happening.

The behavior of e^{At} as $t \to \infty$ depends on the eigenvalues of A.

Our next major step is to understand how e^{At} behaves. We have already seen how e^{At} behaves when A is a diagonal matrix. What should we do when A is not diagonal? Well, diagonalize it, of course!

Assuming that A diagonalizes, we can write

$$A = S\Lambda S^{-1},$$

and therefore

$$At = (S\Lambda S^{-1})t = S(t\Lambda)S^{-1}.$$

We substitute this expression for At into the definition of e^{At} and we compute

$$e^{At} = \exp(At)$$

$$= \exp\left(S(t\Lambda)S^{-1}\right)$$

$$= I + \left(S(t\Lambda)S^{-1}\right) + \frac{1}{2!}\left(S(t\Lambda)S^{-1}\right)^2 + \frac{1}{3!}\left(S(t\Lambda)S^{-1}\right)^3 + \cdots .$$

Because the terms in this expression are of the form

$$\frac{1}{k!}\left(S(t\Lambda)S^{-1}\right)^k$$

we can write it out as

$$\frac{1}{k!}\left(S(t\Lambda)S^{-1}S(t\Lambda)S^{-1}\cdots S(t\Lambda)S^{-1}\right),$$

and we watch the S^{-1} terms annihilate the S terms, leaving

$$\frac{1}{k!}\left(S(t\Lambda)^k S^{-1}\right).$$

We continue our computation of e^{At}:

$$e^{At} = I + \left(S(t\Lambda)S^{-1}\right) + \frac{1}{2!}\left(S(t\Lambda)S^{-1}\right)^2 + \frac{1}{3!}\left(S(t\Lambda)S^{-1}\right)^3 + \cdots$$

$$= SIS^{-1} + \left(S(t\Lambda)S^{-1}\right) + \frac{1}{2!}\left(S(t\Lambda)^2 S^{-1}\right) + \frac{1}{3!}\left(S(t\Lambda)^3 S^{-1}\right) + \cdots$$

$$= S\left[I + (t\Lambda) + \frac{1}{2!}(t\Lambda)^2 + \frac{1}{3!}(t\Lambda)^3 + \cdots\right]S^{-1}$$

$$= S\exp(t\Lambda)S^{-1}.$$

In summary,

$$e^{At} = Se^{\Lambda t}S^{-1}, \tag{2.10}$$

where Λ is a diagonal matrix of A's eigenvalues, and the columns of S are A's eigenvectors.

Let's do an example. Suppose $A = \begin{bmatrix} 2 & 1 \\ 1 & 2 \end{bmatrix}$. Then A's characteristic

An example of e^A where A is not a diagonal matrix.

polynomial is

$$\det(xI - A) = \det \begin{bmatrix} x-2 & 1 \\ 1 & x-2 \end{bmatrix}$$

$$= (x-2)(x-2) - 1$$

$$= x^2 - 4x + 3$$

$$= (x-3)(x-1).$$

Thus A's eigenvalues are 3 and 1. Observe that $\begin{bmatrix} 1 \\ 1 \end{bmatrix}$ is an eigenvector of A associated with 3:

$$\begin{bmatrix} 2 & 1 \\ 1 & 2 \end{bmatrix} \begin{bmatrix} 1 \\ 1 \end{bmatrix} = \begin{bmatrix} 3 \\ 3 \end{bmatrix} = 3 \begin{bmatrix} 1 \\ 1 \end{bmatrix},$$

and $\begin{bmatrix} 1 \\ -1 \end{bmatrix}$ is an eigenvector of A associated with 1:

$$\begin{bmatrix} 2 & 1 \\ 1 & 2 \end{bmatrix} \begin{bmatrix} 1 \\ -1 \end{bmatrix} = 1 \begin{bmatrix} 1 \\ -1 \end{bmatrix}.$$

Thus we let $S = \begin{bmatrix} 1 & 1 \\ 1 & -1 \end{bmatrix}$ (and therefore $S^{-1} = \begin{bmatrix} \frac{1}{2} & \frac{1}{2} \\ \frac{1}{2} & -\frac{1}{2} \end{bmatrix}$) and

$\Lambda = \begin{bmatrix} 3 & 0 \\ 0 & 1 \end{bmatrix}$. As a quick check we compute (do it!)

$$S\Lambda S^{-1} = \begin{bmatrix} 1 & 1 \\ 1 & -1 \end{bmatrix} \begin{bmatrix} 3 & 0 \\ 0 & 1 \end{bmatrix} \begin{bmatrix} \frac{1}{2} & \frac{1}{2} \\ \frac{1}{2} & -\frac{1}{2} \end{bmatrix} = \begin{bmatrix} 2 & 1 \\ 1 & 2 \end{bmatrix} = A.$$

An example of e^{At}.　　Now we compute e^{At}:

$$\exp(At) = S \exp(\Lambda t) S^{-1}$$

$$= \begin{bmatrix} 1 & 1 \\ 1 & -1 \end{bmatrix} \exp \left(t \begin{bmatrix} 3 & 0 \\ 0 & 1 \end{bmatrix} \right) \begin{bmatrix} \frac{1}{2} & \frac{1}{2} \\ \frac{1}{2} & -\frac{1}{2} \end{bmatrix}$$

$$= \begin{bmatrix} 1 & 1 \\ 1 & -1 \end{bmatrix} \begin{bmatrix} e^{3t} & 0 \\ 0 & e^t \end{bmatrix} \begin{bmatrix} \frac{1}{2} & \frac{1}{2} \\ \frac{1}{2} & -\frac{1}{2} \end{bmatrix}$$

$$= \begin{bmatrix} 1 & 1 \\ 1 & -1 \end{bmatrix} \begin{bmatrix} e^{3t}/2 & e^{3t}/2 \\ e^t/2 & -e^t/2 \end{bmatrix}$$

$$= \begin{bmatrix} \frac{1}{2}(e^{3t} + e^t) & \frac{1}{2}(e^{3t} - e^t) \\ \frac{1}{2}(e^{3t} - e^t) & \frac{1}{2}(e^{3t} + e^t) \end{bmatrix}.$$

Finally, we know that $\mathbf{x}(t) = e^{At}\mathbf{x}_0$; in this example, $\mathbf{x}_0 = \begin{bmatrix} 1 \\ 0 \end{bmatrix}$:

$$\mathbf{x}(t) = \begin{bmatrix} x_1(t) \\ x_2(t) \end{bmatrix} = \begin{bmatrix} \frac{1}{2}(e^{3t} + e^t) & \frac{1}{2}(e^{3t} - e^t) \\ \frac{1}{2}(e^{3t} - e^t) & \frac{1}{2}(e^{3t} + e^t) \end{bmatrix} \begin{bmatrix} 1 \\ 0 \end{bmatrix} = \begin{bmatrix} \frac{1}{2}(e^{3t} + e^t) \\ \frac{1}{2}(e^{3t} - e^t) \end{bmatrix}.$$

Thus we arrive at the formulas

$$x_1(t) = \frac{e^{3t} + e^t}{2} \qquad \text{and} \qquad x_2(t) = \frac{e^{3t} - e^t}{2}.$$

A quick substitution of $t = 0$ verifies that $x_1(0) = 1$ and $x_2(0) = 0$. Furthermore,

$$\begin{aligned} x_1'(t) &= \frac{d}{dt}\left(\frac{e^{3t} + e^t}{2}\right) \\ &= \frac{3e^{3t} + e^t}{2} \\ &= 2\left(\frac{e^{3t} + e^t}{2}\right) + \left(\frac{e^{3t} - e^t}{2}\right) \\ &= 2x_1(t) + x_2(t), \end{aligned}$$

and

$$\begin{aligned} x_2'(t) &= \frac{d}{dt}\left(\frac{e^{3t} - e^t}{2}\right) \\ &= \frac{3e^{3t} - e^t}{2} \\ &= \left(\frac{e^{3t} + e^t}{2}\right) + 2\left(\frac{e^{3t} - e^t}{2}\right) \\ &= x_1(t) + 2x_2(t). \end{aligned}$$

Thus we have verified that $\begin{bmatrix} x_1 \\ x_2 \end{bmatrix}' = \begin{bmatrix} 2 & 1 \\ 1 & 2 \end{bmatrix} \begin{bmatrix} x_1 \\ x_2 \end{bmatrix}$, as required.

We do another quick check: Earlier, we used MATLAB to find $\mathbf{x}(2) \approx \begin{bmatrix} 205.4089 \\ 198.0199 \end{bmatrix}$. Use your calculator to verify

$$x_1(2) = \frac{e^{3\cdot 2} + e^2}{2} \approx 205.4089, \qquad \text{and}$$

$$x_2(2) = \frac{e^{3\cdot 2} - e^2}{2} \approx 198.0199.$$

It is now clear that since $e^{3t} \to \infty$ and even $e^{3t} - e^t \to \infty$ as $t \to \infty$, that $|\mathbf{x}(t)| \to \infty$, as we suspected from the numerical evidence.

Recapping the case $\mathbf{x}' = A\mathbf{x}$.   This was a lot of work (and there are some more bumps and twists ahead—sigh!), but let's focus on the highlights of what we have witnessed in this example:

1. We began with the system $\mathbf{x}' = A\mathbf{x}$, where $A = \begin{bmatrix} 2 & 1 \\ 1 & 2 \end{bmatrix}$.

2. The eigenvalues of A are 3 and 1.

3. The answer (i.e., the functions $x_1(t)$ and $x_2(t)$) consists of linear combinations of e^{3t} and e^{1t}.

These observations hold true for any system of the form $\mathbf{x}' = A\mathbf{x}$. Although we can succinctly express the solution as $\mathbf{x}(t) = e^{At}\mathbf{x}_0$, the individual components of the answer (the $x_1(t), \ldots, x_n(t)$) are linear combinations of $e^{\lambda_1 t}, e^{\lambda_2 t}, \ldots, e^{\lambda_n t}$, where $\lambda_1, \lambda_2, \ldots, \lambda_n$ are the eigenvalues of A.

For example, if our matrix had been $A = \begin{bmatrix} 2 & 3 \\ 6 & -8 \end{bmatrix}$, then the solution to the system $\mathbf{x}' = A\mathbf{x}$ would be some linear combination of $e^{\lambda_1 t}$ and $e^{\lambda_2 t}$, where λ_1, λ_2 are the eigenvalues of A. What are these eigenvalues? We could work them out in the usual way,[6] or we might be satisfied with just knowing their numerical values. MATLAB easily gives us the eigenvalues:

```
a =
     2     3
     6    -8

>>eig(a)

ans =
    3.5574
   -9.5574
```

So now we know that $x_1(t)$ and $x_2(t)$ involve terms of the form $e^{3.5574t}$ and $e^{-9.5574t}$. What happens as $t \to \infty$? We note that while $e^{-9.5574t} \to 0$, we also have $e^{3.5574t} \to \infty$. Thus for typical $\mathbf{x}_0$, we have $|\mathbf{x}(t)| \to \infty$.

---

[6] ... and we would find that A's eigenvalues are $-3 + \sqrt{43}$ and $-3 - \sqrt{43}$.

We are on the verge of a general principle: If the eigenvalues are all negative, then $\mathbf{x}(t) \to \mathbf{0}$ as $t \to \infty$. However, if some of A's eigenvalues are positive, then typically $|\mathbf{x}(t)| \to \infty$.

You may be wondering (or worrying) about how to handle the case when A has *complex* eigenvalues. It is still true that $A = S\Lambda S^{-1}$, that $\exp(At) = S\exp(\Lambda t)S^{-1}$, and that $\exp(\Lambda t)$ is a diagonal matrix whose diagonal entries are $e^{\lambda_1 t}, e^{\lambda_2 t}, \ldots, e^{\lambda_n t}$. The worry is, how do we handle terms like $e^{\lambda t}$ when λ is complex? This exact matter is discussed in §A.2. (Yes, please spend some time there now.) Suppose $\lambda = a + bi$. Then $e^{\lambda t} = e^{at+bti} = e^{at}e^{bti}$. The e^{at} part is just a real number whereas the e^{bti} equals (by Euler's formula, equation (A.4) on page 348) $\cos bt + i \sin bt$. Summarizing, we have

> What if there are *complex* eigenvalues?

$$e^{\lambda t} = e^{(a+bi)t} = e^{at}e^{(bt)i} = e^{at}(\cos bt + i \sin bt).$$

Of particular note is that $|e^{(a+bi)t}| = |e^{at}|$. Thus if a, the real part of λ, denoted by $\Re\lambda$, is positive, then $|e^{\lambda t}| \to \infty$ as $t \to \infty$. If $a < 0$ ($\Re\lambda$ is negative) then $e^{\lambda t} \to 0$ as $t \to \infty$. Finally, if $a = 0$ (λ is purely imaginary), then $|e^{\lambda t}| = 1$ for all t and (unless $b = 0$) $e^{\lambda t}$ spins around the origin for all time.

> $\Re\lambda$ stands for the real part of the complex number λ.

Thus our general principle becomes: If the real parts of all of A's eigenvalues are negative, then $\mathbf{x}(t) \to \mathbf{0}$ as $t \to \infty$; if some eigenvalue has positive real part, then typically $|\mathbf{x}(t)| \to \infty$; and if $\Re\lambda \leq 0$ for all λ, but $\Re\lambda = 0$ for some λ, then $\mathbf{x}(t)$ neither settles down to any specific value nor does it blow up.

What is the nature of individual functions (the $x_j(t)$'s) in $\mathbf{x}$? Since the starting vector $\mathbf{x}_0$ and the changes $\mathbf{x}' = A\mathbf{x}$ are all real, we expect all the $x_j(t)$ functions to be real-valued. Well, indeed they are! If $\lambda = a + bi$ is an eigenvalue of A, then, necessarily, so is $\bar{\lambda} = a - bi$. Now,

> With complex numbers everywhere, how do we know the final answer we get will be real?

$$e^{\lambda t} = e^{(a+bi)t} = e^{at}(\cos bt + i \sin bt), \quad \text{and}$$

$$e^{\bar{\lambda} t} = e^{(a-bi)t} = e^{at}(\cos bt - i \sin bt).$$

In other words, $e^{\lambda t}$ and $e^{\bar{\lambda} t}$ can be expressed as linear combinations of $e^{at}\cos bt$ and $e^{at}\sin bt$.

Suppose $\lambda = a + bi$ is complex ($b \neq 0$) and is an eigenvalue of A. Our theory tells us that the $x_j(t)$'s contain terms of the form $e^{\lambda t}$ and $e^{\bar{\lambda} t}$. However, we can replace those terms by terms involving $e^{at}\cos bt$ and $e^{at}\sin bt$. When we do this, all the terms involving i will disappear. Hard to believe? Let us see this in action by revisiting the oscillator example of Chapter 1.

Mass-and-spring, again.   In §1.2.1 on page 8 we considered an ideal system consisting of a block
sliding on a frictionless surface attached to a wall by an ideal spring. (In
§1.2.2 on page 10 we observed that a resistance-free RLC circuit behaves
in the same manner as the frictionless mass and spring.) If we take the
mass m of the block and the spring constant k both equal to 1, we derive
equation (1.8) on page 9 which we repeat here:

$$\begin{bmatrix} x \\ v \end{bmatrix}' = \begin{bmatrix} 0 & 1 \\ -1 & 0 \end{bmatrix} \begin{bmatrix} x \\ v \end{bmatrix}.$$

This system is of the form $\mathbf{y}' = A\mathbf{y}$, where $\mathbf{y} = \begin{bmatrix} x \\ v \end{bmatrix}$ and $A = \begin{bmatrix} 0 & 1 \\ -1 & 0 \end{bmatrix}$.

We take $\mathbf{y}_0 = \mathbf{y}(0) = \begin{bmatrix} 1 \\ 0 \end{bmatrix}$.

We know that the solution to this system is $\mathbf{y}(t) = e^{At}\mathbf{y}_0$, so we need
to compute e^{At}.

First, we find A's eigenvalues. The characteristic equation is

$$\det(\lambda I - A) = \lambda^2 + 1,$$

so the eigenvalues are i and $-i$. Notice that $\begin{bmatrix} 1 \\ i \end{bmatrix}$ is an eigenvector associ-

ated with $\lambda = i$ and that $\begin{bmatrix} 1 \\ -i \end{bmatrix}$ is an eigenvector associated with $\lambda = -i$.

Put $S = \begin{bmatrix} 1 & 1 \\ i & -i \end{bmatrix}$ (and so $S^{-1} = \begin{bmatrix} \frac{1}{2} & -\frac{i}{2} \\ \frac{1}{2} & \frac{i}{2} \end{bmatrix}$) and $\Lambda = \begin{bmatrix} i & 0 \\ 0 & -i \end{bmatrix}$.

Therefore,

$$e^{At} = \exp\left(t\begin{bmatrix} 0 & 1 \\ -1 & 0 \end{bmatrix}\right)$$

$$= S\exp(t\Lambda)S^{-1}$$

$$= \begin{bmatrix} 1 & 1 \\ i & -i \end{bmatrix} \begin{bmatrix} e^{it} & 0 \\ 0 & e^{-it} \end{bmatrix} \begin{bmatrix} \frac{1}{2} & -\frac{i}{2} \\ \frac{1}{2} & \frac{i}{2} \end{bmatrix}$$

$$= \begin{bmatrix} 1 & 1 \\ i & -i \end{bmatrix} \begin{bmatrix} \frac{1}{2}e^{it} & -\frac{i}{2}e^{it} \\ \frac{1}{2}e^{-it} & \frac{i}{2}e^{-it} \end{bmatrix}$$

$$= \begin{bmatrix} \frac{1}{2}(e^{it} + e^{-it}) & \frac{1}{2}(-ie^{it} + ie^{-it}) \\ \frac{1}{2}(ie^{it} - ie^{-it}) & \frac{1}{2}(e^{it} + e^{-it}) \end{bmatrix}.$$

Using the complex form (see §A.2) for the sine and cosine functions,

$$\cos t = \frac{e^{it} + e^{-it}}{2} \quad \text{and} \quad \sin t = \frac{e^{it} - e^{-it}}{2i},$$

and recognizing that

$$\frac{-ie^{it} + ie^{-it}}{2} = \frac{e^{it} - e^{-it}}{2i} = \sin t,$$

we finish our computation of e^{At}:

$$e^{At} = \begin{bmatrix} \frac{1}{2}(e^{it} + e^{-it}) & \frac{1}{2}(-ie^{it} + ie^{-it}) \\ \frac{1}{2}(ie^{it} - ie^{-it}) & \frac{1}{2}(e^{it} + e^{-it}) \end{bmatrix}$$

$$= \begin{bmatrix} \cos t & \sin t \\ -\sin t & \cos t \end{bmatrix}.$$

Finally, we know that $\mathbf{y}(t) = e^{At}\mathbf{y}_0$, so

$$\mathbf{y} = \begin{bmatrix} x \\ v \end{bmatrix} = \begin{bmatrix} \cos t & \sin t \\ -\sin t & \cos t \end{bmatrix} \begin{bmatrix} 1 \\ 0 \end{bmatrix} = \begin{bmatrix} \cos t \\ -\sin t \end{bmatrix}.$$

Hence $x(t) = \cos t$, as we saw in §1.2.1.

Now it is time to add some friction. If the block is moving at a modest speed through air, the air pushes back on the block with a force proportional to the speed of the block. Symbolically, the force on the block due to air resistance is $-\mu v$; the minus sign indicates that the direction of the force is opposite the direction of motion. The number μ is a positive constant. Thus the total force on the block (from the spring and air resistance) is $-kx - \mu v$. If we take (as before) $m = k = 1$, then the system is

Adding friction and resistance.

$$\begin{bmatrix} x \\ v \end{bmatrix}' = \begin{bmatrix} 0 & 1 \\ -1 & -\mu \end{bmatrix} \begin{bmatrix} x \\ v \end{bmatrix}. \tag{2.11}$$

Now we find the eigenvalues of $A = \begin{bmatrix} 0 & 1 \\ -1 & -\mu \end{bmatrix}$. The characteristic equation is

$$\det(\lambda I - A) = \det \begin{bmatrix} \lambda & -1 \\ 1 & \lambda + \mu \end{bmatrix}$$

$$= \lambda(\lambda + \mu) + 1$$

$$= \lambda^2 + \mu\lambda + 1.$$

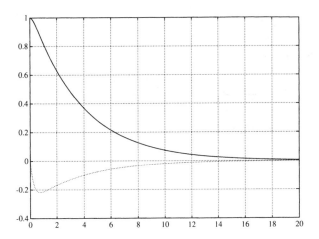

Figure 2.11. Mass-and-spring oscillator with $\mu = 4$. Solid curve is position x, and dotted curve is velocity v.

Hence A's eigenvalues are

$$\frac{-\mu \pm \sqrt{\mu^2 - 4}}{2}.$$

This formula suggests three cases to consider: (1) $\mu > 2$, (2) $0 < \mu < 2$, and (3) $\mu = 2$.

The case $\mu > 2$: overdamping.

In case (1), with $\mu > 2$, both eigenvalues are real. Clearly, $(-\mu - \sqrt{\mu^2 - 4})/2$ is negative. What about the other eigenvalue? Notice that

$$\sqrt{\mu^2 - 4} < \sqrt{\mu^2} = \mu,$$

hence $-\mu + \sqrt{\mu^2 - 4} < 0$. Thus, both eigenvalues are negative. It follows then, that as time progresses, both x and v go to 0. Moreover, the resistance is so heavy that no oscillation can take place. We know this because the formulas for $x(t)$ and $v(t)$ are both linear combinations of $e^{\lambda_1 t}$ and $e^{\lambda_2 t}$, where both λ's are negative real numbers. Figure 2.11 is a plot of the behavior of the system with $\mathbf{y}_0 = \begin{bmatrix} 1 \\ 0 \end{bmatrix}$ and $\mu = 4$.

The case $\mu < 2$: underdamping.

In case (2), when $\mu < 2$, we see that the two eigenvalues, which are $(-\mu \pm \sqrt{\mu^2 - 4})/2$, are complex numbers (since $\mu^2 - 4 < 0$). The real parts of these eigenvalues are both $-\mu/2$, which is negative. Thus we again

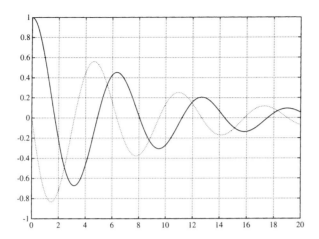

Figure 2.12. Mass-and-spring oscillator with $\mu = \frac{1}{4}$. Solid curve is position x, and dotted curve is velocity v.

expect to see x and v tending to 0 as $t \to \infty$. However, in this case we expect to see oscillation, since the functions $x(t)$ and $v(t)$ will involve sines and cosines. In particular, they will be linear combinations of

$$e^{-(\mu/2)t} \sin\left(t\sqrt{4 - \mu^2}\right) \quad \text{and} \quad e^{-(\mu/2)t} \cos\left(t\sqrt{4 - \mu^2}\right).$$

Figure 2.12 shows the behavior of $x(t)$ and $v(t)$ when $\mu = \frac{1}{4}$.

Finally we consider case (3), in which $\mu = 2$. In this case we have a repeated eigenvalue, $\lambda = -1$. Unfortunately, in this case the matrix $A = \begin{bmatrix} 0 & 1 \\ -1 & -\mu \end{bmatrix} = \begin{bmatrix} 0 & 1 \\ -1 & -2 \end{bmatrix}$ is *not* diagonalizable! In §2.2.3 we discuss the behavior of systems in which the matrix is nondiagonalizable. We will learn that $x(t)$ and $v(t)$ are linear combinations of e^{-t} and te^{-t}. Figure 2.13 shows the behavior of $x(t)$ and $v(t)$ when $\mu = 2$. Notice that the system does not oscillate; there are no sines or cosines in the formulas. Interestingly, the system settles down *faster* when $\mu = 2$ than when $\mu = 4$.

The case $\mu = 2$: critical damping.

(If we consider the RLC circuit from Figure 1.2 on page 10 in which $L = C = 1$, but R is not 0, we have (see equation (1.11) on page 11)

RLC circuits behave just like the mass-and-spring system.

$$\begin{bmatrix} V \\ I \end{bmatrix}' = \begin{bmatrix} 0 & -1 \\ 1 & -R \end{bmatrix} \begin{bmatrix} V \\ I \end{bmatrix}.$$

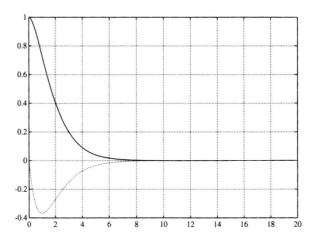

Figure 2.13. Mass-and-spring oscillator with $\mu = 2$. Solid curve is position x, and dotted curve is velocity v.

This equation is nearly identical with equation (2.11) on page 69 for the mass-and-spring system with air resistance. Indeed, the eigenvalues of the matrix $\begin{bmatrix} 0 & -1 \\ 1 & -R \end{bmatrix}$ are the same as those of the matrix $\begin{bmatrix} 0 & 1 \\ -1 & -\mu \end{bmatrix}$ if we put $\mu = R$. The mass-and-spring system and the RLC circuit behave in the same manner.)

Reviewing $\mathbf{x}' = A\mathbf{x}$.

Let us summarize the important points of what we have learned about the case $\mathbf{x}' = A\mathbf{x}$ with $\mathbf{x}(0) = \mathbf{x}_0$.

- The solution can be written as $\mathbf{x}(t) = e^{At}\mathbf{x}_0$.

- If A diagonalizes, $A = S\Lambda S^{-1}$, then $e^{At} = Se^{\Lambda t}S^{-1}$.

- It follows, therefore, that the individual components of $\mathbf{x}$ (i.e., $x_1(t)$ through $x_n(t)$) are linear combinations of $e^{\lambda_1 t}, \ldots, e^{\lambda_n t}$.

- If $a \pm bi$ are complex eigenvalues of A, then $e^{(a \pm bi)t}$ can be expressed using the functions $e^{at}\sin bt$ and $e^{at}\cos bt$.

- Finally:

  - If all of A's eigenvalues have negative real part, then $\mathbf{x}(t) \to \mathbf{0}$ as $t \to \infty$.

- If some of A's eigenvalues have positive real part, then, typically, $|\mathbf{x}(t)| \to \infty$ as $t \to \infty$.

- If all of A's eigenvalues have negative or zero real part, and some have zero real part, then, typically, $\mathbf{x}(t)$ neither tends toward $\mathbf{0}$ nor explodes.

We are ready for the full case: The full system: $\mathbf{x}' = A\mathbf{x} + \mathbf{b}$.

$$\mathbf{x}' = A\mathbf{x} + \mathbf{b} \qquad \text{with} \qquad \mathbf{x}(0) = \mathbf{x}_0. \qquad (2.12)$$

We are still assuming that the matrix A diagonalizes, so $A = S\Lambda S^{-1}$. If we replace A with $S\Lambda S^{-1}$ in $\mathbf{x}' = A\mathbf{x} + \mathbf{b}$, we get

$$\mathbf{x}' = \left(S\Lambda S^{-1}\right)\mathbf{x} + \mathbf{b}.$$

We now multiply on the left by S^{-1}

$$S^{-1}\left[\quad \mathbf{x}' = S\Lambda S^{-1}\mathbf{x} + \mathbf{b} \quad\right]$$

$$\Rightarrow \ S^{-1}\mathbf{x}' = \Lambda S^{-1}\mathbf{x} + S^{-1}\mathbf{b}.$$

Now, $S^{-1}\mathbf{x}' = (S^{-1}\mathbf{x})'$ (see problem 8 on page 89), so we make the substitution $\mathbf{u} = S^{-1}\mathbf{x}$ (and therefore $\mathbf{u}' = S^{-1}\mathbf{x}'$) and arrive at

$$\mathbf{u}' = \Lambda\mathbf{u} + \mathbf{c} \qquad \text{with} \qquad \mathbf{u}(0) = \mathbf{u}_0, \qquad (2.13)$$

where $\mathbf{c} = S^{-1}\mathbf{b}$ and $\mathbf{u}_0 = S^{-1}\mathbf{x}_0$. The matrix equation (2.13) can now be written out as

$$u_1' = \lambda_1 u_1 + c_1,$$

$$u_2' = \lambda_2 u_2 + c_2,$$

$$\vdots$$

$$u_n' = \lambda_n u_n + c_n.$$

Notice that each of these differential equations involves only *one* of the u_j's at a time! Thus we can use the methods of §2.1.2 (in particular, see equation (2.6) on page 46) and we have that

$$u_j(t) = \begin{cases} e^{\lambda_j t}\left(u_j(0) + \frac{c_j}{\lambda_j}\right) - \frac{c_j}{\lambda_j} & \text{when } \lambda_j \neq 0, \text{ and} \\ c_j t + u_j(0) & \text{when } \lambda_j = 0, \end{cases}$$

which gives us explicit formulas for $u_1(t), \ldots, u_n(t)$. To work out the formulas for the x_j's we recall that $\mathbf{u} = S^{-1}\mathbf{x}$, and so $\mathbf{x} = S\mathbf{u}$.

A full example of a system of the form $\mathbf{x}' = A\mathbf{x} + \mathbf{b}$.

Let's do an example. Suppose the system is

$$\begin{bmatrix} x_1 \\ x_2 \\ x_3 \end{bmatrix}' = \begin{bmatrix} 2 & 1 & 1 \\ 1 & 2 & 1 \\ 3 & 3 & 2 \end{bmatrix} \begin{bmatrix} x_1 \\ x_2 \\ x_3 \end{bmatrix} + \begin{bmatrix} -3 \\ 0 \\ 2 \end{bmatrix}$$

with $\mathbf{x}_0 = \begin{bmatrix} 1 \\ 1 \\ 1 \end{bmatrix}$.

We begin by computing the eigenvalues of $A = \begin{bmatrix} 2 & 1 & 1 \\ 1 & 2 & 1 \\ 3 & 3 & 2 \end{bmatrix}$. (You may work them out by hand or with computer software.) They are 5, 1, and 0. Corresponding to these we have eigenvectors

$$\begin{bmatrix} 1 \\ 1 \\ 2 \end{bmatrix}, \quad \begin{bmatrix} 1 \\ -1 \\ 0 \end{bmatrix}, \quad \text{and} \quad \begin{bmatrix} 1 \\ 1 \\ -3 \end{bmatrix},$$

respectively. Thus S and S^{-1} are

$$S = \begin{bmatrix} 1 & 1 & 1 \\ 1 & -1 & 1 \\ 2 & 0 & -3 \end{bmatrix} \quad \text{and} \quad S^{-1} = \frac{1}{10} \begin{bmatrix} 3 & 3 & 2 \\ 5 & -5 & 0 \\ 2 & 2 & -2 \end{bmatrix}.$$

Let $\mathbf{u} = S^{-1}\mathbf{x}$ and

$$\mathbf{c} = S^{-1}\mathbf{b} = \frac{1}{10} \begin{bmatrix} 3 & 3 & 2 \\ 5 & -5 & 0 \\ 2 & 2 & -2 \end{bmatrix} \begin{bmatrix} -3 \\ 0 \\ 2 \end{bmatrix} = \begin{bmatrix} -1/2 \\ -3/2 \\ -1 \end{bmatrix}.$$

Thus the system becomes $\mathbf{u}' = \begin{bmatrix} 5 & 0 & 0 \\ 0 & 1 & 0 \\ 0 & 0 & 0 \end{bmatrix} \mathbf{u} + \begin{bmatrix} -1/2 \\ -3/2 \\ -1 \end{bmatrix}$, with $\mathbf{u}_0 =$

$$S^{-1}\mathbf{x}_0 = \frac{1}{10} \begin{bmatrix} 3 & 3 & 2 \\ 5 & -5 & 0 \\ 2 & 2 & -2 \end{bmatrix} \begin{bmatrix} 1 \\ 1 \\ 1 \end{bmatrix} = \begin{bmatrix} 8/10 \\ 0 \\ 2/10 \end{bmatrix}.$$ In long notation, the

problem is

$$\begin{array}{llll} u_1'(t) & = & 5u_1(t) - 1/2, & u_1(0) & = & 8/10, \\ u_2'(t) & = & u_2(t) - 3/2, & u_2(0) & = & 0, \\ u_3'(t) & = & -1, & u_3(0) & = & 2/10. \end{array}$$

What makes this system tractable is that each equation involves only one $u_j(t)$. We can solve them separately. Recall that the solution to the one-dimensional equation $x' = ax + b$ is $x(t) = e^{at}\left(x_0 + \frac{b}{a}\right) - \frac{b}{a}$ (see equation (2.6) on page 46). Applying that solution to this case, we get

$$u_1(t) = e^{5t}\left(\frac{8}{10} - \frac{1}{10}\right) + \frac{1}{10} = \frac{7e^{5t} + 1}{10},$$

$$u_2(t) = e^t\left(0 - \frac{3}{2}\right) + \frac{3}{2} = \frac{-3e^t + 3}{2},$$

$$u_3(t) = -t + 2/10.$$

Finally, we really want to know $\mathbf{x}$, which equals $S\mathbf{u}$, hence

$$\mathbf{x} = \begin{bmatrix} x_1 \\ x_2 \\ x_3 \end{bmatrix} = \begin{bmatrix} 1 & 1 & 1 \\ 1 & -1 & 1 \\ 2 & 0 & -3 \end{bmatrix} \begin{bmatrix} (7e^{5t} + 1)/10 \\ (-3e^t + 3)/2 \\ -t + 2/10 \end{bmatrix}$$

$$= \frac{1}{10} \begin{bmatrix} 18 - 15e^t + 7e^{5t} - 10t \\ -12 + 15e^t + 7e^{5t} - 10t \\ -4 + 14e^{5t} + 30t \end{bmatrix},$$

or in long notation,

$$x_1(t) = \left(18 - 15e^t + 7e^{5t} - 10t\right)/10,$$

$$x_2(t) = \left(-12 + 15e^t + 7e^{5t} - 10t\right)/10,$$

$$x_3(t) = \left(-4 + 14e^{5t} + 30t\right)/10.$$

What happens as $t \to \infty$? It is clear that the e^{5t} term dominates all others in the preceding formulas. Letting $t \to \infty$, we see that

$$\mathbf{x}(t) \approx \frac{7e^{5t}}{10} \begin{bmatrix} 1 \\ 1 \\ 2 \end{bmatrix},$$

that is, $\mathbf{x}(t)$ is exploding in the direction of the eigenvector associated with the largest eigenvalue ($\lambda = 5$).

We can now describe the behavior of multivariable, linear continuous time dynamical systems $\mathbf{x}' = A\mathbf{x} + \mathbf{b}$ (assuming that A diagonalizes). By changing variables to $\mathbf{u} = S^{-1}\mathbf{x}$, we decouple the system, i.e., we have

A full description of the system $\mathbf{x}' = A\mathbf{x} + \mathbf{b}$.

Continuous: $\mathbf{x}'(t) = A\mathbf{x}(t) + \mathbf{b}$ with $\mathbf{x}(0) = \mathbf{x}_0$	
Conditions on eigenvalues of A	Behavior of $\mathbf{x}(t)$
All have $\Re\lambda < 0$	converges to $\tilde{\mathbf{x}} = -A^{-1}\mathbf{b}$
Some have $\Re\lambda > 0$	typically, $|\mathbf{x}(t)| \to \infty$
All have $\Re\lambda \leq 0$; and some have $\Re\lambda = 0$	stays near but does not approach $\tilde{\mathbf{x}}$, or $|\mathbf{x}(t)| \to \infty$

Table 2.4. The possible behaviors of multidimensional continuous time linear systems.

$\mathbf{u}' = \Lambda\mathbf{u} + \mathbf{c}$, where Λ is a diagonal matrix. Thus we have converted our original problem with n equations in n intertwined variables into a new system consisting of n equations in n separated variables. We solve the easy system for $\mathbf{u}$ and convert back, knowing that $\mathbf{x} = S\mathbf{u}$.

If there are complex eigenvalues, then S will be a complex matrix and $\mathbf{u}$ may have complex entries. However, when we convert back to $\mathbf{x}$ we are able to convert the $e^{(a\pm bi)t}$ terms into $e^{at}\sin bt$ and $e^{at}\cos bt$ terms. (See problem 12 on page 89.)

Although it is nice to be able to compute explicitly the solutions to these linear systems, we are mostly interested in their long-term behavior. We see that the critical issue is the sign of the real part of the eigenvalues. Table 2.4 summarizes what we have learned.

Geometry

All eigenvalues in the left half-plane for stability; an eigenvalue in the right half-plane implies instability.

We have seen that the behavior of the continuous time system $\mathbf{x}' = A\mathbf{x} + \mathbf{b}$ is driven by the real part of the eigenvalues of A. The three cases discussed above are illustrated in Figure 2.14. These cases can be described geometrically as follows. In case (a), all the eigenvalues are points lying strictly to the left of the y-axis. In case (b), some points (eigenvalues) lie strictly to the right of the y-axis. And in case (c), we have some points on, and the remaining points left of, the y-axis.

Now we focus on the two-dimensional case (i.e., $\mathbf{x} \in \mathbf{R}^2$ and A is a 2×2 matrix), as we can plot interesting diagrams in the plane. Let $f(\mathbf{x}) = A\mathbf{x} + \mathbf{b}$, where A is a 2×2 matrix and $\mathbf{b} \in \mathbf{R}^2$. For several points

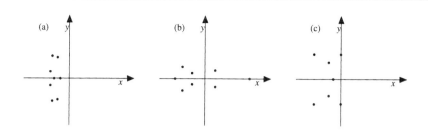

Figure 2.14. The behavior of continuous linear systems. Three cases are shown: (a) all $\Re\lambda < 0$, (b) some $\Re\lambda > 0$, and (c) all $\Re\lambda \leq 0$ and some $\Re\lambda = 0$.

x in the plane, we plot an arrow anchored at **x** pointing in the direction of $f(\mathbf{x})$. Such a diagram depicts a *vector field*: every point in the plane **x** is assigned[7] a vector $f(\mathbf{x})$.

As you examine the figures, follow the arrows in your mind to see how the system evolves. We have plotted some trajectories of the system in the pictures as well.

Let's begin with Figure 2.15, which illustrates the system

$$\left[\begin{array}{c} x_1 \\ x_2 \end{array}\right]' = \left[\begin{array}{cc} \frac{5}{2} & \frac{1}{8} \\ 0 & 1 \end{array}\right] \left[\begin{array}{c} x_1 \\ x_2 \end{array}\right].$$

Two real, positive eigenvalues.

The eigenvalues of A are $\frac{5}{2}$ and 1, so since both are positive, we expect to see the system explode. Indeed, the eigenvector corresponding to $\frac{5}{2}$ is $\left[\begin{array}{c} 1 \\ 0 \end{array}\right]$, and as $t \to \infty$ we see that the system is heading off toward infinity parallel to the x-axis.

A similar system is illustrated in Figure 2.16. In this case the system is

$$\left[\begin{array}{c} x_1 \\ x_2 \end{array}\right]' = \left[\begin{array}{cc} 0 & 1 \\ -5 & 2 \end{array}\right] \left[\begin{array}{c} x_1 \\ x_2 \end{array}\right].$$

Complex eigenvalues with positive real parts. Notice the swirling.

The eigenvalues of A are $1 \pm 2i$, which both have positive real part. Thus we observe that $|\mathbf{x}(t)| \to \infty$ as $t \to \infty$. However, in this case $\mathbf{x}(t)$ is not exploding in a straight-line direction; rather, the trajectory circles around and around in an ever widening spiral, corresponding to the sine and cosine terms generated by the imaginary part of the eigenvalues.

---

[7]For clarity in the figures, the arrows we draw all have the same length. We show just the direction of the vector $f(\mathbf{x})$.

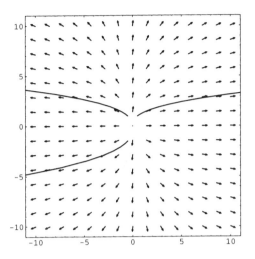

Figure 2.15. A dynamical system $\mathbf{x}' = A\mathbf{x}$ where the eigenvalues of A are $5/2$ and 1.

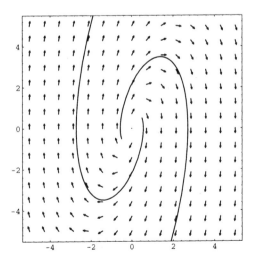

Figure 2.16. A dynamical system $\mathbf{x}' = A\mathbf{x}$ where the eigenvalues of A are $1 \pm 2i$.

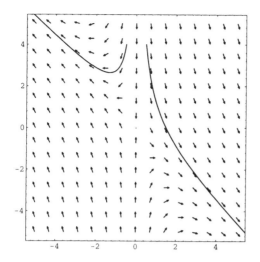

Figure 2.17. A dynamical system $\mathbf{x}' = A\mathbf{x}$ where the eigenvalues of A are ± 1.

Figure 2.17 is another example of a dynamical system in divergence. In this case, the eigenvalues of A are ± 1. Since there is a positive eigenvalue, most $\mathbf{x}_0$ will cause the system to explode. In this system, however, if we choose $\mathbf{x}_0$ on the y-axis, we would have $\mathbf{x}(t) \to \mathbf{0}$ as $t \to \infty$.

One positive, one negative.

Figure 2.18 illustrates a system in which the eigenvalues are both negative. In this example, $A = \begin{bmatrix} \frac{1}{8} & \frac{1}{4} \\ -\frac{5}{6} & -\frac{3}{4} \end{bmatrix}$, whose eigenvalues are $-\frac{1}{4}$ and $-\frac{1}{3}$. Notice that all trajectories lead to the origin, $\mathbf{0}$.

Both eigenvalues negative.

Figure 2.19 illustrates a system in which the eigenvalues both have negative real part. In this example, $A = \begin{bmatrix} 0 & 1 \\ -\frac{5}{4} & -1 \end{bmatrix}$, whose eigenvalues are $-\frac{1}{2} \pm i$. In this system, all trajectories spiral in to the origin, $\mathbf{0}$. The swirling is caused by the sine and cosine terms.

Complex eigenvalues with negative real parts.

We now consider cases in which we have one zero and one negative eigenvalue. Figure 2.20 illustrates the system $\mathbf{x}' = A\mathbf{x}$ where $A = \begin{bmatrix} -1 & \frac{1}{2} \\ 2 & -1 \end{bmatrix}$. The eigenvalues of A are -2 and 0. This system has an entire line of fixed points (see the dots in the figure). All states *not* on this line of fixed points gravitate toward some fixed point. These fixed points are marginally stable.

Two examples with zero and negative eigenvalues.

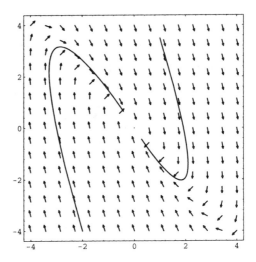

Figure 2.18. A dynamical system $\mathbf{x}' = A\mathbf{x}$ where the eigenvalues of A are $-1/4$ and $-1/3$.

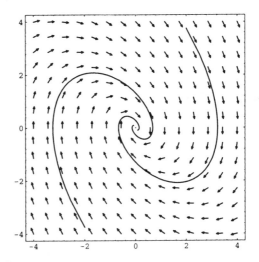

Figure 2.19. A dynamical system $\mathbf{x}' = A\mathbf{x}$ where the eigenvalues of A are $-\frac{1}{2} \pm i$.

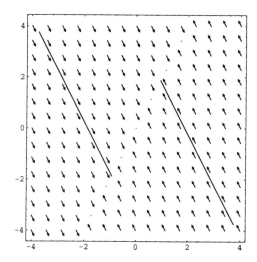

Figure 2.20. A dynamical system $\mathbf{x}' = A\mathbf{x}$ where the eigenvalues of A are 0 and -2.

In Figure 2.21 we have a system of the form $\mathbf{x}' = A\mathbf{x} + \mathbf{b}$ where again the eigenvalues of A are -2 and 0. Notice that $\mathbf{x}(t)$ goes to infinity as $t \to \infty$. In this system there are *no* fixed points and the trajectories all go off to infinity.

Finally, consider Figure 2.22. In this system the eigenvalues of A are pure imaginary numbers (real part is 0). The trajectories of this system are all closed curves (ellipses, in fact) about the fixed point $\mathbf{0}$. The origin is a marginally stable fixed point.

Purely imaginary eigenvalues.

2.2.3   The nondiagonalizable case*

We have been studying linear systems of the form

$$\begin{aligned} \mathbf{x}' &= A\mathbf{x} + \mathbf{b} & \text{(continuous time)} \\ \mathbf{x}(k+1) &= A\mathbf{x}(k) + \mathbf{b} & \text{(discrete time)} \end{aligned}$$

Nondiagonalizable systems have the same general behavior as diagonalizable systems.

where A is a *diagonalizable* matrix. We have seen two types of behavior: gravitation to a fixed point $\tilde{\mathbf{x}}$ or explosion to infinity for most $\mathbf{x}_0$ depending on either the real part or the absolute value of the eigenvalues of A.

Since the behavior of a linear system depends on its eigenvalues, it seems natural to worry that nondiagonalizable systems might behave rather

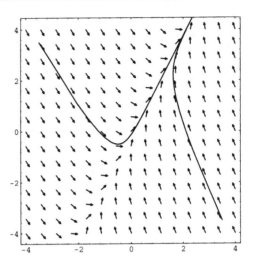

Figure 2.21. A dynamical system $\mathbf{x}' = A\mathbf{x} + \mathbf{b}$ where the eigenvalues of A are 0 and -2.

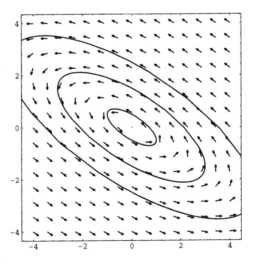

Figure 2.22. A dynamical system $\mathbf{x}' = A\mathbf{x}$ where the eigenvalues of A are $\pm i$.

differently. Here's the good news: Although the *exact* formulas for nondiagonalizable linear dynamical systems are a little different from the exact
formulas for the diagonalizable systems, the rules governing their gross
behavior are the same. In discrete time, if all eigenvalues have absolute
value less than 1, then the system gravitates toward a fixed point; if some
eigenvalue has absolute value greater than 1, then the system typically explodes. In continuous systems, if the real parts of the eigenvalues are all
negative, then the system tends to a fixed point; if some eigenvalue has
positive real part, we can expect the system to blow up.

So the question is, Can we skip this section? The answer is, Yes, for
now. It is safe to assume that nondiagonalizable systems behave just like
their diagonalizable cousins. The purpose of this section is to sketch the
theoretical underpinnings of this claim.

Are you still with us? Good, but before going on, please read Appendix A, especially §A.1.5 about the Jordan canonical form of a matrix.

We limit our discussion to systems in which $\mathbf{b} = \mathbf{0}$, that is, systems of
the form $\mathbf{x}' = A\mathbf{x}$ (continuous) or $\mathbf{x}(k+1) = A\mathbf{x}(k)$ (discrete). The general
solutions to these systems are

System	Solution
$\mathbf{x}' = A\mathbf{x}$	$\mathbf{x}(t) = e^{At}\mathbf{x}_0$
$\mathbf{x}(k+1) = A\mathbf{x}(k)$	$\mathbf{x}(k) = A^k\mathbf{x}_0$

The difficulty is knowing how e^{At} and A^k behave. Earlier, we deduced the
behavior by diagonalization; we no longer have that tool available. Instead,
we have the next best thing: the Jordan canonical form. If A is *any* square
matrix, then we can write $A = SJS^{-1}$, where J is in Jordan form (see
§A.1.5). Let's see how we can use this to study e^{At} and A^k.

The behavior of e^{At} and A^k
for nondiagonalizable A.

If we substitute SJS^{-1} for A in the formula for e^{At}, we get

$$e^{At} = I + tA + \frac{t^2}{2!}A^2 + \frac{t^3}{3!}A^3 + \cdots$$

$$= I + tSJS^{-1} + \frac{t^2}{2!}(SJS^{-1})^2 + \frac{t^3}{3!}(SJS^{-1})^3 + \cdots$$

$$= SS^{-1} + S(tJ)S^{-1} + S(\frac{t^2}{2!}J^2)S^{-1} + S(\frac{t^3}{3!}J^3)S^{-1} + \cdots$$

$$= S\left(I + tJ + \frac{t^2}{2!}J^2 + \frac{t^3}{3!}J^3 + \cdots\right)S^{-1}$$

$$= S e^{Jt} S^{-1}.$$

So far, this has been very similar to the case where A is diagonalizable. Now $e^{\Lambda t}$, where Λ is a diagonal matrix, is relatively easy to compute. To see what happens in the case e^{Jt} where J is a Jordan matrix, we consider an example. Suppose J is a 4×4 Jordan block, i.e.,

$$J = \begin{bmatrix} \lambda & 1 & 0 & 0 \\ 0 & \lambda & 1 & 0 \\ 0 & 0 & \lambda & 1 \\ 0 & 0 & 0 & \lambda \end{bmatrix}.$$

To compute e^{Jt} we first want to compute the powers of J. We get

$$J^0 = \begin{bmatrix} 1 & 0 & 0 & 0 \\ 0 & 1 & 0 & 0 \\ 0 & 0 & 1 & 0 \\ 0 & 0 & 0 & 1 \end{bmatrix} \quad \text{(identity matrix)},$$

$$J^1 = \begin{bmatrix} \lambda & 1 & 0 & 0 \\ 0 & \lambda & 1 & 0 \\ 0 & 0 & \lambda & 1 \\ 0 & 0 & 0 & \lambda \end{bmatrix} \quad \text{(J itself)},$$

$$J^2 = \begin{bmatrix} \lambda^2 & 2\lambda & 1 & 0 \\ 0 & \lambda^2 & 2\lambda & 1 \\ 0 & 0 & \lambda^2 & 2\lambda \\ 0 & 0 & 0 & \lambda^2 \end{bmatrix},$$

$$J^3 = \begin{bmatrix} \lambda^3 & 3\lambda^2 & 3\lambda & 1 \\ 0 & \lambda^3 & 3\lambda^2 & 3\lambda \\ 0 & 0 & \lambda^3 & 3\lambda^2 \\ 0 & 0 & 0 & \lambda^3 \end{bmatrix},$$

$$J^4 = \begin{bmatrix} \lambda^4 & 4\lambda^3 & 6\lambda^2 & 4\lambda \\ 0 & \lambda^4 & 4\lambda^3 & 6\lambda^2 \\ 0 & 0 & \lambda^4 & 4\lambda^3 \\ 0 & 0 & 0 & \lambda^4 \end{bmatrix},$$

$$J^5 = \begin{bmatrix} \lambda^5 & 5\lambda^4 & 10\lambda^3 & 10\lambda^2 \\ 0 & \lambda^5 & 5\lambda^4 & 10\lambda^3 \\ 0 & 0 & \lambda^5 & 5\lambda^4 \\ 0 & 0 & 0 & \lambda^5 \end{bmatrix}.$$

Do the first rows of these matrices look familiar? Yes? Good! What we see is the first few terms in the expansion of $(\lambda + 1)^k$ using the binomial theorem. For example,

$$(\lambda + 1)^5 = \underbrace{\lambda^5 + 5\lambda^4 + 10\lambda^3 + 10\lambda^2} + 5\lambda + 1.$$

Using binomial coefficient notation,[8] we have the first row of J^k:

$$\begin{bmatrix} \lambda^k & \binom{k}{1}\lambda^{k-1} & \binom{k}{2}\lambda^{k-2} & \binom{k}{3}\lambda^{k-3} \end{bmatrix}$$

and form the successive rows by pushing in a zero on the left. Now we can work on computing e^{Jt}.

Exponentiating a Jordan block.

The entries below the main diagonal of e^{Jt} are obviously all zero.

The main diagonal entries of e^{Jt} are the next simplest. Since the main diagonal entries of J^k are simply λ^k, the main diagonal entries of e^{Jt} are just

$$1 + \lambda t + \frac{1}{2!}\lambda^2 t^2 + \frac{1}{2!}\lambda^2 t^2 + \cdots = e^{\lambda t}.$$

Let us consider the diagonal just above the main diagonal. The entries in J^k are $k\lambda^{k-1}$, so the corresponding entries in e^{Jt} are

$$0 + t + \frac{1}{2!}t^2 2\lambda + \frac{1}{3!}t^3 3\lambda^2 + \frac{1}{4!}t^4 4\lambda^3 + \cdots + \frac{1}{k!}t^k k\lambda^{k-1} + \cdots,$$

from which we can cancel the k in the numerator with the k of the $k!$ in the denominator, and we get

$$t\left(1 + t\lambda + \frac{1}{2!}t^2\lambda^2 + \frac{1}{3!}t^3\lambda^3 + \cdots\right) = te^{\lambda t}.$$

So far we know that e^{Jt} has the form

$$\begin{bmatrix} e^{\lambda t} & te^{\lambda t} & ? & ?? \\ 0 & e^{\lambda t} & te^{\lambda t} & ? \\ 0 & 0 & e^{\lambda t} & te^{\lambda t} \\ 0 & 0 & 0 & e^{\lambda t} \end{bmatrix}.$$

Let's move up to the next diagonal (the '?' entries above). The entries in J^k are $\binom{k}{2}\lambda^{k-2}$, so the corresponding entries in e^{Jt} are

$$0+0+\frac{1}{2!}t^2+\frac{1}{3!}t^3 3\lambda+\frac{1}{4!}t^4 6\lambda^2+\frac{1}{5!}t^5 10\lambda^3+\cdots+\frac{1}{k!}t^k\frac{k(k-1)}{2}\lambda^{k-2}+\cdots.$$

---

[8] The expression $\binom{k}{r}$ stands for the coefficient of x^r in the expansion of $(x+1)^k$. It is also the number of ways to select an r-element subset from a fixed k-element set. Numerically, $\binom{k}{r}$ equals $\frac{k!}{r!(k-r)!}$. This can also be written as $\binom{k}{r} = \frac{1}{r!}k(k-1)(k-2)\cdots(k-r+1)$. This form suggests we think of $\binom{k}{r}$ as a polynomial in k of degree r.

As before, we cancel $k(k-1)$ upstairs with the first two terms of the $k!$ downstairs (and factor out a $t^2/2$) to get

$$\frac{t^2}{2}\left(1 + t\lambda + \frac{1}{2!}t^2\lambda^2 + \frac{1}{3!}t^3\lambda^3 + \cdots\right) = \frac{t^2}{2}e^{\lambda t}.$$

Finally, we consider the upper right corner (the '??' from before). The upper right entry of J^k is $\binom{k}{3}\lambda^{k-3}$, so the upper right entry of e^{Jt} is

$$0+0+0+\frac{1}{3!}t^3+\frac{1}{4!}t^4 4\lambda+\frac{1}{5!}t^5 10\lambda^2+\cdots+\frac{1}{k!}t^k\frac{k(k-1)(k-2)}{3!}\lambda^{k-3}+\cdots.$$

Factoring out $\frac{t^3}{3!}$ and canceling the $k(k-1)(k-2)$ from the $k!$, we get

$$\frac{t^3}{3!}\left(1 + t\lambda + \frac{1}{2!}t^2\lambda^2 + \frac{1}{3!}t^3\lambda^3 + \cdots\right) = \frac{t^3}{3!}e^{\lambda t}.$$

We have computed all the entries in e^{Jt}, and finally we have

$$e^{Jt} = \begin{bmatrix} e^{\lambda t} & te^{\lambda t} & \frac{t}{2!}e^{\lambda t} & \frac{t^2}{3!}e^{\lambda t} \\ 0 & e^{\lambda t} & te^{\lambda t} & \frac{t}{2!}e^{\lambda t} \\ 0 & 0 & e^{\lambda t} & te^{\lambda t} \\ 0 & 0 & 0 & e^{\lambda t} \end{bmatrix}.$$

These computations are for a 4×4 Jordan block. Had J been a Jordan block of a different size, the same pattern would have emerged. The first row of e^{Jt} is

$$\begin{bmatrix} e^{\lambda t} & te^{\lambda t} & \frac{t^2}{2!}e^{\lambda t} & \frac{t^3}{3!}e^{\lambda t} & \cdots \end{bmatrix},$$

and the successive rows are formed by pushing in zeros from the left.

What if J is not a simple Jordan block but is in Jordan canonical form? That is, suppose

$$J = \begin{bmatrix} J_1 & 0 & 0 & \cdots & 0 \\ 0 & J_2 & 0 & \cdots & 0 \\ 0 & 0 & J_3 & \cdots & 0 \\ \vdots & \vdots & \vdots & \ddots & \vdots \\ 0 & 0 & 0 & \cdots & J_m \end{bmatrix},$$

where each J_ℓ is a Jordan block (and the 0's denote rectangular blocks of zeros). It is a simple matter to verify that

$$e^{Jt} = J = \begin{bmatrix} e^{J_1 t} & 0 & 0 & \cdots & 0 \\ 0 & e^{J_2 t} & 0 & \cdots & 0 \\ 0 & 0 & e^{J_3 t} & \cdots & 0 \\ \vdots & \vdots & \vdots & \ddots & \vdots \\ 0 & 0 & 0 & \cdots & e^{J_m t} \end{bmatrix}.$$

Recapping, we know that any square matrix A can be written as $A = SJS^{-1}$, where J is in Jordan canonical form. We therefore have $e^{At} = Se^{Jt}S^{-1}$. The entries in e^{Jt} look like $e^{\lambda t}$ multiplied by polynomials in t.

We can now justify what we said earlier. If all A's eigenvalues have negative real part, then as $t \to \infty$, all entries in $e^{At} = Se^{Jt}S^{-1}$ tend to 0 and therefore $\mathbf{x}(t) \to \mathbf{0}$. However, if some eigenvalues have positive real part, then some entries in e^{At} will explode, and then, typically, $|\mathbf{x}(t)| \to \infty$.

The preceding discussion handles the continuous systems of the form $\mathbf{x}' = A\mathbf{x}$ for *any* square matrix A. Now we consider discrete time systems $\mathbf{x}(k + 1) = A\mathbf{x}(k)$.

We know that $\mathbf{x}(k) = A^k \mathbf{x}_0$. The question is, How does A^k behave? Writing $A = SJS^{-1}$, we have

$$\begin{aligned} A^k &= \left(SJS^{-1}\right)^k \\ &= \left(SJS^{-1}\right)\left(SJS^{-1}\right)\cdots\left(SJS^{-1}\right) \\ &= SJ^k S^{-1}. \end{aligned}$$

Thus the behavior of A^k is controlled by the behavior of J^k. Fortunately, we have just considered how J^k behaves! The entries in J^k are of the following form: a polynomial in k times λ^k, where λ is an eigenvalue. Thus if $|\lambda| < 1$, these terms vanish. If $|\lambda| > 1$, the terms explode. Finally, if $|\lambda| = 1$, then the term λ^k neither explodes nor vanishes, but the terms of the form $k\lambda^k$, $\frac{1}{2}k^2\lambda^k$, etc. all explode.

Hence, if all A's eigenvalues have absolute value less than 1, then A^k tends to the zero matrix. However, if some eigenvalue of A has absolute value greater than 1, then entries in A^k will explode.

In summary, the general behavior of linear systems does not depend on the diagonalizability of the matrix A.

Problems for §2.2

◆1.   Find the eigenvalues and eigenvectors of the following matrices:

(a) $\begin{bmatrix} 1 & 2 \\ 0 & 3 \end{bmatrix}$.

(b) $\begin{bmatrix} 0 & 1 \\ 1 & 0 \end{bmatrix}$.

(c) $\begin{bmatrix} -1 & 1 \\ 1 & -1 \end{bmatrix}$.

(d) $\begin{bmatrix} 1 & 1 \\ 1 & 0 \end{bmatrix}$.

◆2.   For each of the matrices in the previous problem, find a general formula for A^k and e^{At}.

The *trace* of a matrix is the sum of its diagonal elements.

◆3.   Let A be a 2×2 matrix. Suppose that the determinant of A is positive and its trace is negative. Show that the system $\mathbf{x}' = A\mathbf{x}$ has $\mathbf{0}$ as a stable fixed point.

◆4.   Consider the discrete time dynamical system $\mathbf{x}(k + 1) = A\mathbf{x}(k) + \mathbf{b}$ with $\mathbf{b} \neq \mathbf{0}$. Find a change of variables which converts this problem into one of the form $\mathbf{z}(k + 1) = A\mathbf{z}(k)$. If you need to make some special assumptions for your change of variables to work, please be sure to state them clearly.

Conclude that if A has an eigenvalue with absolute value greater than 1, then $|\mathbf{x}(k)| \to \infty$ for typical $\mathbf{x}_0$.

◆5.   Give an example of a system of the form $\mathbf{x}(k+1) = A\mathbf{x}(k)+\mathbf{b}$ in which one of A's eigenvalues has absolute value bigger than 1, and a starting vector $\mathbf{x}_0$ for which $|\mathbf{x}(k)|$ does *not* tend to infinity.

Try to find such a $\mathbf{x}_0$ which is *not* a fixed point of the system.

◆6.   The points in figure 2.10 are symmetric about the x-axis. Why?

◆7.   The power method. Let A be a matrix for which there is a unique eigenvalue of maximum absolute value. In other words, there is an eigenvalue λ_{max} with the property that $|\lambda_{max}| > |\lambda|$ for all other eigenvalues λ (and λ_{max} is of multiplicity 1).

Now, consider the following algorithm.

(a) Let $\mathbf{v}$ be a randomly chosen vector. (For example, choose each of the components in $\mathbf{v}$ to be a number in the range $[0, 1]$ chosen uniformly at random.)

(b) Do the following steps several times:

(a) Let $\mathbf{v} \leftarrow \mathbf{v}/|\mathbf{v}|$ (this makes $\mathbf{v}$ a unit vector).

(b) Let $\mathbf{v} \leftarrow A\mathbf{v}$.

(c) Observe that $A\mathbf{v}$ is (very nearly) a scalar multiple of $\mathbf{v}$; output that scalar.

Explain why the output of the algorithm is $_{\max}$.

[Notes: (1) You may assume that A diagonalizes, although this is not essential for the method to work. (2) The vector $\mathbf{v}$ at the end of the algorithm is (nearly) an eigenvector associated with $_{\max}$. (3) The "divide by $|\mathbf{v}|$" step is not necessary in *theory*. It serves only to prevent the computer arithmetic from under- or overflow.]

◆8. Let $A = \begin{bmatrix} 1 & 2 & 3 \\ 4 & 5 & 6 \\ 7 & 8 & 9 \end{bmatrix}$ and let $\mathbf{v} = \begin{bmatrix} x \\ y \\ z \end{bmatrix}$, where x, y, and z are functions of t. First compute $A(\mathbf{v}')$ and then work out $(A\mathbf{v})'$. (Note that differentiation is with respect to t in both cases. Write the derivative of x as x'.) Note that your two answers are the same. Explain why $(A\mathbf{v})' = A(\mathbf{v}')$ for a general $n \times n$ matrix of constants A and an n-vector of functions $\mathbf{v}$.

◆9. Let $A = \begin{bmatrix} 3 & 1 \\ 1 & 3 \end{bmatrix}$, $\mathbf{b} = \begin{bmatrix} 1 \\ 2 \end{bmatrix}$, and $\mathbf{x}_0 = \begin{bmatrix} -1 \\ 2 \end{bmatrix}$. Consider the discrete time dynamical system $\mathbf{x}(k+1) = A\mathbf{x}(k) + \mathbf{b}$. Find an exact formula for $\mathbf{x}(k)$.

◆10. Let $A = \begin{bmatrix} 2 & 1 \\ 1 & 2 \end{bmatrix}$, $\mathbf{b} = \begin{bmatrix} 1 \\ 2 \end{bmatrix}$, and $\mathbf{x}_0 = \begin{bmatrix} -1 \\ 2 \end{bmatrix}$. Consider the discrete time dynamical system $\mathbf{x}(k+1) = A\mathbf{x}(k) + \mathbf{b}$. Find an exact formula for $\mathbf{x}(k)$.

[Note: This problem is virtually identical to the one before it but more challenging (why?). You will need to review the methods from the text to find the exact formulas.]

◆11. Let $A = \begin{bmatrix} 2 & 1 \\ 1 & 2 \end{bmatrix}$, $\mathbf{b} = \begin{bmatrix} 1 \\ 2 \end{bmatrix}$, and $\mathbf{x}_0 = \begin{bmatrix} -1 \\ 2 \end{bmatrix}$. Consider the continuous time dynamical system $\mathbf{x}' = A\mathbf{x} + \mathbf{b}$. Find an exact formula for $\mathbf{x}(t)$.

◆12. Let $A = \begin{bmatrix} 2 & 1 \\ -1 & 2 \end{bmatrix}$, $\mathbf{b} = \begin{bmatrix} 1 \\ 2 \end{bmatrix}$, and $\mathbf{x}_0 = \begin{bmatrix} -1 \\ 2 \end{bmatrix}$. Consider the continuous time dynamical system $\mathbf{x}' = A\mathbf{x} + \mathbf{b}$. Find an exact formula for $\mathbf{x}(t)$. Please be sure your answer does not involve imaginary numbers.

◆13. Let $A = \begin{bmatrix} -1 & 1 \\ 1 & -1 \end{bmatrix}$ and $\mathbf{b} = \begin{bmatrix} -1 \\ 2 \end{bmatrix}$. What is the long-term behavior of the dynamical system $\mathbf{x}' = A\mathbf{x}$?

What is the long-term behavior of the system $\mathbf{x}' = A\mathbf{x} + \mathbf{b}$?

◆14. Let $A = \begin{bmatrix} -1 & 0 & 0 \\ 0 & 0 & 1 \\ 0 & -1 & 0 \end{bmatrix}$ and $\mathbf{b} = \begin{bmatrix} 2 \\ 1 \\ 2 \end{bmatrix}$. What is the long-term behavior of the dynamical system $\mathbf{x}' = A\mathbf{x}$?

What is the long-term behavior of the system $\mathbf{x}' = A\mathbf{x} + \mathbf{b}$? Try various starting values.

◆15. Let $A = \begin{bmatrix} 0 & 1 & 0 \\ 0 & 0 & 1 \\ 0 & -1 & 0 \end{bmatrix}$ and $\mathbf{b} = \begin{bmatrix} 2 \\ 1 \\ 2 \end{bmatrix}$. What is the long-term behavior of the dynamical system $\mathbf{x}' = A\mathbf{x}$?

What is the long-term behavior of the system $\mathbf{x}' = A\mathbf{x} + \mathbf{b}$? Try various starting values.

◆16. Let $A = \begin{bmatrix} 0 & 1 \\ -1 & 0 \end{bmatrix}$ and $\mathbf{b} = \begin{bmatrix} 2 \\ 1 \end{bmatrix}$. What is the long-term behavior of the dynamical system $\mathbf{x}(k + 1) = A\mathbf{x}(k)$?

What is the long-term behavior of the dynamical system $\mathbf{x}(k+1) = A\mathbf{x}(k) + \mathbf{b}$?

◆17. Let $A = \begin{bmatrix} 1 & 0 \\ 0 & -1 \end{bmatrix}$ and $\mathbf{b} = \begin{bmatrix} 2 \\ 1 \end{bmatrix}$. What is the long-term behavior of the dynamical system $\mathbf{x}(k + 1) = A\mathbf{x}(k)$?

What is the long-term behavior of the dynamical system $\mathbf{x}(k+1) = A\mathbf{x}(k) + \mathbf{b}$?

◆18. Let $A = \begin{bmatrix} 1 & 0 & 0 \\ 0 & 0.8 & 0.2 \\ 0 & 0 & -0.2 \end{bmatrix}$ and $\mathbf{b} = \begin{bmatrix} 2 \\ 1 \\ 2 \end{bmatrix}$. What is the long-term behavior of the dynamical system $\mathbf{x}(k + 1) = A\mathbf{x}(k)$?

What is the long-term behavior of the dynamical system $\mathbf{x}(k+1) = A\mathbf{x}(k) + \mathbf{b}$?

◆19. Let $A = \begin{bmatrix} 1 & 0 & 0 \\ 0 & 0.8 & 0.2 \\ 0 & 0 & -0.2 \end{bmatrix}$ and $\mathbf{b} = \begin{bmatrix} 0 \\ 1 \\ 2 \end{bmatrix}$. What is the long-term behavior of the dynamical system $\mathbf{x}(k + 1) = A\mathbf{x}(k)$?

What is the long-term behavior of the dynamical system $\mathbf{x}(k+1) = A\mathbf{x}(k) + \mathbf{b}$?

◆20. Let $A = \begin{bmatrix} 0.2 & 0.3 \\ -0.4 & 0.5 \end{bmatrix}$ and $\mathbf{b} = \begin{bmatrix} 0.3 \\ -0.1 \end{bmatrix}$. Consider the discrete time system $\mathbf{x}(k + 1) = A\mathbf{x}(k) + \mathbf{b}$. Plot the points $\mathbf{x}(0), \mathbf{x}(1), \mathbf{x}(2),\ldots$ for various starting values $\mathbf{x}(0) = \mathbf{x}_0$. You might like to color code your dots depending on the starting value.

The various dot paths should spiral in toward a common point in the plane. Why? What is the significance of that point?

2.3 Examplification: Markov chains

Thus far in our discussion we have considered systems which are *deter-ministic*; once we know the state of the system, the future of the system is fixed. It's not hard to imagine, however, a system where the next state is determined by a random mechanism. For example, consider the position of a token in a game such as Monopoly. In a simplified model of this game, the *state* of the token is its position on the board (one of 40 possible squares).[9] At each discrete time period (the player's turn) the piece moves to another square between 2 and 12 steps ahead depending on the roll of the dice.

Introducing the idea of a random dynamical system.

Although such a random system may seem antithetical to the notion of a dynamical system, we shall see that it can be viewed as an example of a linear dynamical (deterministic!) system.

The study of Markov chains is a course unto itself; the brief introduction we give hardly does the topic justice. Our motivation is to show how Markov chains can be viewed as dynamical systems and to utilize the ideas we have learned from linear systems to understand them.

2.3.1 Introduction

A *Markov chain* consists of two parts:

1. A set of possible states—called the *state space*—the system might be in. We only deal with Markov chains with finitely many possible states.

2. A *transition rule* which tells us the probability of moving from one state to another. Specifically, for any pair of states i and j, we have a nonnegative number $p(i \rightarrow j)$, called the *transition probability*, which tells us that if we are in state i, then the probability we are in state j at the next time period is $p(i \rightarrow j)$. If we hold i constant and sum $p(i \rightarrow j)$ for all possible states j, the result must be 1.

Consider our Monopoly example. There are 40 possible states corresponding to the 40 squares on the Monopoly board. We can refer to these states using the numbers 1 through 40, with 1 corresponding to the "Go" square and 40 corresponding to Boardwalk. Thus the state space for Monopoly is the set $\{1, 2, \ldots, 40\}$. Next we need a transition rule. If we are in state i we can move to any of the states $i + 2$ though $i + 12$ (with

---

[9]To make this more accurate we would have to take into account the "Go to Jail" square, the fact that rolling doubles three times in a row ends up in Jail, the rules for leaving Jail, etc.

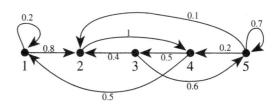

Figure 2.23. A directed graph representation of a Markov chain.

addition mod 40, i.e., state "41" is really state 1). No other move is possible. The probability that we will roll a 2 is $1/36$, a 3 is $2/36$, a 4 is $3/36$, ..., a 12 is $1/36$. Thus we have

$$
\begin{aligned}
p(i \rightarrow i + 2) &= 1/36, & p(i \rightarrow i + 3) &= 2/36, \\
p(i \rightarrow i + 4) &= 3/36, & p(i \rightarrow i + 5) &= 4/36, \\
p(i \rightarrow i + 6) &= 5/36, & p(i \rightarrow i + 7) &= 6/36, \\
p(i \rightarrow i + 8) &= 5/36, & p(i \rightarrow i + 9) &= 4/36, \\
p(i \rightarrow i + 10) &= 3/36, & p(i \rightarrow i + 11) &= 2/36, \\
p(i \rightarrow i + 12) &= 2/36, & p(i \rightarrow j) &= 0 \text{ otherwise.}
\end{aligned}
$$

A pictorial representation of a Markov chain.

We can draw a directed graph to give a picture of a Markov chain. A *directed graph* consists of a collection of dots called *vertices*. These vertices are joined by arrows called *arcs*. An arc from vertex i to vertex j means it is possible to move from vertex i to vertex j in one step. We label the arc from i to j with the number $p(i \rightarrow j)$ to show what the probability of this transition is. [If $p(i \rightarrow j) = 0$, we simply don't draw an arc.] See Figure 2.23. In this figure the arc from 1 to 2 is labeled 0.8; this means that when we are at state 1, there is an 80% chance we will be at state 2 in the next time period, i.e., $p(1 \rightarrow 2) = 0.8$. There is also an arc from vertex 1 to itself labeled 0.2. This means there is a 20% chance that if we are currently at state 1, we will still be there in the next time period, i.e., $p(1 \rightarrow 1) = 0.2$. Notice there is no arc from vertex 2 to vertex 1; this is because it is impossible to move from state 2 to state 1 in a single step, i.e., $p(2 \rightarrow 1) = 0$. Indeed, from state 2, the only possible next state is state 4, since $p(2 \rightarrow 4) = 1$.

Representing a Markov chain in a matrix.

We can condense all the information from the directed graph in Fig-

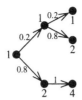

Figure 2.24. Taking two steps from state 1 of the Markov chain in Figure 2.23.

ure 2.23 into a matrix P whose ij entry is $p(i \rightarrow j)$, namely,

$$P = \begin{bmatrix} 0.2 & 0.8 & 0 & 0 & 0 \\ 0 & 0 & 0 & 1 & 0 \\ 0 & 0.4 & 0 & 0 & 0.6 \\ 0.5 & 0 & 0.5 & 0 & 0 \\ 0 & 0.1 & 0 & 0.2 & 0.7 \end{bmatrix}.$$

2.3.2 Markov chains as linear systems

Let us continue our discussion of the Markov chain depicted in Figure 2.23. Suppose at time $k = 0$ we are in state 1. In which state will we be at time $k = 1$? The answer, of course, is, we don't know. There is a 20% chance we will still be in state 1 and an 80% chance we will be in state 2. The probability we are in any other state is 0.

Where will we be at time $k = 2$? Again, we don't know, but we can work out the probabilities. On the first step, we can either stay at state 1 (20%) or move to state 2 (80%). In the first case, we again stay at state 1 or move to state 2. In the second case, we must move to state 4. We summarize these options in Figure 2.24. The three possible routes the Markov chain might follow, and their respective probabilities are:

Multistep transition probabilities.

$$p(1 \rightarrow 1 \rightarrow 1) = p(1 \rightarrow 1)p(1 \rightarrow 1) = 0.2^2 = 0.04,$$

$$p(1 \rightarrow 1 \rightarrow 2) = p(1 \rightarrow 1)p(1 \rightarrow 2) = 0.2 \times 0.8 = 0.16,$$

$$p(1 \rightarrow 2 \rightarrow 4) = p(1 \rightarrow 2)p(2 \rightarrow 4) = 0.8 \times 1 = 0.8.$$

Let us write $p(i \xrightarrow{2} j)$ to denote the probability that if the Markov chain is currently in state i then two steps later it will be in state j. Thus

$$p(1 \xrightarrow{2} 1) = 0.04,$$

$$p(1 \xrightarrow{2} 2) = 0.16,$$

$$p(1 \xrightarrow{2} 4) = 0.8, \quad \text{and}$$

$$p(1 \xrightarrow{2} j) = 0 \quad \text{otherwise.}$$

We can collect this information into a matrix whose (i, j)-entry is $p(i \xrightarrow{2} j)$ with the following result:

$$\begin{bmatrix} 0.04 & 0.16 & 0 & 0.8 & 0 \\ 0.5 & 0 & 0.5 & 0 & 0 \\ 0 & 0.06 & 0 & 0.52 & 0.42 \\ 0.1 & 0.6 & 0 & 0 & 0.3 \\ 0.1 & 0.07 & 0.1 & 0.24 & 0.49 \end{bmatrix}.$$

The entries in the ith row of this matrix are the two-step probabilities starting from state i.

The two step transition probabilities are the entries in the matrix P^2.

You might be wondering if there is a simple relation between the preceding matrix (whose (i, j)-entry is $p(i \xrightarrow{2} j)$) and the matrix P (whose (i, j)-entry is $p(i \to j)$). Indeed there is. The former matrix is simply P^2. [Run, do not walk, to your nearest computer to test this out!] This is not a coincidence. Let's see why this works by taking a detailed look at how we compute $p(i \xrightarrow{2} j)$.

The probability $p(i \xrightarrow{2} j)$ is actually a sum of probabilities depending on what the intermediate state between i and j might be. We can express this as

$$p(i \xrightarrow{2} j) = p(i \to 1 \to j) + p(i \to 2 \to j) + \cdots + p(i \to n \to j),$$

where n is the number of states in the Markov chain. The general term in this sum is $p(i \to k \to j)$: the probability that we move from state i to state k in the first step and then move from state k to state j in the second. This we compute as $p(i \to k \to j) = p(i \to k)p(k \to j)$. Thus we can rewrite the expression above as

$$p(i \xrightarrow{2} j) = p(i \to 1)p(1 \to j) + p(i \to 2)p(2 \to j) + \cdots +$$
$$+ p(i \to n)p(n \to j),$$

or using summation notation,

$$p(i \xrightarrow{2} j) = \sum_{k=1}^{n} p(i \to k)p(k \to j) = \sum_{k=1}^{n} P_{ik} P_{kj}.$$

This is exactly the formula for matrix multiplication! Thus P^2 is the matrix whose (i, j)-entry is $p(i \xrightarrow{2} j)$. Would you like to guess the matrix whose (i, j)-entry is $p(i \xrightarrow{3} j)$? You've got it! It is, of course, P^3; let's be sure we understand why. We can think of the three-step transition $i \xrightarrow{3} j$ as consisting of first a *single* step from i to some state k, followed by a *double* step from k to state j. We can therefore write

$$p(i \xrightarrow{3} j) = p(i \to 1)p(1 \xrightarrow{2} j) + p(i \to 2)p(2 \xrightarrow{2} j) + \cdots +$$
$$+ p(i \to n)p(n \xrightarrow{2} j).$$

Rewriting this in matrix notation, we have

$$p(i \xrightarrow{3} j) = \sum_{k=0}^{n} P_{ik}(P^2)_{kj},$$

which is precisely the formula for the matrix multiplication $P \times P^2$.

In this way we see that the (i, j)-entry of the matrix P^m is the m-step transition probability $p(i \xrightarrow{m} j)$.

The m-step transition matrix is P^m.

If the Markov chain starts in state 1, where will it be after m steps? We can't know *exactly* where the system will be, but we *do* know the probability with which it assumes any given state; these numbers are the first row of P^m. If the Markov chain begins in state 2, then the numbers in the second row of P^m give the probabilities of where the system might be after m steps, and so on. The ith row of P^m gives the probability distribution of the state of the Markov chain after m steps, given that we began in state i.

There is a simple matrix multiplication trick we can use to extract the first (or any other) row of P^m. We multiply P^m on the left by the row vector $[1, 0, 0, \ldots, 0]$. The result is a row vector consisting of the first row of P^m.

For example, in the Markov chain from Figure 2.23 we have

$$[1, 0, 0, 0, 0]P^5 = [0.0483, 0.1853, 0.0160, 0.5344, 0.2160].$$

We learn from this computation that after five steps the Markov chain might be in any state, but it's most likely to be in state 4 (with probability about 53%).

A row vector of nonnegative numbers which sum to 1 is called a *probability vector*. We can think of the initial state of the Markov chain as the probability vector $[1, 0, 0, 0, 0]$, which says that with probability 1 (i.e., certainly) we are in state 1, and with probability 0 we are in any other

Probability vectors are row vectors of nonnegative numbers which sum to 1.

state. Let us write $\mathbf{p}(m)$ to denote the probability vector after m steps of the system. The jth component of $\mathbf{p}(m)$ is the probability we are in state j after m steps. Thus we can write

$$\mathbf{p}(m) = \mathbf{p}(0) P^m,$$

where $\mathbf{p}(0) = [1, 0, \ldots, 0]$ (assuming we started in state 1).

Now it is conceivable that we don't know precisely in which state the system begins. For example, the system may have be running for a while before we start, and we know only the probability of where the system begins. Or suppose we flip a fair coin and start the system either in state 1 or state 5 (each with probability $\frac{1}{2}$). We can write this as $\mathbf{p}(0) = [0.5, 0, 0, 0, 0.5]$. Where will the system be after m steps? You should work out for yourself that the answer is again $\mathbf{p}(0) P^m$.

We can summarize all this by the single equation,

$$\mathbf{p}(m + 1) = \mathbf{p}(m) P,$$

where $\mathbf{p}(\cdot)$ is a probability n-row vector and P is an $n \times n$ nonnegative matrix whose rows all sum to 1. (Every row of P is a probability vector; such matrices are called *stochastic matrices*.) In this way we can think of the row vector $\mathbf{p}(m)$ as being the *state* of the Markov chain at time m.

This is almost exactly the form of a linear discrete time dynamical system. The only difference is in the notation: Our state vector is a row (instead of a column).[10]

2.3.3   The long term

It now makes sense to ask typical dynamical system questions about Markov chains, especially, What is the fate of these systems in the long term? The example in Figure 2.23 is small enough to try out by computer. We start by computing P^m for a large value of m (say 1000) and we get

$$P^{1000} = \begin{bmatrix} 0.1592 & 0.2038 & 0.1274 & 0.2548 & 0.2548 \\ 0.1592 & 0.2038 & 0.1274 & 0.2548 & 0.2548 \\ 0.1592 & 0.2038 & 0.1274 & 0.2548 & 0.2548 \\ 0.1592 & 0.2038 & 0.1274 & 0.2548 & 0.2548 \\ 0.1592 & 0.2038 & 0.1274 & 0.2548 & 0.2548 \end{bmatrix}.$$

The remarkable feature about this matrix is that all five rows are the same. Starting in *any* state, after 1000 iterations of the Markov chain, there is a

---

[10]We could fix this by writing $\mathbf{p}$ as a column and replacing P with P^T.

15.92% chance of being in state 1, a 20.38% chance of being in state 2, etc. Indeed, we can compute P^{1001} or P^{99999} and get the same result. In short, regardless of $\mathbf{p}(0)$ we have

$$\mathbf{p}(m) \to [0.1592, 0.2038, 0.1274, 0.2548, 0.2548]$$

as $m \to \infty$.

Why does this happen? To understand, we look at the eigenvalues and eigenvectors of P. First, we note that 1 must be an eigenvalue of P. Since all the rows in P sum to 1, then we must have $P\mathbf{1} = \mathbf{1}$, where $\mathbf{1}$ is a column vector of all 1's. We can use a computer to find all of P's eigenvalues, namely,

```
-0.3724 + 0.7414i
-0.3724 - 0.7414i
 1.0000
 0.0541
 0.5907
```

This means that 1 is the unique eigenvalue of P with the largest absolute value (since the absolute value of $-0.3724 \pm 0.7414i$ is 0.8296).

The successive iterates of the system, $\mathbf{p}(m)$, are now *row* vectors, not column vectors as we had before. So it is not right to think about the eigenvalues/vectors of P, but rather of P^T. However, since the eigenvalues of a matrix and its transpose are the same (see problem 1 on page 98), we note that the list above also gives the eigenvalues of P^T. Hence there is an eigenvector of P^T corresponding to the eigenvalue 1. It is *not* simply a row vector of all 1's, since

$$[1, 1, 1, 1, 1] \cdot \begin{bmatrix} 0.2 & 0.8 & 0 & 0 & 0 \\ 0 & 0 & 0 & 1 & 0 \\ 0 & 0.4 & 0 & 0 & 0.6 \\ 0.5 & 0 & 0.5 & 0 & 0 \\ 0 & 0.1 & 0 & 0.2 & 0.7 \end{bmatrix} = [0.7, 1.3, 0.5, 1.2, 1.3000].$$

Instead, we find (using the computer to find the eigenvectors of P^T) that the row vector $[25, 32, 20, 40, 40]$ satisfies

$$[25, 32, 20, 40, 40] \cdot \begin{bmatrix} 0.2 & 0.8 & 0 & 0 & 0 \\ 0 & 0 & 0 & 1 & 0 \\ 0 & 0.4 & 0 & 0 & 0.6 \\ 0.5 & 0 & 0.5 & 0 & 0 \\ 0 & 0.1 & 0 & 0.2 & 0.7 \end{bmatrix} = [25, 32, 20, 40, 40].$$

If we rescale this row vector by multiplying by $1/(25+32+20+40+40) = 1/157$, we get

$$\mathbf{p}^* = [0.1592, 0.2038, 0.1274, 0.2548, 0.2548],$$

which is precisely the row we saw repeated five times in P^{1000}.

The Markov chain we have been studying is a bit contrived. After many iterations, the probability vector $\mathbf{p}(m)$ settles down to a fixed values. There are other behaviors which Markov chains can exhibit. Some of the possibilities are explored in the problems.

Problems for §2.3

◆1.　Let A be a square matrix. Recall that the determinant of A and its transpose, A^T, are the same. Use this fact to explain why the eigenvalues of A and A^T are the same.

Although the eigenvalues of A and A^T must be the same, their eigenvectors need not be the same. Give an example of a matrix A for which the eigenvectors of A and A^T are different.

◆2.　For each of the following Markov chain transition matrices, draw the corresponding directed graph representation (as in Figure 2.23 on page 92).

1.
$$\begin{bmatrix} 0.6 & 0 & 0.2 & 0 & 0.2 & 0 \\ 0.7 & 0.1 & 0 & 0.2 & 0 & 0 \\ 0 & 0.4 & 0 & 0.6 & 0 & 0 \\ 0 & 0 & 0 & 0 & 0.1 & 0.9 \\ 0 & 0 & 0 & 0 & 0 & 1 \\ 0 & 0 & 0.6 & 0.4 & 0 & 0 \end{bmatrix}.$$

2.
$$\begin{bmatrix} 0 & 0 & 0 & 0.5 & 0.5 \\ 0.7 & 0 & 0.3 & 0 & 0 \\ 0 & 0.5 & 0 & 0 & 0.5 \\ 0 & 0 & 1 & 0 & 0 \\ 0.9 & 0 & 0.1 & 0 & 0 \end{bmatrix}.$$

3.
$$\begin{bmatrix} 0.7 & 0 & 0.2 & 0 & 0 & 0 & 0.1 & 0 \\ 0 & 0 & 0 & 0 & 1 & 0 & 0 & 0 \\ 0.8 & 0 & 0 & 0.2 & 0 & 0 & 0 & 0 \\ 0 & 0 & 0 & 0 & 0 & 0.8 & 0 & 0.2 \\ 0 & 0.4 & 0 & 0 & 0.6 & 0 & 0 & 0 \\ 0 & 0 & 0 & 0.3 & 0 & 0.3 & 0 & 0.4 \\ 0 & 0.5 & 0.4 & 0 & 0 & 0 & 0 & 0.1 \\ 0 & 0 & 0 & 0 & 0 & 0.7 & 0 & 0.3 \end{bmatrix}.$$

◆3.　Calculate P^{512} and P^{513} for each of the matrices in problem 2.

◆4. Write a computer program to simulate each of the above Markov chains, starting in state 1, through 1000 steps. Count the number of times each state is entered. Divide by 1000 to approximate the average fraction of time the chain spends in each state. Repeat this several times. Better yet, repeat for 10,000 steps.

◆5. For each matrix P in problem 2, find the eigenvector(s) of P^T corresponding to the eigenvalue $\lambda = 1$.

Please multiply each vector you find by a scalar so that the eigenvector's entries sum to 1.

Compare your results with what you found in problems 3 and 4. Comment.

◆6. Based on what you have learned from problems 2 through 5, give a qualitative description of the behavior of each of the above Markov chains.

◆7. Let P be the transition matrix for a Markov chain. Show that 1 is an eigenvalue for P.

◆8. (Previous problem continued.) Suppose that for all other eigenvalues λ of P we have $|\lambda| < 1$. Explain why the state vector $\mathbf{p}(m)$ settles down to a fixed vector $\mathbf{p}^*$, i.e., why $\mathbf{p}(m) \to \mathbf{p}^*$ as $m \to \infty$.

(Make any reasonable assumptions you wish to simplify this problem. For example, you may assume P diagonalizes.)

Chapter 3

Nonlinear Systems 1: Fixed Points

The general forms for dynamical systems are

$$\begin{aligned} \mathbf{x}' &= f(\mathbf{x}) & \text{continuous time, and} \\ \mathbf{x}(k+1) &= f(\mathbf{x}(k)) & \text{discrete time.} \end{aligned}$$

We have closely examined the case when f is linear. In that case, we can answer nearly any question we might consider. We can work out exact formulas for the behavior of $\mathbf{x}(t)$ (or $\mathbf{x}(k)$) and deduce from them the long-term behavior of the system. There are two main behaviors: (1) the system gravitates toward a fixed point, or (2) the system blows up. There are some marginal behaviors as well.

Now we begin our study of more general systems in which f can be virtually any function. However, we do make the following assumption:

> Throughout this chapter, we assume f is differentiable with continuous derivative.

Will this broad generality make our work more complicated? Yes and no:

Nonlinear systems are more complicated; we seek qualitative descriptions in place of exact formulas.

Yes: Nonlinear functions can present insurmountable problems. Typically, it is impossible to find exact formulas for $\mathbf{x}$. Further, the range of behaviors available to nonlinear systems is much greater than that for linear systems (but that's why nonlinear systems are more interesting).

No: Because it can be terribly difficult to find exact solutions to nonlinear systems, we have a valid excuse for not even trying! Instead, we settle for a more modest goal: determine the long-term behavior of the system. This is often feasible even when finding an exact solution is not.

In this chapter we focus on the notion of a *fixed point* (sometimes called an *equilibrium point*) of a dynamical system. We discuss how to find fixed points and then to determine if they are *stable* or *unstable*. Often, understanding the fixed points of a dynamical system can tell us much about the global behavior of the system.

Our study of nonlinear systems continues in Chapter 4, where we examine other behaviors nonlinear systems exhibit.

3.1 Fixed points

3.1.1 What is a fixed point?

A state vector that doesn't change.

The vector **x** is the state of the dynamical system, and the function f tells us how the system moves. In special circumstances, however, the system does *not* move. The system can be stuck (we'll say *fixed*) in a special state; we call these states *fixed points* of the dynamical system.

For example, consider the nonlinear discrete time system

$$x(k + 1) = [x(k)]^2 - 6.$$

Suppose the system is in the state $x(k) = 3$; where will it be in the next instant? This is easy to compute:

$$x(k + 1) = x(k)^2 - 6 = 3^2 - 6 = 3.$$

Aha! The system is again at state $x = 3$. Where will it be in the next time period? Of course, still in state 3. The value $\tilde{x} = 3$ is a fixed point of the system $x(k + 1) = x(k)^2 - 6$, since if we are ever in state 3 we remain there for all time. (This system has another fixed point; try to find it.)

Let's consider a continuous time example:

$$x' = x^3 - 8.$$

What happens if $x(t) = 2$? We compute that dx/dt equals $x^3 - 8 = 2^3 - 8 = 0$. Thus $x(t)$ is neither increasing nor decreasing; in other words, it's stuck at 2. Thus $\tilde{x} = 2$ is a fixed point of this system. (This system has no other fixed points; try to figure out why.)

Thus a *fixed point* of a dynamical system is a state vector $\tilde{\mathbf{x}}$ with the property that if the system is ever in the state $\tilde{\mathbf{x}}$, it will remain in that state for all time.

3.1.2   Finding fixed points

In the preceding examples, the fixed points were handed to us on a silver platter. You may be wondering, Given a dynamical system, how do I find its fixed points? In principle this is easy; in practice, however, it can present some challenges.

If the system is discrete, $\mathbf{x}(k+1) = f(\mathbf{x}(k))$, we want a value $\tilde{\mathbf{x}}$ so that $\tilde{\mathbf{x}} = f(\tilde{\mathbf{x}})$. In other words, we need to solve the equation $\mathbf{x} = f(\mathbf{x})$.

Solve $f(\mathbf{x}) = \mathbf{x}$ to find fixed points of discrete time systems.

For example, suppose the system is

$$\begin{bmatrix} x_1(k+1) \\ x_2(k+1) \end{bmatrix} = \begin{bmatrix} x_1(k)^2 + x_2(k) \\ x_1(k) + x_2(k) - 2 \end{bmatrix}.$$

We may write this as $\mathbf{x}(k+1) = f(\mathbf{x}(k))$, where $f\begin{bmatrix} u \\ v \end{bmatrix} = \begin{bmatrix} u^2 + v \\ u + v - 2 \end{bmatrix}$.

To find a point $\tilde{\mathbf{x}}$ with the property that $\tilde{\mathbf{x}} = f(\tilde{\mathbf{x}})$, we solve

$$u = u^2 + v,$$

$$v = u + v - 2.$$

Notice that v drops out of the second equation and we find that $u = 2$. Substituting $u = 2$ into the first equation, we have $2 = 2^2 + v$, so $v = -2$. Thus we have

$$f\begin{bmatrix} 2 \\ -2 \end{bmatrix} = \begin{bmatrix} 2^2 + (-2) \\ 2 + (-2) - 2 \end{bmatrix} = \begin{bmatrix} 2 \\ -2 \end{bmatrix},$$

and therefore $\begin{bmatrix} 2 \\ -2 \end{bmatrix}$ is a fixed point (indeed the only one) of the system.

Next let us consider continuous time systems ($\mathbf{x}' = f(\mathbf{x})$). We want a state vector $\tilde{\mathbf{x}}$ with the property that if the system is in state $\tilde{\mathbf{x}}$, it stays put. In other words, it doesn't change with time. Since $\mathbf{x}$ does not change as time marches on, its derivative, $\mathbf{x}'$, must be $\mathbf{0}$ at this state. Thus to find a fixed point of $\mathbf{x}' = f(\mathbf{x})$, we solve the equation $f(\mathbf{x}) = \mathbf{0}$ for $\mathbf{x}$. For example, suppose the system is

Solve $f(\mathbf{x}) = \mathbf{0}$ to find fixed points of continuous time systems.

$$\begin{bmatrix} x_1' \\ x_2' \end{bmatrix} = \begin{bmatrix} x_1^2 + x_2^2 - 25 \\ x_1 + x_2 + 1 \end{bmatrix}.$$

Finding fixed points		
Time	System	To find $\tilde{\mathbf{x}}$
Continuous	$\mathbf{x}' = f(\mathbf{x})$	solve $f(\mathbf{x}) = \mathbf{0}$
Discrete	$\mathbf{x}(k+1) = f(\mathbf{x}(k))$	solve $f(\mathbf{x}) = \mathbf{x}$

Table 3.1. How to find fixed points $\tilde{\mathbf{x}}$ of continuous and discrete time dynamical systems.

Thus $f\begin{bmatrix} u \\ v \end{bmatrix} = \begin{bmatrix} u^2 + v^2 - 25 \\ u + v + 1 \end{bmatrix}$. To find the fixed points of this system, we solve $f(\mathbf{x}) = \mathbf{0}$, i.e., we solve the system of equations

$$u^2 + v^2 - 25 = 0,$$

$$u + v + 1 = 0.$$

We can solve for u in the second equation ($u = -1 - v$) and substitute this expression into the first equation:

$$(-1 - v)^2 + v^2 - 25 = 0 \quad \Rightarrow \quad 2v^2 + 2v - 24 = 0.$$

This quadratic equation has two roots: $v = 3$ and $v = -4$, which yield (since $u = -1 - v$) $u = -4$ and $u = 3$ respectively. Thus the fixed points of this system are $\begin{bmatrix} -4 \\ 3 \end{bmatrix}$ and $\begin{bmatrix} 3 \\ -4 \end{bmatrix}$.

Finding fixed points reduces to solving equations.

Finding fixed points of dynamical systems does *not* require us to find exact formulas for $\mathbf{x}(k)$ [or $\mathbf{x}(t)$]. All we have to do is solve some equations. Of course, solving systems of equations can be difficult, but it is at least comforting to know that this is the only issue involved. The equations we solve depend on f and whether the system is in discrete or continuous time. The methods at hand to find fixed points are recounted in Table 3.1.

3.1.3 Stability

Not all fixed points are the same. We call some *stable* and others *unstable*. We begin by illustrating these concepts with an example.

Let $f(x) = x^2$ and consider the discrete time dynamical system

$$x(k+1) = f(x(k)) = [x(k)]^2.$$

In other words, we are interested in seeing what happens when we iterate the square function.

The system has two fixed points: 0 and 1 (these are the solutions to $f(x) = x$, i.e., $x^2 = x$). If you enter either 0 or 1 into your calculator and start pressing the $\boxed{x^2}$ button, you will notice something very boring: nothing happens. Both 0 and 1 are fixed points, and the x^2 function just leaves them alone.

Now, let's put other numbers into our calculator and see what happens. First, let us start with a number which is close to (but not equal) 0, say 0.1. If we iterate x^2, we see

$$0.1 \mapsto 0.01 \mapsto 0.0001 \mapsto 0.0000001 \mapsto \cdots.$$

Clearly $x(k) \to 0$ as $k \to \infty$. It's not hard to see why this works. If we begin with any number x_0, our iterations go

$$x_0 \mapsto x_0^2 \mapsto x_0^4 \mapsto x_0^8 \mapsto \cdots \mapsto x_0^{2^k} \mapsto \cdots.$$

Thus if x_0 is near 0, then, clearly, $x(k) \to 0$ as $k \to \infty$. (How near zero must we be? All we need is $|x_0| < 1$.) We say that x_0 is a *stable* or an *attractive* fixed point of the system $x(k+1) = f(x(k))$ because if we start the system near x_0, then the system gravitates toward x_0.

Now let's examine the other fixed point, 1. What happens if we put a number near (but not equal to) 1 in our calculator and start iterating the x^2 function. If $x_0 = 1.1$, we see

$$1.1 \mapsto 1.21 \mapsto 1.4641 \mapsto 2.1436 \mapsto 4.5950 \mapsto 21.1138 \mapsto 445.7916 \mapsto \cdots.$$

Clearly, $x(k) \to \infty$. If we take $x_0 = 0.9$, we see

$$0.9 \mapsto 0.81 \mapsto 0.6561 \mapsto 0.4305 \mapsto 0.1853 \mapsto 0.0343 \mapsto 0.0012 \mapsto \cdots.$$

Clearly $x(k) \to 0$. In any case, starting points near (but not equal to) 1 tend to iterate away from 1. We call 1 an *unstable* fixed point of the system.

Let us now describe three types of fixed points a system may possess.

First, a fixed point $\tilde{x}$ is called *stable* provided the following is true: For all starting values x_0 near $\tilde{x}$, the system not only stays near $\tilde{x}$ but also $x(t) \to \tilde{x}$ as $t \to \infty$ [or $x(k) \to \tilde{x}$ as $k \to \infty$ in discrete time].[1]

Stable fixed point.

Second, a fixed point $\tilde{x}$ is called *marginally stable* or *neutral* provided the following: For all starting values x_0 near $\tilde{x}$, the system stays near $\tilde{x}$ but does *not* converge to $\tilde{x}$.

Marginally stable (or neutral) fixed point.

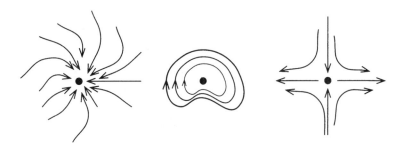

Figure 3.1. Fixed points with three different types of stability. The fixed point on the left is stable. The fixed point in the center is marginally stable. The fixed point on the right is unstable.

Unstable fixed point.

Third, a fixed point $\tilde{\mathbf{x}}$ is called *unstable* if it is neither stable nor marginally stable. In other words, there are starting values $\mathbf{x}_0$ very near $\tilde{\mathbf{x}}$ so that the system moves far away from $\tilde{\mathbf{x}}$.

Figure 3.1 illustrates each of these possibilities. The fixed point on the left of the figure is stable; all trajectories which begin near $\tilde{\mathbf{x}}$ remain near, and converge to, $\tilde{\mathbf{x}}$. The fixed point in the center of the figure is marginally stable (neutral). Trajectories which begin near $\tilde{\mathbf{x}}$ stay nearby but never converge to $\tilde{\mathbf{x}}$. Finally, the fixed point on the right of the figure is unstable. There are trajectories which start near $\tilde{\mathbf{x}}$ and move far away from $\tilde{\mathbf{x}}$.

Let's look at a few explicit examples to reinforce these ideas. We consider four examples, all of the form $\mathbf{x}' = A\mathbf{x}$ where A is a simple 2×2 matrix. In every case, the fixed point of $\mathbf{x}' = A\mathbf{x}$ is the origin, $\tilde{\mathbf{x}} = \mathbf{0}$.

A system with a stable fixed point.

1. $A = \begin{bmatrix} -2 & 0 \\ 0 & -1 \end{bmatrix}$. We observe that

---

[1]What does *near* mean? This footnote explains the definition of stability more precisely.

The first part of the definition says "for all starting values $\mathbf{x}_0$ near $\tilde{\mathbf{x}}$, the system stays near $\tilde{\mathbf{x}}$." The precise meaning is the following. For every positive number ε, one can find a positive number δ with the following property: If $\mathbf{x}_0$ is within distance δ of $\tilde{\mathbf{x}}$, then $\mathbf{x}(t)$ is within distance ε of $\tilde{\mathbf{x}}$ for all $t \geq 0$.

The definition also requires "for all starting values $\mathbf{x}_0$ near $\tilde{\mathbf{x}}$ we have $\mathbf{x}(t) \to \tilde{\mathbf{x}}$ as $t \to \infty$." This means there is a positive number δ so that for any $\mathbf{x}_0$ within distance δ of $\tilde{\mathbf{x}}$ the following is true: For every $\varepsilon > 0$ there is a $T > 0$ so that if $t \geq T$, then $\mathbf{x}(t)$ is within distance ε of $\tilde{\mathbf{x}}$.

$$\begin{bmatrix} x_1(t) \\ x_2(t) \end{bmatrix} = \begin{bmatrix} e^{-2t} & 0 \\ 0 & e^{-t} \end{bmatrix} \mathbf{x}_0,$$

and since e^{-2t} and e^{-t} both tend to 0 as $t \to \infty$, we see that $\mathbf{0}$ is a stable fixed point. If we start this system at any vector $\mathbf{x}_0$ near $\mathbf{0}$, the system will converge to $\mathbf{0}$. [Indeed, we do not need to start *near* $\mathbf{x}_0$ to converge to $\mathbf{0}$; no matter where we start the system, it converges to $\mathbf{x}_0$. We call such a fixed point *globally stable*.

2. $A = \begin{bmatrix} 2 & 0 \\ 0 & 1 \end{bmatrix}$. We observe that

A system with an unstable fixed point.

$$\begin{bmatrix} x_1(t) \\ x_2(t) \end{bmatrix} = \begin{bmatrix} e^{2t} & 0 \\ 0 & e^t \end{bmatrix} \mathbf{x}_0,$$

and since e^{2t} and e^t both tend to infinity as $t \to \infty$, we see that $\mathbf{0}$ is an unstable fixed point. No matter how close to $\mathbf{0}$ we begin, the system moves away from $\mathbf{0}$.

3. $A = \begin{bmatrix} 2 & 0 \\ 0 & -1 \end{bmatrix}$. We observe that

Another system with an unstable fixed point.

$$\begin{bmatrix} x_1(t) \\ x_2(t) \end{bmatrix} = \begin{bmatrix} e^{2t} & 0 \\ 0 & e^{-t} \end{bmatrix} \mathbf{x}_0.$$

We claim that $\mathbf{0}$ is, again, an unstable fixed point of the system; let's see why. First, consider two starting positions: $\mathbf{x}_0 = \begin{bmatrix} a \\ 0 \end{bmatrix}$, and $\mathbf{x}_0 = \begin{bmatrix} 0 \\ a \end{bmatrix}$ where a is some number (not 0). For the first position, observe that as $t \to \infty$ we have $|\mathbf{x}(t)| \to \infty$. For the second, we have $\mathbf{x}(t) \to \mathbf{0}$ as $t \to \infty$. Thus some points gravitate toward $\mathbf{0}$ while others get blown away. This fixed point is *unstable* because there are trajectories which begin very close to $\mathbf{0}$ and which go far away. Even though for *some* starting points near $\mathbf{0}$ the system converges to $\mathbf{0}$, we still call this fixed point unstable.

4. $A = \begin{bmatrix} 0 & 1 \\ -1 & 0 \end{bmatrix}$. We check that

A system with a marginally stable fixed point.

$$e^{At} = \begin{bmatrix} \cos t & \sin t \\ -\sin t & \cos t \end{bmatrix}.$$

Is the fixed point $\mathbf{0}$ stable or unstable? The answer is, neither. To see that it is not stable, consider any point $\mathbf{x}_0$ near (but not equal to) $\mathbf{0}$. As $t \to \infty$ the system never approaches $\mathbf{0}$. Further, $\mathbf{0}$ is not unstable. To be unstable, points near $\mathbf{0}$ must be sent "far" away from $\mathbf{0}$. Clearly, if we start at certain distance from $\mathbf{0}$ the system does not get any farther away.

Thus $\mathbf{0}$ for this system is *marginally stable*.

Summary

Stable fixed points give excellent information about the fate of a dynamical system. We know how to find fixed points: In discrete time we solve $\mathbf{x} = f(\mathbf{x})$, and in continuous time we solve $f(\mathbf{x}) = \mathbf{0}$. The question remains, Once we have found our fixed points, how can we tell whether they are stable or unstable? The next sections will give us some tools for making this determination. The most important tool (see §3.2) is linearization: We approximate our system near its fixed points by linear functions. This method doesn't always work, in which case (see §3.3) we resort to our emergency backup method: Lyapunov functions.

Problems for §3.1

◆1.   Find all fixed points of the following discrete time systems $x(k + 1) = f(x(k))$.

    (a) $f(x) = x^2 - 2$.

    (b) $f(x) = \sin x$.

    (c) $f(x) = 1/x$.

    (d) $f(x) = \sqrt[3]{x^2}$.

    (e) $f\begin{bmatrix} x \\ y \end{bmatrix} = \begin{bmatrix} x^2 + y^2 \\ x + y - 1 \end{bmatrix}$.

◆2.   Do numerical experiments near each of the fixed points you found in the previous problem to determine their stability.

◆3.   Find all fixed points of the following continuous time systems $x' = f(x)$.

    (a) $f(x) = x^2 - x - 1$.

    (b) $f(x) = \sin x$.

    (c) $f(x) = e^x - 1$.

    (d) $f(x) = \log(x^2)$.

    (e) $f(x) = x/(1 - x)$.

(f) $f \begin{bmatrix} x \\ y \end{bmatrix} = \begin{bmatrix} x - y^2 \\ x + y - 2 \end{bmatrix}.$

◆4. Show that $[sA/(d + \rho)]^2$ is a fixed point of the dynamical system from the economic growth model (see §1.2.5, especially equation (1.22) on page 19).

◆5. Explain why it is impossible for a linear system (either discrete or continuous) to have exactly two fixed points.

◆6. Use graphical analysis to show that iterating $\cos x$ from any starting value x_0 always leads to the same answer: the unique fixed point of $x(k + 1) = \cos x(k)$.

◆7. Consider the function $f(\theta) = 3\theta$ where θ is an *angle*. Thus θ may take values only in the range $[0, 2\pi)$. For example, $f(5\pi/4) = 7\pi/4$ (since $15\pi/4 = 7\pi/4$ for angles).

   Find all the fixed points of the system $\theta(k + 1) = f(\theta(k))$.

   Generalize this problem by letting $f(\theta) = t\theta$, where t is a positive integer.

◆8. Exact solutions for one-dimensional continuous time systems. There is a technique (called *separation of variables*) which often can yield exact formulas for one-dimensional continuous time systems. This is how it works.

   Separation of variables.

   You are given the system $dx/dt = f(x)$.

   - Step 1: Rewrite this as $f(x) \, dx = dt$.
   - Step 2: Integrate both sides: $\int f(x) \, dx = \int dt$ which yields $F(x) = t + C$.
   - Step 3: Substitute $t = 0$ and $x = x_0$ to find C.
   - Step 4: Solve for x in terms of t.

   For example, suppose the system is $x' = x^2$ with $x(0) = 1$. The first step is to write the system as $dx/x^2 = dt$. The second step is to integrate both sides:

$$\int \frac{dx}{x^2} = \int dt \quad \Longrightarrow \quad \frac{-1}{x} = t + C.$$

   Step 3: Substitute $t = 0$ and $x = x_0 = 1$ to yield $-1 = 0 + C$, so $C = -1$. Finally (step 4): Solve for x to yield $x(t) = 1/(1 - t)$.

   Use this method to find exact formulas for $x(t)$ for each of the following systems.

   (a) $x' = -3x, x(0) = 2$.
   (b) $x' = 2/x, x(0) = 1$.
   (c) $x' = x + 1, x(0) = 1$.
   (d) $x' = x + 1, x(0) = -1$.
   (e) $x' = x^2 + 1, x(0) = 0$.
   (f) $x' = \sqrt{x}, x(0) = 1$.

3.2 Linearization

The purpose of this section is to provide a method to tell whether a fixed point $\tilde{\mathbf{x}}$ of a dynamical system [either discrete $\mathbf{x}(k + 1) = f(\mathbf{x}(k))$ or continuous $\mathbf{x}' = f(\mathbf{x})$] is stable or unstable. If the function f is linear, i.e., of the form $f(\mathbf{x}) = A\mathbf{x} + \mathbf{b}$, the answer is relatively easy: We check the eigenvalues of A (either their absolute values or their real parts, depending on the nature of time). The idea we present here is to *approximate* f near its fixed point $\tilde{\mathbf{x}}$ by a linear function.

To help us to build our intuition, we begin with one-dimensional systems (f is a function of just one variable). We then generalize to the higher dimensional setting.

3.2.1 One dimension

Derivatives: why and how?

How to approximate a nonlinear function with a linear function.

We use the derivative of f to approximate f. Two questions arise, Why the derivative? And what about the derivative?

We want to approximate f near a point $\tilde{x}$ (a fixed point of our system, but for the moment, that's irrelevant). In other words, if x is near $\tilde{x}$, we want

$$f(x) \approx a(x - \tilde{x}) + f(\tilde{x}), \qquad (3.1)$$

where a is a constant.[2] The right-hand side of equation (3.1) is the equation of a straight line through the point $(\tilde{x}, f(\tilde{x}))$ (to see why, substitute $\tilde{x}$ for x in equation (3.1)). Another way to write equation (3.1) is

$$f(x) = a(x - \tilde{x}) + f(\tilde{x}) + \text{error}(x - \tilde{x}), \qquad (3.2)$$

where $\text{error}(x - \tilde{x})$ measures how far off our answer is. What is the best number a to use in equation (3.2)? Certainly we want $\text{error}(x - \tilde{x}) \to 0$ as $x \to \tilde{x}$, but this will happen no matter what value for a we choose! We can ask for much more: We want $\text{error}(x - \tilde{x})$ to be very much smaller than the difference between x and $\tilde{x}$, i.e., we want

$$\frac{\text{error}(x - \tilde{x})}{x - \tilde{x}} \to 0 \qquad \text{as } x \to \tilde{x}.$$

---

[2]We often write linear functions in the form $y = mx + b$, where m is the slope and b is the y-intercept. The right-hand side of equation (3.1) is a variant of this. We wrote $a(x - \tilde{x}) + f(\tilde{x})$, which can be rewritten $ax + b$, where $b = -a\tilde{x} + f(\tilde{x})$ is a constant. The advantage to the form $a(x - \tilde{x}) + f(\tilde{x})$ is that we can easily see that when $x = \tilde{x}$, the result is $f(\tilde{x})$.

Can we really get such a wonderful approximation? Yes—here's how: We divide both sides of equation (3.2) by $x - \tilde{x}$ to get

$$\frac{f(x) - f(\tilde{x})}{x - \tilde{x}} - a = \frac{\text{error}(x - \tilde{x})}{x - \tilde{x}}. \tag{3.3}$$

Notice that if we choose $a = f'(\tilde{x})$, then automatically we will have

$$\frac{\text{error}(x - \tilde{x})}{x - \tilde{x}} \to 0 \qquad \text{as } x \to \tilde{x}.$$

Thus taking $a = f'(\tilde{x})$ gives us the very best approximation of f near $\tilde{x}$ by a linear function.

In conclusion, the best linear approximation to $f(x)$ for x near $\tilde{x}$ is given by

$$f(x) \approx f'(x)(x - \tilde{x}) + f(\tilde{x}).$$

How does this approximation enable us to determine the stability of fixed points? The answer depends on whether our system is continuous or discrete.

Continuous time

Let $x' = f(x) = x^2 - 1$. The fixed points of this continuous time system are ± 1. A graph of f is shown in Figure 3.2. What is the nature of each of these fixed points? Suppose the system begins at a value x_0 just greater than 1. Since $f(x_0) > 0$, we see that $x(t)$ is increasing; indeed, it will continue to increase forever, moving farther and farther away from 1. On the other side, if x_0 is slightly less than 1, then $f(x_0) < 0$, and so $x(t)$ is decreasing, moving farther away from 1 (but in the opposite direction). Thus 1 is an unstable fixed point.

Next let us consider the fixed point -1. If x_0 is slightly greater than -1, we see that $f(x_0) < 0$, and therefore $x(t)$ is decreasing *down toward* -1. On the other side, if x_0 is slightly less than -1, then $f(x_0) > 0$, and thus $x(t)$ is increasing *up toward* -1. Thus we see that -1 is a stable fixed point.

What about the derivative? Since $f(x) = x^2 - 1$, we know that $f'(x) = 2x$. Near $\tilde{x} = 1$ (the unstable fixed point) we have $f(x) \approx 2x - 2$ (see Figure 3.2). Thus f is very well approximated by a line of slope $+2$, and near $\tilde{x}$ we know that $f(x)$ and $2x - 2$ are nearly the same. Were f a linear function $(2x - 2)$, then since $2 > 0$ we know that $\tilde{x} = 1$ is an unstable fixed point and the system would explode away from 1.

A continuous time system with two fixed points: the solutions to $f(x) = 0$.

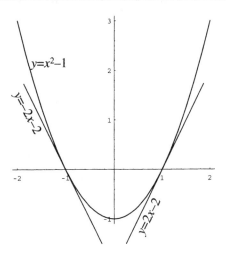

Figure 3.2. The graph of $f(x) = x^2 - 1$. The system $x' = f(x)$ has a stable fixed point at -1 and an unstable fixed point at $+1$.

Near $\tilde{x} = -1$ (the stable fixed point) we have $f(x) \approx -2x - 2$ (see the figure). Now f is approximated by a line with slope -2. Since $-2 < 0$, the linear system (as well as the true system it approximates) gravitates toward $\tilde{x} = -1$.

More generally, let $\tilde{x}$ be a fixed point of the continuous time system $x' = f(x)$. Since $\tilde{x}$ is a fixed point, we know that $f(\tilde{x}) = 0$.

Why $f'(\tilde{x}) > 0$ implies $\tilde{x}$ is unstable.

Suppose first that $f'(\tilde{x}) > 0$. Since f' is continuous, $f'(x) > 0$ for all values of x near $\tilde{x}$. Thus, near $\tilde{x}$, we know that f is a strictly increasing function. Since $f(\tilde{x}) = 0$, we know that for x near (but less than) $\tilde{x}$ we have $f(x) < 0$, while for x slightly larger than $\tilde{x}$ we have $f(x) > 0$. This means that for x just below $\tilde{x}$ we have $x' = f(x) < 0$, so the system is decreasing away from $\tilde{x}$. Likewise, for x just above $\tilde{x}$ we have $x' = f(x) > 0$, so the system is increasing away from $\tilde{x}$. Thus $\tilde{x}$ is an unstable fixed point.

Why $f'(\tilde{x}) < 0$ implies $\tilde{x}$ is stable.

Suppose now that $f'(\tilde{x}) < 0$. Since f' is continuous, we know that $f'(x) < 0$ for all x near $\tilde{x}$. This means that near $\tilde{x}$, f is a strictly decreasing function. Thus, for x just below $\tilde{x}$ we have $x' = f(x) > 0$, and for x just above $\tilde{x}$ we have $x' = f(x) < 0$. This implies that for x near $\tilde{x}$ the system tends to $\tilde{x}$. In other words, $\tilde{x}$ is a stable fixed point.

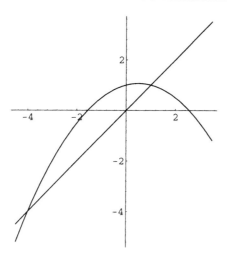

Figure 3.3. A graph of the function $y = f(x) = \frac{1}{4}(4 + x - x^2)$ and the line $y = x$. The points of intersection ($x = -4$ and $x = 1$) give the fixed points of the system $x(k + 1) = f(x(k))$.

Discrete time

Let $f(x) = \frac{1}{4}\left(4 + x - x^2\right)$ and, let's consider the system $x(k + 1) = f(x(k))$; in other words, we want to know what happens as we iterate f.

First, we locate the fixed points of f by solving $f(x) = x$, i.e., we solve

$$\frac{4 + x - x^2}{4} = x \quad \Rightarrow \quad x^2 - 3x + 4 = 0.$$

A discrete time system with two fixed points: the solutions to $f(x) = x$.

The roots of the latter equation are -4 and 1. Figure 3.3 shows a plot of the curve $y = f(x)$ and the line $y = x$. The intersection of the line and the curve gives the roots of the equation $x = f(x)$, i.e., at -4 and 1. Now we want to know how the system behaves near these fixed points. Let's zoom in on each.

Figure 3.4 shows a close-up of the graph $y = f(x)$ near the fixed point -4. I hope your reaction is, The curve $y = f(x)$ looks like a straight line! Good! That's the reason we treat it like a straight line (at least locally).

Returning to the graph, we consider a value x_0 just less than $\tilde{x} = -4$. We see that as we iterate f, we move farther and farther to the left, away from -4. We also consider an initial value just greater than -4. As we iterate f, the values of $x(k)$ get larger and larger, again, moving away from

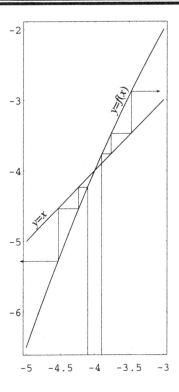

Figure 3.4. The graph of the function $y = f(x) = \frac{1}{4}(4 + x - x^2)$ near the fixed point -4.

-4. Clearly, -4 is an unstable fixed point.

Another way to see this is to consider the derivative of f at -4. Now, $f'(x) = (1 - 2x)/4$, and at -4 we have $f'(-4) = \frac{9}{4}$. Thus near $\tilde{x} = -4$ we have $f(x) \approx \frac{9}{4}x + 5$. Since the slope of this line ($\frac{9}{4}$) is greater than 1, we know we are in an unstable situation.

Now let's consider the other fixed point, 1. Figure 3.5 shows a close-up of the graph $y = f(x)$ near the fixed point 1. The figure shows what happens as we iterate f starting at values near 0.8 and 1.2. In both cases, the iterates move closer and closer to the fixed point, $\tilde{x} = 1$. Clearly, this fixed point is stable.

Let's examine this situation analytically. Since $f'(x) = (1 - 2x)/4$, we have $f'(1) = -\frac{1}{4}$. So, near $\tilde{x} = 1$ we have $f(x) \approx -\frac{1}{4}x + \frac{5}{4}$. Thus, very near to $\tilde{x}$, f behaves just like a linear function with slope $-\frac{1}{4}$. Since

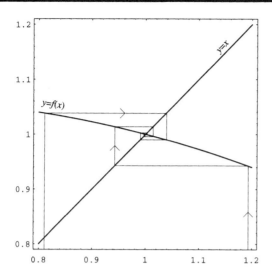

Figure 3.5. The graph of the function $y = f(x) = \frac{1}{4}(4 + x - x^2)$ near the fixed point 1.

$|-\frac{1}{4}| < 1$, we know that $\tilde{x} = 1$ is a stable fixed point.

We can use the mean value theorem (see §A.3.1 on page 349) to understand why $|f'(\tilde{x})| < 1$ implies stability and why $|f'(\tilde{x})| > 1$ implies instability.

Let $\tilde{x}$ be a fixed point of a discrete time system $x(k + 1) = f(x(k))$. Thus $f(\tilde{x}) = \tilde{x}$.

First we consider the case $|f'(\tilde{x})| < 1$. Since f' is continuous, we know that $|f'(x)| < 1$ for all numbers x near $\tilde{x}$. More precisely, for any x in the interval $[\tilde{x} - a, \tilde{x} + a]$ we have $|f'(x)| \leq b$, where a and b are positive numbers and $b < 1$. We work to show that for any $x \in [\tilde{x} - a, \tilde{x} + a]$ we have $f^k(x) \to \tilde{x}$ as $k \to \infty$. *Why $|f'(\tilde{x})| < 1$ implies stability.*

Let $x \in [\tilde{x} - a, \tilde{x} + a]$ (but $x \neq \tilde{x}$). We want to compare the distance between x and $\tilde{x}$ with the distance between $f(x)$ and $\tilde{x}$; let us estimate

$$\frac{|f(x) - \tilde{x}|}{|x - \tilde{x}|}. \tag{3.4}$$

We know that $\tilde{x} = f(\tilde{x})$ (because $\tilde{x}$ is a fixed point). We can rewrite the fraction in equation (3.4) by substituting $f(\tilde{x})$ for $\tilde{x}$

$$\frac{|f(x) - \tilde{x}|}{|x - \tilde{x}|} = \left|\frac{f(x) - f(\tilde{x})}{x - \tilde{x}}\right|. \tag{3.5}$$

We can now apply the mean value theorem to the fraction on the right side of equation (3.5): There must be a number c between x and $\tilde{x}$ for which

$$f'(c) = \frac{f(x) - f(\tilde{x})}{x - \tilde{x}}. \tag{3.6}$$

Since $x \in [\tilde{x} - a, \tilde{x} + a]$ and c is between x and $\tilde{x}$, we know that $c \in [\tilde{x} - a, \tilde{x} + a]$ and therefore $|f'(c)| \le b < 1$. Thus equation (3.6) becomes

$$\left|\frac{f(x) - f(\tilde{x})}{x - \tilde{x}}\right| \le b, \tag{3.7}$$

which can also be written as

$$|f(x) - \tilde{x}| = |f(x) - f(\tilde{x})| \le b|x - \tilde{x}|. \tag{3.8}$$

Let us examine equation (3.8) closely. It says: The distance between $f(x)$ and $\tilde{x}$ is at most b times as great as the distance between x and $\tilde{x}$. In other words, in moving from x to $f(x)$ we have shrunk the distance to $\tilde{x}$ by a factor of b (or better). Thus $f(x)$ is closer to $\tilde{x}$ than x is. Further, consider $f^2(x) = f[f(x)]$. Since $f(x) \in [\tilde{x} - a, \tilde{x} + a]$, we have (using equation (3.8) twice)

$$|f^2(x) - \tilde{x}| \le b|f(x) - \tilde{x}| \le b^2|x - \tilde{x}|.$$

Using equation (3.8) repeatedly, we have

$$|f^k(x) - \tilde{x}| \le b^k|x - \tilde{x}|.$$

Since $0 < b < 1$, we know that $b^k \to 0$ and therefore the distance between $f^k(x)$ and $\tilde{x}$ shrinks to 0 as $k \to \infty$. Thus we have shown that $f^k(x) \to \tilde{x}$, proving that $\tilde{x}$ is stable.

Why $|f'(\tilde{x})| > 1$ implies instability.

The case when $|f'(\tilde{x})| > 1$ is quite similar. Since f' is continuous, we know that for any x in the interval $[\tilde{x} - a, \tilde{x} + a]$ we have $|f'(x)| \ge b$ where a and b are constants and $b > 1$. Working as we did before (see equations (3.4) through (3.8)), we arrive at

$$|f(x) - \tilde{x}| \ge b|x - \tilde{x}|,$$

from which it follows that

$$|f^k(x) - \tilde{x}| \ge b^k|x - \tilde{x}|.$$

Since $b > 1$, we know that b^k gets larger and larger as k grows. Thus for any x (except $x = \tilde{x}$) within distance a of $\tilde{x}$, after several iterations we will have $f^k(x) \notin [\tilde{x} - a, \tilde{x} + a]$; therefore, $\tilde{x}$ is unstable.

Classification of fixed points: one dimension		
Time	Derivative at $\tilde{x}$	Fixed point is
Continuous	$f'(\tilde{x}) < 0$	stable
	$f'(\tilde{x}) > 0$	unstable
	$f'(\tilde{x}) = 0$	test fails
Discrete	$|f'(\tilde{x})| < 1$	stable
	$|f'(\tilde{x})| > 1$	unstable
	$|f'(\tilde{x})| = 1$	test fails

Table 3.2. Linearization test for one-dimensional nonlinear systems.

What we have learned

We considered two examples—one discrete and the other continuous—of nonlinear dynamical systems. We found the fixed points and then observed that we may approximate the systems near their fixed points by straight lines. We then brought what we learned from Chapter 2 to bear on analyzing these fixed points. Here are our conclusions:

Continuous time. Let $\tilde{x}$ be a fixed point of the continuous time dynamical system $x' = f(x)$. If $f'(\tilde{x}) < 0$, then $\tilde{x}$ is a stable fixed point. If $f'(\tilde{x}) > 0$, then $\tilde{x}$ is an unstable fixed point.

Discrete time. Let $\tilde{x}$ be a fixed point of the discrete time dynamical system $x(k+1) = f(x(k))$. If $|f'(\tilde{x})| < 1$, then $\tilde{x}$ is a stable fixed point. If $|f'(\tilde{x})| > 1$, then $\tilde{x}$ is an unstable fixed point.

Wait a minute! The preceding cases do not cover all the possibilities. We might have $f'(\tilde{x}) = 0$ (in continuous time) or $|f'(\tilde{x})| = 1$ (in discrete time). What then? The answer is, I don't know—better yet, we can't know from the information we are given. The fixed point may or may not be stable. Exercises 6 and 7 (page 128) lead you through some examples of these borderline cases; see also problem 8.

Table 3.2 summarizes what we have learned about the stability of fixed points of one-dimensional dynamical systems.

> Summarizing the linearization technique for one-dimensional systems.

3.2.2   Two and more dimensions

Derivatives again

We are ready to examine higher dimensional nonlinear systems. Again, we wish to approximate our function f by a linear function in the neighborhood

> Approximating a nonlinear function of several variables by a linear function.

of a fixed point $\tilde{\mathbf{x}}$. Since f is a function from $\mathbf{R}^n$ to $\mathbf{R}^n$, the approximation we seek is of the form

$$f(\mathbf{x}) \approx A(\mathbf{x} - \tilde{\mathbf{x}}) + f(\tilde{\mathbf{x}}),$$

where A is an $n \times n$ matrix which gives the best approximation.[3] In particular, if we write

$$f(\mathbf{x}) = A(\mathbf{x} - \tilde{\mathbf{x}}) + f(\tilde{\mathbf{x}}) + \text{error}(\mathbf{x} - \tilde{\mathbf{x}}),$$

then we want

$$\frac{|\text{error}(\mathbf{x} - \tilde{\mathbf{x}})|}{|\mathbf{x} - \tilde{\mathbf{x}}|} \to 0$$

as $\mathbf{x} \to \tilde{\mathbf{x}}$. The matrix A which does this job is the *Jacobian matrix* of f, which we denote by Df.

> The Jacobian matrix of f is the matrix of its partial derivatives.

Let's review the Jacobian matrix. Recall that f is a vector-valued function of several values. Thus we can write

$$f(\mathbf{x}) = f \begin{bmatrix} x_1 \\ x_2 \\ \vdots \\ x_n \end{bmatrix} = \begin{bmatrix} f_1(\mathbf{x}) \\ f_2(\mathbf{x}) \\ \vdots \\ f_n(\mathbf{x}) \end{bmatrix},$$

where each f_j is a scalar-valued function of n variables. For example, if

$$f(\mathbf{x}) = f \begin{bmatrix} x_1 \\ x_2 \end{bmatrix} = \begin{bmatrix} e^{x_1} \cos x_2 \\ x_1 - x_2 \end{bmatrix}$$

then $f_1(\mathbf{x}) = e^{x_1} \cos x_2$, and $f_2(\mathbf{x}) = x_1 - x_2$. The Jacobian, Df, of f is the matrix of partial derivatives of the f_j's:

$$Df = \begin{bmatrix} \frac{\partial f_1}{\partial x_1} & \frac{\partial f_1}{\partial x_2} & \cdots & \frac{\partial f_1}{\partial x_n} \\ \frac{\partial f_2}{\partial x_1} & \frac{\partial f_2}{\partial x_2} & \cdots & \frac{\partial f_2}{\partial x_n} \\ \vdots & \vdots & \ddots & \vdots \\ \frac{\partial f_n}{\partial x_1} & \frac{\partial f_n}{\partial x_2} & \cdots & \frac{\partial f_n}{\partial x_n} \end{bmatrix}.$$

For the preceding example,

$$\frac{\partial f_1}{\partial x_1} = e^{x_1} \cos x_2 \qquad \frac{\partial f_1}{\partial x_2} = -e^{x_1} \sin x_2$$

$$\frac{\partial f_2}{\partial x_1} = 1 \qquad \frac{\partial f_2}{\partial x_2} = -1$$

---

[3] As in the one-dimensional case, we seek a linear (affine) function of the form $A\mathbf{x} + \mathbf{b}$. Note that $A(\mathbf{x} - \tilde{\mathbf{x}}) + f(\tilde{\mathbf{x}})$ can be rewritten as $A\mathbf{x} + \mathbf{b}$, with $\mathbf{b} = -A\tilde{\mathbf{x}} + f(\tilde{\mathbf{x}})$.

hence

$$Df = \begin{bmatrix} e^{x_1} \cos x_2 & -e^{x_1} \sin x_2 \\ 1 & -1 \end{bmatrix}.$$

Continuous time

Let's consider a specific example, the continuous time system

$$x_1' = x_1^2 + x_2^2 - 1,$$
$$x_2' = x_1 - x_2.$$

In our usual notation, $\mathbf{x}' = f(\mathbf{x})$, where

$$f \begin{bmatrix} x_1 \\ x_2 \end{bmatrix} = \begin{bmatrix} x_1^2 + x_2^2 - 1 \\ x_1 - x_2 \end{bmatrix}. \tag{3.9}$$

To find the fixed points of the system, we solve $f(\mathbf{x}) = \mathbf{0}$, i.e., we solve

$$x_1^2 + x_2^2 - 1 = 0,$$
$$x_1 - x_2 = 0.$$

We find two fixed points: $\begin{bmatrix} 1/\sqrt{2} \\ 1/\sqrt{2} \end{bmatrix}$ and $\begin{bmatrix} -1/\sqrt{2} \\ -1/\sqrt{2} \end{bmatrix}$.

Now we ask, Which (if either) is stable? ... is unstable?

Let's look at a picture of f to attempt to understand how it behaves. Figure 3.6 shows such an overview of the function f. At each point $\mathbf{x}$ of the plane ($\mathbf{R}^2$) we draw an arrow in the direction of $f(\mathbf{x})$.

In the vicinity of the first fixed point, $\begin{bmatrix} 1/\sqrt{2} \\ 1/\sqrt{2} \end{bmatrix} \approx \begin{bmatrix} 0.7 \\ 0.7 \end{bmatrix}$ (upper right), the arrows appear to be moving away, and in the vicinity of the second fixed point, $\begin{bmatrix} -1/\sqrt{2} \\ -1/\sqrt{2} \end{bmatrix} \approx \begin{bmatrix} -0.7 \\ -0.7 \end{bmatrix}$ (lower left), they appear to be swirling in. Let's have a closer look.

Figure 3.7 focuses on the region around $\begin{bmatrix} 1/\sqrt{2} \\ 1/\sqrt{2} \end{bmatrix}$. Although some arrows are pointing inward, it is clear that a typical starting value near this fixed point is swept away from the fixed point. In this sense it is similar to the example of Figure 2.17 (page 79), in which there is a linear system with one positive and one negative eigenvalue.

Next we consider Figure 3.8, which illustrates the vector field near

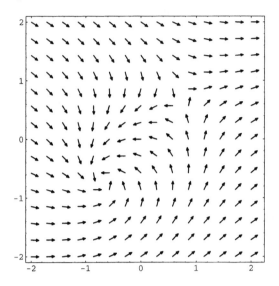

Figure 3.6. An overview of the behavior of the function f from equation (3.9).

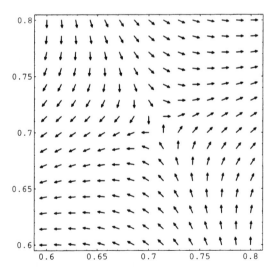

Figure 3.7. The behavior of f, from equation (3.9), near one of its fixed points.

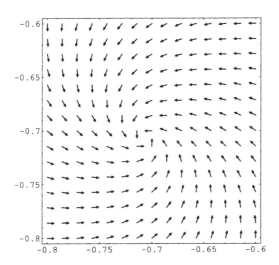

Figure 3.8. The behavior of f, from equation (3.9), near another fixed point.

the fixed point $\begin{bmatrix} -1/\sqrt{2} \\ -1/\sqrt{2} \end{bmatrix}$. Notice that the arrows are spiraling inward. This fixed point is stable. The picture is similar to the vector field of a system with complex eigenvalues with negative real parts; see Figure 2.19 (page 80).

The geometric pictures tell us the story of this system quite clearly. However, they are time consuming to compute and do not help us for systems in many variables. We can glean the same information more efficiently by approximating f near its fixed points by a linear function.

To do this, we approximate f near each of its fixed points $\tilde{\mathbf{x}}$ using the formula
$$f(\mathbf{x}) \approx Df(\tilde{\mathbf{x}})[\mathbf{x} - \tilde{\mathbf{x}}] + f(\tilde{\mathbf{x}}) = Df(\tilde{\mathbf{x}})[\mathbf{x} - \tilde{\mathbf{x}}].$$
We need to compute the Jacobian of f:
$$\partial f_1/\partial x_1 = 2x_1,$$
$$\partial f_1/\partial x_2 = 2x_2,$$
$$\partial f_2/\partial x_1 = 1,$$
$$\partial f_2/\partial x_2 = -1,$$

and therefore

$$Df = \begin{bmatrix} 2x_1 & 2x_2 \\ 1 & -1 \end{bmatrix}.$$

Near the first fixed point, $\tilde{\mathbf{x}}_1 = \begin{bmatrix} 1/\sqrt{2} \\ 1/\sqrt{2} \end{bmatrix}$, we have $Df(\tilde{\mathbf{x}}_1) = \begin{bmatrix} \sqrt{2} & \sqrt{2} \\ 1 & -1 \end{bmatrix}.$

The eigenvalues of this matrix are (approximately) 1.9016 and -1.4874. Aha! Near $\tilde{\mathbf{x}}_1$ the system behaves just like a linear system with one positive and one negative eigenvalue (just as we observed geometrically). This fixed point is *unstable*.

Let's linearize near the second fixed point, $\tilde{\mathbf{x}}_2 = \begin{bmatrix} -1/\sqrt{2} \\ -1/\sqrt{2} \end{bmatrix}.$ Here

the Jacobian equals $Df(\tilde{\mathbf{x}}_2) = \begin{bmatrix} -\sqrt{2} & -\sqrt{2} \\ 1 & -1 \end{bmatrix}.$ The eigenvalues of this

matrix are (approximately) $-1.2071 \pm 1.1710i$, complex eigenvalues with negative real parts. Thus we confirm that $\tilde{\mathbf{x}}_2$ is a *stable* fixed point, and near $\tilde{\mathbf{x}}_2$ the system behaves like a linear system with complex eigenvalues with negative real parts.

Summary for continuous time

In a continuous time system, the stability of a fixed point $\tilde{\mathbf{x}}$ can be judged by the signs of the real parts of the eigenvalues of the Jacobian $Df(\mathbf{x})$.

Let's review what we have learned: Let $\tilde{\mathbf{x}}$ be a fixed point of a continuous time dynamical system $\mathbf{x}' = f(\mathbf{x})$. If the eigenvalues of the Jacobian $Df(\tilde{\mathbf{x}})$ all have negative real part, then $\tilde{\mathbf{x}}$ is a stable fixed point. If some eigenvalues of $Df(\tilde{\mathbf{x}})$ have positive real part, then $\tilde{\mathbf{x}}$ is an unstable fixed point. Otherwise (all eigenvalues have nonpositive real part and some have zero real part) we cannot judge the stability of the fixed point.

Discrete time

Next we consider a discrete time nonlinear dynamical system $\mathbf{x}(k + 1) = f(\mathbf{x}(k))$ where

$$f(\mathbf{x}) = f \begin{bmatrix} x_1 \\ x_2 \end{bmatrix} = \begin{bmatrix} (x_1 + x_2)/2 \\ 4 \cos x_1 \end{bmatrix}. \tag{3.10}$$

First, we need to find the fixed points of this system, i.e., to solve the equation $\mathbf{x} = f(\mathbf{x})$. We need to solve

$$x_1 = (x_1 + x_2)/2,$$

$$x_2 = 4 \cos x_1.$$

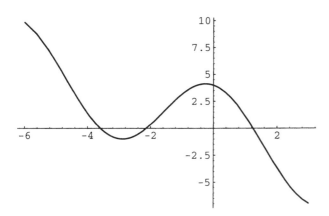

Figure 3.9. A plot of the function $4 \cos x - x$. The curve crosses the x-axis three times, so the equation $4 \cos x = x$ has three roots.

Solving this pair of equations is not as bad as it might first seem. The first equation rearranges to $x_1 = x_2$; then we just need to solve the equation $4 \cos x_2 = x_2$. There is no simple solution we can give for this equation, so we proceed numerically. It is helpful to plot a graph of $4 \cos x - x$ and see where it crosses the x-axis. Figure 3.9 shows such a plot. Notice that the curve crosses the x-axis three times, at approximately 1.2, -2.1, and -3.6. Using numerical methods (such as Newton's method), we can pin down these values. More precisely, the fixed points of dynamical system (3.10) are

$$\tilde{\mathbf{x}}_1 = \begin{bmatrix} 1.25235 \\ 1.25235 \end{bmatrix}, \quad \tilde{\mathbf{x}}_2 = \begin{bmatrix} -2.1333 \\ -2.1333 \end{bmatrix}, \quad \text{and} \quad \tilde{\mathbf{x}}_3 = \begin{bmatrix} -3.5953 \\ -3.5953 \end{bmatrix}.$$

Which of these fixed points are stable? To find out, we linearize.

The Jacobian matrix for f is

$$Df(\mathbf{x}) = \begin{bmatrix} 1/2 & 1/2 \\ -4 \sin x_1 & 0 \end{bmatrix}.$$

We can now compute the eigenvalues (numerically) near each fixed point.

- At $\tilde{\mathbf{x}}_1 = \begin{bmatrix} 1.25235 \\ 1.25235 \end{bmatrix}$ we have $Df(\tilde{\mathbf{x}}_1) = \begin{bmatrix} 0.5 & 0.5 \\ -3.7989 & 0 \end{bmatrix}$,
  whose eigenvalues are $0.25 \pm 1.35534i$. The absolute value of these eigenvalues is 1.3782, which is greater than 1. Hence $\tilde{\mathbf{x}}_1$ is *unstable*.

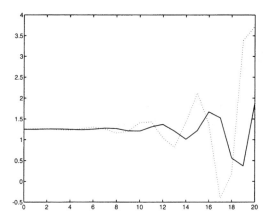

Figure 3.10. Twenty iterations of the system in equation (3.10), starting near $\tilde{\mathbf{x}}_1$.

- At $\tilde{\mathbf{x}}_2 = \begin{bmatrix} -2.1333 \\ -2.1333 \end{bmatrix}$ we have $Df(\tilde{\mathbf{x}}_2) = \begin{bmatrix} 0.5 & 0.5 \\ 3.3836 & 0 \end{bmatrix}$, whose eigenvalues are 1.5745 and -1.0745. Both have absolute value greater than 1, hence $\tilde{\mathbf{x}}_2$ is *unstable*.

- Finally, we consider $\tilde{\mathbf{x}}_3 = \begin{bmatrix} -3.5953 \\ -3.5953 \end{bmatrix}$. In this case we have

$Df(\tilde{\mathbf{x}}_2) = \begin{bmatrix} 0.5 & 0.5 \\ -1.7532 & 0 \end{bmatrix}$, whose eigenvalues are $0.25 \pm 0.902281i$.

The absolute value of these eigenvalues is 0.936275, which is less than 1. Hence $\tilde{\mathbf{x}}_3$ is *stable*.

Let's see the results in action. We choose starting vectors $\mathbf{x}_0$ near each of the three fixed points. Figures 3.10, 3.11, and 3.12 illustrate several iterations of the system (3.10), starting at

$$\begin{bmatrix} 1.25 \\ 1.25 \end{bmatrix}, \quad \begin{bmatrix} -2.13 \\ -2.13 \end{bmatrix}, \quad \text{and} \quad \begin{bmatrix} -4 \\ -3.5 \end{bmatrix},$$

respectively.

In the graphs the solid line represents the successive values of x_1, and the dotted line the values of x_2. Notice that in the first two figures the values are clearly moving away from their fixed points. However in the third figure the iterations are moving closer progressively to $\tilde{\mathbf{x}}_3$.

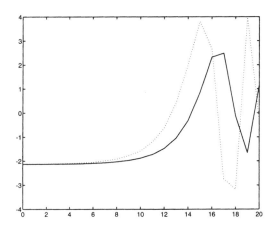

Figure 3.11. Twenty iterations of the system in equation (3.10), starting near $\tilde{\mathbf{x}}_2$.

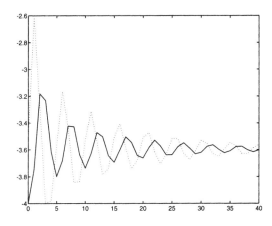

Figure 3.12. Forty iterations of the system in equation (3.10), starting near $\tilde{\mathbf{x}}_3$.

Classification of fixed points $\tilde{\mathbf{x}}$ in several dimensions		
Time	Eigenvalues of $Df(\tilde{\mathbf{x}})$	Fixed point is
Continuous	all $\Re\lambda < 0$	stable
	some $\Re\lambda > 0$	unstable
	all $\Re\lambda \leq 0$, some $\Re\lambda = 0$	test fails
Discrete	all $|\lambda| < 1$	stable
	some $|\lambda| > 1$	unstable
	all $|\lambda| \leq 1$, some $|\lambda| = 1$	test fails

Table 3.3. Linearization test for multidimensional nonlinear systems.

What we have learned

We can use linearization to understand the nature of fixed points of multidimensional nonlinear systems. Near a fixed point $\tilde{\mathbf{x}}$ we can approximate $f(\mathbf{x})$ by the linear function $Df(\tilde{\mathbf{x}})[x - \tilde{\mathbf{x}}] + f(\tilde{\mathbf{x}})$. We then inspect the eigenvalues of the Jacobian matrix $Df(\tilde{\mathbf{x}})$ and apply what we learned in Chapter 2 about linear systems.

The linearization tests for continuous and discrete time systems.

    Continuous time. Let $\tilde{\mathbf{x}}$ be a fixed point of the system $\mathbf{x}' = f(\mathbf{x})$, that is, $f(\tilde{\mathbf{x}}) = \mathbf{0}$. Compute the Jacobian evaluated at $\tilde{\mathbf{x}}$, i.e., find $Df(\tilde{\mathbf{x}})$. If the eigenvalues of the Jacobian all have negative real part, then $\tilde{\mathbf{x}}$ is a stable fixed point. If some eigenvalue of the Jacobian has positive real part, then $\tilde{\mathbf{x}}$ is an unstable fixed point.

    Discrete time. Let $\tilde{\mathbf{x}}$ be a fixed point of the system $\mathbf{x}(k+1) = f(\mathbf{x}(k))$, that is, $f(\tilde{\mathbf{x}}) = \mathbf{x}$. Compute the Jacobian evaluated at $\tilde{\mathbf{x}}$, i.e., find $Df(\tilde{\mathbf{x}})$. If the eigenvalues of the Jacobian all have absolute value less than 1, then $\tilde{\mathbf{x}}$ is a stable fixed point. If some eigenvalue of the Jacobian has absolute value greater than 1, then $\tilde{\mathbf{x}}$ is an unstable fixed point.

    These results are summarized in Table 3.3.

Problems for §3.2

◆1.   For each of the following discrete time systems $x(k+1) = f(x(k))$, find all fixed points and determine their stability.

    (a) $f(x) = \cos x$.

    (b) $f(x) = -x^3 - 2$.

    (c) $f(x) = x^2 - x + \frac{1}{4}$.

    (d) $f(x) = e^{x/2} - 1$.

   (e) $f(x) = \frac{3}{2}\sin x$.

   (f) $f(x) = \frac{3}{2}\cos x$.

   (g) $f(x) = e^{\cos x}$.

   (h) $f(x) = \frac{1}{2}e^{\sin x}$.

   (i) $f(x) = e^x$.

◆2.   For each of the following continuous time systems $x' = f(x)$, find all fixed
      points and determine their stability.

      (a) $f(x) = 1 - e^x$.

      (b) $f(x) = e^x - 1$.

      (c) $f(x) = x^2 + 3x - 2$.

      (d) $f(x) = x^3 - 3x^2 + 2x$.

      (e) $f(x) = x^6 + 4$.

      (f) $f(x) = 2^x - 3^x$.

      (g) $f(x) = |x - 1| - 1$.

      (h) $f(x) = x + 1/x - 4$.

      (i) $f(x) = 1 + \sin x$.

◆3.   For each of the following discrete time systems $\mathbf{x}(k + 1) = f(\mathbf{x}(k))$, find
      all fixed points and determine their stability.

      (a) $f\begin{bmatrix} x \\ y \end{bmatrix} = \begin{bmatrix} x + y^2 \\ x + 2y \end{bmatrix}$.

      (b) $f\begin{bmatrix} x \\ y \end{bmatrix} = \begin{bmatrix} x + y - x^2 \\ 2x + 3y \end{bmatrix}$.

      (c) $f\begin{bmatrix} x \\ y \end{bmatrix} = \begin{bmatrix} x^2 + y^2 - \frac{1}{4} \\ x/2 + y \end{bmatrix}$.

      (d) $f\begin{bmatrix} x \\ y \end{bmatrix} = \frac{1}{3}\begin{bmatrix} y^2 - x - 3 \\ x + y \end{bmatrix}$.

      (e) $f\begin{bmatrix} x \\ y \end{bmatrix} = \begin{bmatrix} e^y \\ x/5 \end{bmatrix}$. [Hint: There are two fixed points.]

◆4.   Show that the fixed point $\tilde{k} = [sA/(d + \rho)]^2$ is a stable fixed point of the
      economic growth system of §1.2.5 (see equation (1.22) on page 19).

◆5.   For each of the following continuous time systems $\mathbf{x}' = f(\mathbf{x})$, find all fixed
      points and determine their stability.

      (a) $f\begin{bmatrix} x \\ y \end{bmatrix} = \begin{bmatrix} (y^2 - x^2 - x - 3)/2 \\ (x + y + 1)/2 \end{bmatrix}$.

(b) $f\begin{bmatrix} x \\ y \end{bmatrix} = \begin{bmatrix} x^3 - y \\ x + y \end{bmatrix}.$

(c) $f\begin{bmatrix} x \\ y \end{bmatrix} = \begin{bmatrix} y^2 - x \\ x^2 - y \end{bmatrix}.$

(d) $f\begin{bmatrix} x \\ y \end{bmatrix} = \begin{bmatrix} \sin y \\ x + y \end{bmatrix}.$

(e) $f\begin{bmatrix} x \\ y \end{bmatrix} = \begin{bmatrix} 3y - e^x \\ 2x - y \end{bmatrix}.$ [Hint: There are two fixed points.]

◆6. In this problem we consider one-dimensional discrete time dynamical systems $x(k + 1) = f(x(k))$ with a fixed point $\tilde{x}$ at which $|f'(\tilde{x})| = 1$. For each of the following systems, discuss the stability of the fixed point $\tilde{x}$.

(a) $f(x) = \sin x, \tilde{x} = 0.$

(b) $f(x) = x^3 + x, \tilde{x} = 0.$

(c) $f(x) = 1 + \log x, \tilde{x} = 1.$

(d) $f(x) = x^2 + \frac{1}{4}, \tilde{x} = \frac{1}{2}.$

(e) $f(x) = \frac{3}{4} - x^2, \tilde{x} = \frac{1}{2}.$

(f) $f(x) = 1/x, \tilde{x} = 1.$

(g) $f(x) = 1/(2x - 2) + 1/(2x + 2), \tilde{x} = 0.$

◆7. In this problem we consider one-dimensional continuous time dynamical systems $x' = f(x)$ with a fixed point $\tilde{x}$ at which $f'(\tilde{x}) = 0$. For each of the following systems, discuss the stability of the fixed point $\tilde{x} = 0$.

(a) $f(x) = x^2.$

(b) $f(x) = -x^2.$

(c) $f(x) = x^3.$

(d) $f(x) = -x^3.$

◆8.* Develop higher derivative tests—for both discrete and continuous time—for stability when $f'(\tilde{x})$ leads to inconclusive results. Your test should be able to determine that 0 is an unstable fixed point of $x' = x^3$ but a stable fixed point of $x' = -x^3$.

3.3 Lyapunov functions

3.3.1 Linearization can fail

When linearization fails to determine stability.

Linearization is a great tool for determining the stability of fixed points of dynamical systems. Unfortunately, there's the nasty "Test fails" possibility. What can we do then? Fortunately, we have an "emergency backup" test (which is more difficult to apply and also isn't guaranteed to work) on which we can call. Let's begin with an example.

Weird friction

In Chapter 1 (see §1.2.1 on page 8) we introduced the simple mass-and-spring dynamical system; the system was not particularly realistic because we omitted the effect of friction. We corrected that problem in Chapter 2 (see page 68), where we considered the air resistance on the mass as exerting a force against the motion of the block in an amount proportional to the block's velocity. If the block were moving through a more viscous medium, then a resistive force proportional to the square of the velocity might be a more realistic model. Let's do something even more dramatic here. Imagine the block is moving through a bizarre medium which exerts a force on the block proportional to the *cube* of its velocity.[4] We can model this situation by the following equations:

$$x' = v,$$
$$v' = -x - \mu v^3,$$

where μ is some positive constant. In vector notation $\mathbf{y}' = f(\mathbf{y})$, where $\mathbf{y} = \begin{bmatrix} x \\ v \end{bmatrix}$ and $f(\mathbf{y}) = f\begin{bmatrix} x \\ v \end{bmatrix} = \begin{bmatrix} v \\ -x - \mu v^3 \end{bmatrix}$. What is the fate of this system?

We readily check that the system has only one fixed point: $\tilde{\mathbf{y}} = \begin{bmatrix} 0 \\ 0 \end{bmatrix}$.

Is this fixed point stable? The Jacobian matrix is

$$Df = \begin{bmatrix} 0 & 1 \\ -1 & -3\mu v^2 \end{bmatrix},$$

The linearization test does not tell us the stability of this fixed point...

hence $Df(\tilde{\mathbf{y}}) = \begin{bmatrix} 0 & 1 \\ -1 & 0 \end{bmatrix}$. The eigenvalues of $Df(\tilde{\mathbf{y}})$ are $\pm i$, which have real part equal to 0. Hence the linearization test fails.

What can we do next? We can try numerical methods. Let us take $\mu = 0.25$ and begin in state $\mathbf{y}_0 = \begin{bmatrix} 2 \\ 0 \end{bmatrix}$. Figure 3.13 shows the phase diagram for this situation. Notice that the system's trajectory is spiraling around the fixed point $\mathbf{0}$, but it is not clear if it will eventually converge to $\mathbf{0}$ or will always remain at a comfortable distance from $\mathbf{0}$ and never get

...and numerical evidence is inconclusive.

[4]There does not appear to be, in nature, a fluid which presents a resistive force precisely proportional to the cube of the velocity. However, there are media, such as thixotropic or dilatant fluids, which do present highly nonlinear resistance to motion. Examples include paints, printing inks, and solutions of corn starch in water.

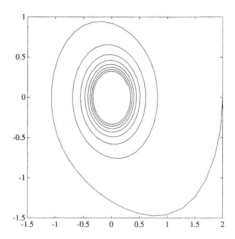

Figure 3.13. Phase diagram for the weird-friction example. The trajectory begins at the far right and spirals inward. Is the system converging to the fixed point?

nearer. In other words, the numerical evidence is too weak to suggest **0**'s stability or lack thereof.

Our linearization method and our numerical experiments have failed us. What next? Let's check our intuition. What should happen when a spring is vibrating in a highly viscous medium? It should stop! Why? Because the friction is bleeding off *energy* from the system.

3.3.2 Energy

Our intuition says, if a system is losing energy then it must eventually grind to a halt. Thus the fixed point **0** of the preceding "weird-friction" example should be stable. Now, energy is a concept from the physical world, and mathematics need not obey the laws of physics. Nonetheless, we can use the idea of energy as motivation for a mathematical method.

Energy as mechanical work.   What is energy? Energy is distance times force. If you lift an object which weighs 1 newton through a distance of 1 meter, you have used 1 joule of energy. Where is that energy now? The law of conservation of energy says it must be somewhere. The answer is, it is in the potential energy of the object being held up at a given height. If we release that object, its potential energy is converted into kinetic (motion) energy. Let's work this out.

You may recall from basic physics that if we lift a mass m to a height h in the presence of a gravitational field of strength g (so the weight of the object is mg), then the potential energy in the mass is $E = mgh$. (We exerted mg units of force through a distance h.)

Potential energy in height.

Now let's allow the mass to fall. It starts with zero velocity and accelerates at a rate of g, i.e., after t seconds, its speed is $v(t) = gt$. Because velocity is the rate of change of position, i.e., $v = dx/dt$, the distance the object has fallen after t seconds is

Kinetic energy.

$$\int_0^t v(\tau)\, d\tau = \int_0^t g\tau\, d\tau = \frac{1}{2}gt^2.$$

Let t be the time it takes the object to return to the position from which we lifted it, i.e., $h = \frac{1}{2}gt^2$. Substituting this value of h back into $E = mgh$, we have $E = mg\left(\frac{1}{2}gt^2\right) = \frac{1}{2}m(gt)^2$. Finally, gt is the velocity at the time the mass reaches its starting position, and we have the well-known formula $E = \frac{1}{2}mv^2$; this is the amount of *kinetic* energy in an object traveling with velocity v.

The weird-friction example includes a spring. When the mass is in its neutral position, the spring contains no energy; however, as the spring is compressed or expanded it stores energy. The spring exerts a varying amount of force depending on how far we compress (or expand) it. Our assumption is that the force the spring exerts equals kx, where x is the distance the spring is compressed (or expanded), and k is a constant (Hooke's law). If we compress a spring through distance x, how much energy have we stored in that spring? The answer is not very simple, because the force changes with the distance. We handle this problem by adding up the amount of energy we need to advance by a tiny distance (ds) when the spring has been compressed a distance s: the force is ks and the distance is ds. We sum over the entire range from 0 to x, i.e., we integrate

Potential energy stored in a spring.

$$E = \int_0^x ks\, ds = \frac{1}{2}kx^2.$$

Thus the energy stored in a spring with constant k and compressed [expanded] a distance x is $\frac{1}{2}kx^2$.

We can now write down how much energy is in the weird-friction system when it's in a given state $\mathbf{y} = \begin{bmatrix} x \\ v \end{bmatrix}$. The energy is the sum of the potential energy in the spring and the kinetic energy in the motion of the mass: $E = \frac{1}{2}kx^2 + \frac{1}{2}mv^2$. As before, we simplify by taking $k = m = 1$, and therefore we can write $E = (x^2 + v^2)/2$.

Now let's justify our intuitive feeling that the system must be losing energy. To this end, we compute dE/dt: the rate at which energy is changing over time. Now, E is a function of two variables, x and v, each of which depends on time. The formula for the derivative of such a function is (see §A.3.2, equation (A.6) on page 350)

$$\frac{dE}{dt} = \frac{\partial E}{\partial x}\frac{dx}{dt} + \frac{\partial E}{\partial v}\frac{dv}{dt}.$$

We compute each part of the right-hand side:

$$
\begin{aligned}
\partial E/\partial x &= x & \text{since } E = (x^2 + v^2)/2, \\
\partial E/\partial v &= v & \text{since } E = (x^2 + v^2)/2, \\
dx/dt &= v & \text{from our weird system,} \quad \text{and} \\
dv/dt &= -x - \mu v^3 & \text{from our weird system.}
\end{aligned}
$$

Plugging these into the formula $\frac{dE}{dt} = \frac{\partial E}{\partial x}\frac{dx}{dt} + \frac{\partial E}{\partial v}\frac{dv}{dt}$, we have

$$\frac{dE}{dt} = xv + v(-x - \mu v^3) = -\mu v^4.$$

The system is losing energy. Notice that dE/dt is *always* negative and therefore the system is always losing energy. Of course, when $v = 0$, we have $dE/dt = 0$. When does $v = 0$? There are two cases. First, if we are at the fixed point **0**, then the system is at rest (in equilibrium) and no energy is being lost. Also, at the instant the spring is maximally compressed or expanded and the block is changing directions, the system is again not losing energy. But this latter case happens for only an instant and then the system starts losing energy again. Thus as time progresses we can never revisit the same state twice because the energy level depends on the state, and it is continually losing energy.

Let us recap what we have learned:

1. We can define a function E on the states **y** of our system.

2. At the fixed point $\tilde{\mathbf{y}}$ we have $E(\tilde{\mathbf{y}}) = 0$; everywhere else $E(\mathbf{y}) > 0$.

3. We have $dE/dt < 0$ at almost all states, and at any state (other than $\tilde{\mathbf{y}}$) where $dE/dt = 0$ the system *immediately* moves to a state where $dE/dt < 0$ again.

The fact that we called E "energy" is actually irrelevant.

3.3.3 Lyapunov's method

By considering the loss of energy in the weird-friction example, we are able to determine that the fixed point **0** is stable: Starting the system near **0** will inevitably lead the system back to **0**.

Abstract energy.

If a dynamical system models a mechanical system, then consideration of energy is appropriate. Further, we can use energy-like ideas to show the stability of fixed points in nonphysical systems. The idea is to make up a function which behaves like the energy. We call such functions *Lyapunov functions*.

Suppose we have a continuous time dynamical system with state vector **x** which has a fixed point $\tilde{\mathbf{x}}$. Let V be a function defined on the states of the space, i.e., to each state **x** we assign a number $V(\mathbf{x})$ (the "energy" of that state). Now suppose V satisfies the following conditions:

- V is a differentiable function with $V(\mathbf{x}) > 0$ for all $\mathbf{x} \neq \tilde{\mathbf{x}}$, and $V(\tilde{\mathbf{x}}) = 0$.

- $dV/dt \leq 0$ at all states **x**. Further, at any state $\mathbf{x} \neq \tilde{\mathbf{x}}$ where $dV/dt = 0$, the system immediately moves to a state where $dV/dt < 0$.

If we can find such a function V (and this can be difficult), then it must be the case that $\tilde{\mathbf{x}}$ is a stable fixed point of the dynamical system. We know that $\tilde{\mathbf{x}}$ is stable because as time progresses "energy" (i.e., V) continually decreases until it bottoms out at the fixed point.

Let's do an example. Consider the system

$$x' = -x^3.$$

This is a one-dimensional system with $f(x) = -x^3$. The only fixed point is $\tilde{x} = 0$. Linearizing, we have $f'(x) = -3x^2$ and $f'(0) = 0$, so the linearization test fails. Thus we don't know if 0 is a stable fixed point.[5]

Now we need to make up a Lyapunov function for this system. There are a few standard tricks for doing this (see page 137), but for now we'll just grab one (seemingly by magic) out of thin air. Let $V(x) = x^2$. Let's see if it satisfies the conditions we set forth above.

First, $V(x)$ is a continuous function defined on the state space. Clearly, $V(x) > 0$ at all x, except that $V(\tilde{x}) = V(0) = 0$. Next we need to compute

---

[5]If you sketch a graph of $y = f(x)$, you should be able to infer that 0 is a stable fixed point. No matter—we use this example to illustrate the method.

dV/dt. [Note that although we write $V(x)$ we see that V is also a function of t since x is a function of t.] By the chain rule,

$$\frac{dV}{dt} = V'(x)\frac{dx}{dt} = 2x\frac{dx}{dt},$$

where dx/dt is, by the definition of our system, equal to $-x^3$. Thus

$$\frac{dV}{dt} = -2x^4.$$

Clearly, $dV/dt < 0$ at all states x (except 0) and therefore satisfies the conditions to be a Lyapunov function. Therefore 0 is a stable fixed point of this system.

Let's consider another example which is a bit more complicated; it also will help us expand the usefulness of this method. The system is

$$x_1' = -x_2,$$
$$x_2' = x_1 + x_2^3 - 3x_2.$$

Expressed in other notation, the system is $\mathbf{x}' = f(\mathbf{x})$, where

$$f\begin{bmatrix} x_1 \\ x_2 \end{bmatrix} = \begin{bmatrix} -x_2 \\ x_1 + x_2^3 - 3x_2 \end{bmatrix}. \tag{3.11}$$

The fixed points of the system are the solutions to $f(\mathbf{x}) = \mathbf{0}$. Please check that $\mathbf{0}$ is the only fixed point of this system. We could use linearization[6] to verify the stability of this fixed point, but we opt instead to use a Lyapunov function to illustrate the method.

Does this system describe a mechanical situation? I don't know. So I can't say what the "energy" of a state is. Instead, we make up a Lyapunov function. A standard guess is $V(\mathbf{x}) = x_1^2 + x_2^2$. Let's see that this is (almost) fine:

First, $V(\mathbf{x}) > 0$ at all states $\mathbf{x}$, except that $V(\tilde{\mathbf{x}}) = V(\mathbf{0}) = 0$.

Second, we need to compute dV/dt:

$$\frac{dV}{dt} = \frac{\partial V}{\partial x_1}\frac{dx_1}{dt} + \frac{\partial V}{\partial x_2}\frac{dx_2}{dt}$$

$$= 2x_1\frac{dx_1}{dt} + 2x_2\frac{dx_2}{dt}$$

---

[6] Yes, that would be easier. You should do it as an exercise.

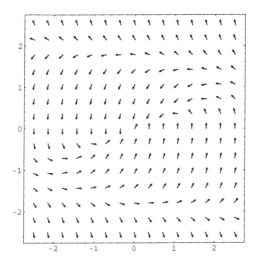

Figure 3.14. An overview of the behavior of the system from equation (3.11).

$$= 2x_1(-x_2) + 2x_2(x_1 + x_2^3 - 3x_2)$$

$$= -2x_1x_2 + 2x_2x_1 + 2x_2^4 - 6x_2^2$$

$$= 2x_2^4 - 6x_2^2$$

$$= 2x_2^2 \left(x_2^2 - 3 \right).$$

Now we'd like to proclaim proudly that $dV/dt < 0$, but this isn't the case. If $|x_2| > \sqrt{3}$, then we have $dV/dt > 0$. We can still use V to explain the stability of $\tilde{\mathbf{x}} = \mathbf{0}$, but we have to be a bit more careful.

Notice that *near* the fixed point $\mathbf{0}$ we *do* have $dV/dt < 0$, since the only points where this goes amiss have $|x_2| \geq \sqrt{3}$. For example, within a circle of radius 1 about $\mathbf{0}$ we are guaranteed to have $dV/dt < 0$. Thus, if our system begins near $\mathbf{0}$, then clearly it must tend toward $\mathbf{0}$. Thus $\mathbf{0}$ is a stable fixed point.

A graphical view might help. Consider Figure 3.14. Observe that arrows within the horizontal strip $-\sqrt{3} < x_2 < \sqrt{3}$ tend to point in toward the origin, while those outside the strip are pointing off toward infinity. This is in consonance with what we see with our "energy" function V. Inside the strip energy is decreasing; outside, it is increasing.

Now let's zoom in on the origin and understand how the function V

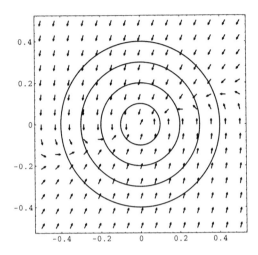

Figure 3.15. A closeup of the behavior of the system from equation (3.11). Level curves of the Lyapunov function V are shown. Notice that all arrows cross these curves in an inward direction.

relates to the vector field of the function f. Figure 3.15 shows the vector field of the function f near the fixed point $\mathbf{0}$. Notice the concentric circles drawn. Recall that $V(\mathbf{x}) = x_1^2 + x_2^2$, so the level curves of V (places where V is constant) are circles about the origin. At the origin, V is zero, and as we move outward we are at greater and greater values of V. Now, notice how the arrows of the vector field relate to the circles. Observe that the arrows all cross the circles *in an inward direction*. This is a direct consequence of the fact that dV/dt is negative near $\mathbf{0}$: V must decrease along any trajectory of the system.

This example shows that we need not have dV/dt negative *everywhere*; it is enough that dV/dt is negative within a fixed radius of the fixed point we are considering.

Summary of the Lyapunov method.

We now summarize the method of Lyapunov functions. Let $\tilde{\mathbf{x}}$ be a fixed point of the the system $\mathbf{x}' = f(\mathbf{x})$. Guess a function V for which:

(1) V is positive except at $\tilde{\mathbf{x}}$, and

(2) $dV/dt < 0$ at "all" points. The word *all* is in quotation marks because there are some exceptions:

– We don't have $dV/dt < 0$ at the fixed point $\tilde{\mathbf{x}}$.

– We need $dV/dt < 0$ only within a fixed positive distance of $\tilde{\mathbf{x}}$.

– We can tolerate some points $\mathbf{x}$ where $dV/dt = 0$ provided the system *immediately* moves to a state where $dV/dt < 0$ again.

Now for the nagging question: Where do I buy a Lyapunov function? There are some standard guesses I can recommend. First, if the dynamical system is a model of a physical system, try computing the energy at each state of the system. Be creative with the definition of *energy*. In an economic system, perhaps it might mean total money. Second, if the state vector is $\mathbf{x}$ and the fixed point is $\mathbf{0}$, let $V(\mathbf{x}) = x_1^2 + x_2^2 + \cdots + x_n^2$; this is the square of the distance from $\mathbf{x}$ to $\mathbf{0}$. (Indeed, this is what we used in the preceding example.) If the fixed point isn't $\mathbf{0}$ but rather $\tilde{\mathbf{x}}$, you can still use the same idea: Let

Some standard guesses for Lyapunov functions.

$$V(\mathbf{x}) = (x_1 - \tilde{x}_1)^2 + (x_2 - \tilde{x}_2)^2 + \cdots + (x_n - \tilde{x}_n)^2.$$

This squared distance idea sometimes won't work, so you can try more complicated ideas. For example, you can try

$$V(\mathbf{x}) = a_1(x_1 - \tilde{x}_1)^2 + a_2(x_2 - \tilde{x}_2)^2 + \cdots + a_n(x_n - \tilde{x}_n)^2,$$

where $a_1, a_2, \ldots, a_n$ are positive numbers. Which positive numbers? Cheat! Work backward from your goal $dV/dt < 0$ to see if you can find appropriate a's. If *that* doesn't work, here's one last desperate suggestion. Try

$$V(\mathbf{x}) = \sum_{i=1}^{n} \sum_{j=1}^{n} a_{ij}(x_i - \tilde{x}_i)(x_j - \tilde{x}_j),$$

where the a_{ij}'s have the following property: The matrix $A = [a_{ij}]$ is symmetric ($A = A^T$, i.e., $a_{ij} = a_{ji}$) and has positive eigenvalues. [In fancy language: $V(\mathbf{x}) = (\mathbf{x} - \tilde{\mathbf{x}})^T A(\mathbf{x} - \tilde{\mathbf{x}})$, where A is a symmetric positive definite matrix.] The positive eigenvalues ensure that $V(\mathbf{x}) > 0$ for all $\mathbf{x} \neq \tilde{\mathbf{x}}$.

3.3.4 Gradient systems

A special class of dynamical system is particularly well suited to the Lyapunov method. These systems arise from the *gradient* of a function. Let us discuss the gradient and how we can use it to build a dynamical system.

Let $h: \mathbf{R}^n \to \mathbf{R}$, i.e., h is a function of n variables which returns a

The gradient of a function.

single number answer. The *gradient* of h, denoted by ∇h, is the vector of h's partial derivatives. For example, if

$$h(\mathbf{x}) = h\begin{bmatrix} x_1 \\ x_2 \end{bmatrix} = \left(x_1^2 + x_2^2\right)^2 = x_1^4 + 2x_1^2 x_2^2 + x_2^4,$$

then the gradient of h is

$$\nabla h = \begin{bmatrix} \partial h/\partial x_1 \\ \partial h/\partial x_2 \end{bmatrix} = \begin{bmatrix} 4x_1^3 + 4x_1 x_2^2 \\ 4x_1^2 x_2 + 4x_2^3 \end{bmatrix}.$$

Notice that ∇h is a function from $\mathbf{R}^n$ to $\mathbf{R}^n$. We can use ∇h to form a dynamical system:

$$\mathbf{x}' = -\nabla h(\mathbf{x}). \qquad (3.12)$$

In other words, $f = -\nabla h$, and we have $\mathbf{x}' = f(\mathbf{x})$. In our example, the system would be

$$\begin{bmatrix} x_1 \\ x_2 \end{bmatrix}' = \begin{bmatrix} -4x_1^3 - 4x_1 x_2^2 \\ -4x_1^2 x_2 - 4x_2^3 \end{bmatrix}.$$

This system has a unique fixed point at $\mathbf{0}$ (please check this yourself). The Jacobian of $f = -\nabla h$ is

$$Df(\mathbf{x}) = \begin{bmatrix} -12x_1^2 - 4x_2^2 & -8x_1 x_2 \\ -8x_1 x_2 & -4x_1^2 - 12x_2^2 \end{bmatrix},$$

which at the fixed point $\tilde{\mathbf{x}} = \mathbf{0}$ is just $\begin{bmatrix} 0 & 0 \\ 0 & 0 \end{bmatrix}$. The eigenvalues of this matrix are both 0, so the linearization test is inconclusive.

We now switch to Lyapunov's method. The usual question is, What should we try for our Lyapunov function V? The answer is embedded in the very way we contrived the problem: We try h as the Lyapunov function, i.e., $V(\mathbf{x}) = h(\mathbf{x}) = \left(x_1^2 + x_2^2\right)^2$. We now calculate dV/dt:

$$\frac{dV}{dt} = \frac{\partial V}{\partial x_1}\frac{dx_1}{dt} + \frac{\partial V}{\partial x_2}\frac{dx_2}{dt} = \begin{bmatrix} \partial V/\partial x_1 \\ \partial V/\partial x_2 \end{bmatrix} \cdot \begin{bmatrix} x_1' \\ x_2' \end{bmatrix} = [\nabla h(\mathbf{x})] \cdot \mathbf{x}'.$$

Now, $\mathbf{x}' = f(\mathbf{x}) = -\nabla h(\mathbf{x})$. We finish our computation of dV/dt and get

$$\frac{dV}{dt} = \nabla h(\mathbf{x}) \cdot [-\nabla h(\mathbf{x})] = -[\nabla h(\mathbf{x}) \cdot \nabla h(\mathbf{x})] = -|\nabla h(\mathbf{x})|^2.$$

This is wonderful! Notice that dV/dt is always negative except when $\nabla h(\mathbf{x})$ equals $\mathbf{0}$. And the great thing is that $\nabla h(\mathbf{x}) = \mathbf{0}$ exactly when $\mathbf{x}$ is a fixed point of the system $\mathbf{x}' = -\nabla h(\mathbf{x})$.

We also have to check that $V(\mathbf{x}) = h(\mathbf{x}) > 0$ for all $\mathbf{x}$ (except $\tilde{\mathbf{x}} = \mathbf{0}$). This is simple for our example, since we chose $h(\mathbf{x}) = \left(x_1^2 + x_2^2 \right)^2$, which is clearly positive except at $\mathbf{0}$.

To summarize, suppose we have a function $h: \mathbf{R}^n \to \mathbf{R}$ which is positive except at a single value $\tilde{\mathbf{x}}$. Let $f(\mathbf{x}) = -\nabla h(\mathbf{x})$; Then the system $\mathbf{x}' = f(\mathbf{x})$ has a stable fixed point at $\tilde{\mathbf{x}}$.

When we are given a dynamical system, we would like to know if it is a gradient system. Consider the following two systems:

How do we recognize gradient systems?

$$\begin{bmatrix} x_1 \\ x_2 \end{bmatrix}' = \begin{bmatrix} -x_1 + x_2 \\ -x_1 - x_2 \end{bmatrix} \tag{3.13}$$

and

$$\begin{bmatrix} x_1 \\ x_2 \end{bmatrix}' = \begin{bmatrix} -2x_1 e^{x_2} \\ -x_1^2 e^{x_2} - 2x_2 \end{bmatrix}. \tag{3.14}$$

Both of these systems have a unique fixed point at $\tilde{\mathbf{x}} = \mathbf{0}$. One of these systems is a gradient system and the other isn't. In other words, in one case we can find a function h so that the system can be rewritten as $\mathbf{x}' = -\nabla h(\mathbf{x})$, and in the other case we can't. How can we tell which is which? How can we find the function h?

Suppose $f(\mathbf{x}) = -\nabla h(\mathbf{x})$. Now f is a vector-valued function; let its components be $f_1(\mathbf{x}), f_2(\mathbf{x}), \ldots, f_n(\mathbf{x})$, where $f_i(\mathbf{x}) = -\partial h(\mathbf{x})/\partial x_i$. If h has continuous second derivatives, then

If $\mathbf{x}' = f(\mathbf{x})$ is a gradient system, then $\partial f_i/\partial x_j = \partial f_j/\partial x_i$.

$$\frac{\partial f_i}{\partial x_j} = -\frac{\partial^2 h}{\partial x_j \partial x_i} = -\frac{\partial^2 h}{\partial x_i \partial x_j} = \frac{\partial f_j}{\partial x_i}.$$

Thus if we can find i and j so that $\partial f_i/\partial x_j \neq \partial f_j/\partial x_i$, then we will know that $\mathbf{x}' = f(\mathbf{x})$ is *not* a gradient system. For example, consider the system in equation (3.13). In this system we have

$$f_1(\mathbf{x}) = -x_1 + x_2 \qquad \text{and} \qquad f_2(\mathbf{x}) = -x_1 - x_2,$$

and therefore

$$\frac{\partial f_1}{\partial x_2} = 1 \qquad \text{and} \qquad \frac{\partial f_2}{\partial x_1} = -1.$$

We see that $\partial f_1/\partial x_2 \neq \partial f_2/\partial x_1$, so the system in equation (3.13) is *not* a gradient system.

On the other hand, consider the system in equation (3.14). Here we have

$$f_1(\mathbf{x}) = -2x_1 e^{x_2} \qquad \text{and} \qquad f_2(\mathbf{x}) = -x_1^2 e^{x_2} - 2x_2.$$

To check if this is a gradient system, we first compute

$$\frac{\partial f_1}{\partial x_2} = -2x_1 e^{x_2} = \frac{\partial f_2}{\partial x_1},$$

so the system in equation (3.14) might be a gradient system. Now we can try to recover the function h so that $f = -\nabla h$.

Recovering h from f.   Suppose $f = -\nabla h$. Then we know that $f_1 = -\partial h/\partial x_1$. If we integrate f_1 with respect to x_1 and hold x_2 constant, we have

$$h(\mathbf{x}) = \int -f_1(\mathbf{x})\, dx_1 = \int 2x_1 e^{x_2}\, dx_1 = x_1^2 e^{x_2} + C(x_2),$$

where $C(x_2)$ is a constant (as far as x_1 is concerned) which depends on x_2. Let's try to figure out what $C(x_2)$ is. We know that

$$\frac{\partial h}{\partial x_2} = -f_2 = x_1^2 e^{x_2} + 2x_2,$$

and since we know that $h(\mathbf{x}) = x_1^2 e^{x_2} + C(x_2)$, we also have

$$\frac{\partial h}{\partial x_2} = x_1^2 e^{x_2} + C'(x_2).$$

Equating these two expressions for $\partial h/\partial x_2$, we learn that

$$C'(x_2) = 2x_2.$$

Finally, integrating both sides of $C'(x_2) = 2x_2$ with respect to x_2, we have that $C(x_2) = x_2^2 + k$, where k is an absolute constant.

We have now learned that $h(\mathbf{x})$ is of the form

$$h(\mathbf{x}) = x_1^2 e^{x_2} + x_2^2 + k.$$

At this point you should compute $\nabla h(\mathbf{x})$ and be sure you get $-f(\mathbf{x})$. We now know that $f = -\nabla h$ regardless of what value we take for k. However, in order for h to serve as a Lyapunov function, we need $h(\tilde{\mathbf{x}}) = 0$ and $h(\mathbf{x}) > 0$ for $\mathbf{x} \neq \tilde{\mathbf{x}}$. Substituting $x_1 = x_2 = 0$ into our formula for h, we get

$$0 = h(\mathbf{0}) = 0^2 e^0 + 0^2 + k = k,$$

so we want to take $k = 0$. Finally, notice that the terms $x_1^2 e^{x_2}$ and x_2^2 can never be negative. Further, if $x_1 \neq 0$, then $x_1^2 e^{x_2} > 0$, and if $x_2 \neq 0$, then $x_2^2 > 0$. Thus for $\mathbf{x} \neq \mathbf{0}$ we have $h(\mathbf{x}) > 0$. We conclude that

$$h(\mathbf{x}) = x_1^2 e^{x_2} + x_2^2$$

is a Lyapunov function for the system of equation (3.14) and therefore $\mathbf{0}$ is a stable fixed point.

To summarize, suppose we are given a system of the form $\mathbf{x}' = f(\mathbf{x})$ with fixed point $\tilde{\mathbf{x}}$. We seek a function $h(\mathbf{x})$ for which:

Recapping how we find a Lyapunov function using the ideas of a gradient system.

(1) $f(\mathbf{x}) = -\nabla h(\mathbf{x})$,

(2) $h(\tilde{\mathbf{x}}) = 0$, and

(3) $h(\mathbf{x}) > 0$ for all $\mathbf{x} \neq \tilde{\mathbf{x}}$.

If such a function exists, then it is a Lyapunov function and we may conclude that $\tilde{\mathbf{x}}$ is stable. In order for condition (1) to hold, we must have $\partial f_i / \partial x_j = \partial f_j / \partial x_i$. If this is the case, we can use integration to try to recover the function h. We adjust arbitrary constants in our formula to make condition (2) true, then we check to see if condition (3) holds.

Geometric view

The stability of fixed points of gradient systems can also be illustrated geometrically.

Let's begin with a one-dimensional example. Let $h: \mathbf{R} \to \mathbf{R}$. The gradient, ∇h, is simply h', the derivative of h. Our gradient system is then

$$\frac{dx}{dt} = -h'(x).$$

In Figure 3.16 we plot a function $y = h(x)$ (solid curve) and its derivative (gradient) $y = h'(x)$ (dashed). Notice that where $y = h(x)$ is sloping downward, $h'(x)$ is negative, and where $h(x)$ is sloping upward, $h'(x)$ is positive. The sign of $h'(x)$ tells us if $h(x)$ is increasing or decreasing. Therefore, in the system $dx/dt = -h'(x)$ we notice that if $h(x)$ is sloping downward, then x is *increasing* (because of the minus sign), while if $h(x)$ is sloping upward, then x is *decreasing*.

Here's a nice way to think about this. Imagine that the state of the system x is represented by a point sitting on the curve $y = h(x)$. As time progresses, the point always moves in a downhill direction.

The state always rolls downhill.

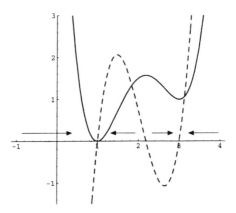

Figure 3.16. The graph of a function $y = h(x)$ (solid curve) and the gradient of h, i.e., $\nabla h = h'$ (dashed curve). The dynamical system $dx/dt = -h'(x)$ moves in the directions of the arrows.

Notice that the system $dx/dt = -h'(x)$ in Figure 3.16 has three fixed points, at 1, 2.2, and 3. Because 1 and 3 are local minima of $h(x)$, these fixed points are stable; however, 2.2 is a local maximum of $h(x)$ and is therefore an unstable fixed point of the system.

Let us consider a two-dimensional system. For example, let

$$h(\mathbf{x}) = h\left[\begin{array}{c} x_1 \\ x_2 \end{array}\right] = 1 - \cos x_1 \cos x_2.$$

A graph of $y = h(\mathbf{x})$ is plotted in three dimensions (since y is a function of x_1 and x_2), and the graph is a surface; see Figure 3.17.

The gradient of h is

$$\nabla h(\mathbf{x}) = \left[\begin{array}{c} \sin x_1 \cos x_2 \\ \cos x_1 \sin x_2 \end{array}\right].$$

Let us analyze the system $\mathbf{x}' = -\nabla h(\mathbf{x})$, i.e.,

$$\left[\begin{array}{c} x_1 \\ x_2 \end{array}\right]' = \left[\begin{array}{c} -\sin x_1 \cos x_2 \\ -\cos x_1 \sin x_2 \end{array}\right]. \tag{3.15}$$

This system has infinitely many fixed points, all of the form $\left[\begin{array}{c} j\pi/2 \\ k\pi/2 \end{array}\right]$, where j and k are integers. We focus on the fixed point $\mathbf{0}$.

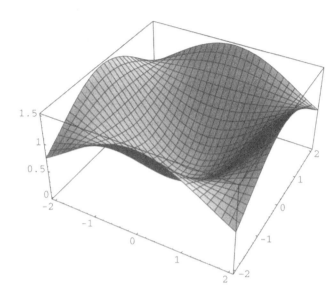

Figure 3.17. The graph $y = h(\mathbf{x})$ where $h(\mathbf{x}) = 1 - \cos x_1 \cos x_2$.

In one dimension, the sign of the gradient of h (i.e., the derivative of h) indicates whether h is increasing or decreasing. In the same way, the direction of the vector $\nabla h(\mathbf{x})$ gives the direction of steepest ascent of the function h. Therefore, the vector $-\nabla h(\mathbf{x})$ is pointing in the direction of steepest *descent*. This means that the system $\mathbf{x}' = -\nabla h(\mathbf{x})$ will lead to a value of $\mathbf{x}$ which minimizes h. Gradients point uphill.

To see this geometrically, consider Figure 3.18. At each point of the plane $\mathbf{x}$ we have plotted the direction of the vector $-\nabla h(\mathbf{x})$; it is quite clear that for any starting point near the origin $\mathbf{0}$ the system of equation (3.15) gravitates to the origin. Indeed, *gravitates* is the right word here. Imagine the state $\mathbf{x}$ as a point sitting on the surface $y = h(\mathbf{x})$ (see Figure 3.17). Because $\mathbf{x}' = -\nabla h(\mathbf{x})$, the point always moves in a downhill direction. Since $\mathbf{0}$ sits at the bottom of a well (a local minimum of h), we conclude that $\mathbf{0}$ must be a stable fixed point.

Let's summarize the behavior of a gradient system $\mathbf{x}' = -\nabla h(\mathbf{x})$. As the state vector $\mathbf{x}$ changes, the value of $h(\mathbf{x})$ decreases. [This is why h can be used as a Lyapunov function.] The fixed points $\tilde{\mathbf{x}}$ of the system are the points where $\nabla h(\mathbf{x}) = \mathbf{0}$. These points include the local minima of h, and

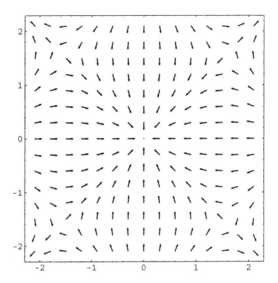

Figure 3.18. The directions of the vectors $-\nabla h(\mathbf{x})$ where $h(\mathbf{x}) = 1 - \cos x_1 \cos x_2$.

these are precisely the stable fixed points of the system.

Problems for §3.3

◆1. For each of the following continuous time dynamical systems we have given a fixed point. Show that the fixed point is stable by means of a Lyapunov function.

(a) $\begin{bmatrix} x \\ y \end{bmatrix}' = \begin{bmatrix} -x^5 - y \\ x \end{bmatrix}$ at $\begin{bmatrix} 0 \\ 0 \end{bmatrix}$.

(b) $\begin{bmatrix} x \\ y \end{bmatrix}' = \begin{bmatrix} y^2 - x - 1 \\ -xy - y \end{bmatrix}$ at $\begin{bmatrix} -1 \\ 0 \end{bmatrix}$.

(c) $x' = -x^2 \sin^{-1} x$ at 0.

◆2. Consider a physical system consisting of two masses arranged as follows. The first mass, m_1, is attached to the ceiling by a spring. The second mass, m_2, is attached to the bottom of the first mass by a second spring. The springs have Hooke's constants k_1 and k_2. Let g denote acceleration due to gravity. See Figure 3.19.

First, model this two-mass, two-spring system as a dynamical system operating in the absence of friction. Second, assume that there is a frictional force, proportional to—but opposite—the velocity. Third, assume that the frictional force is "weird"—proportional to the *cube* of the velocity.

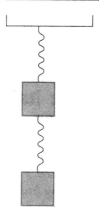

Figure 3.19. A two-mass two-spring system.

In each case, find the fixed point of the system and determine its stability.

◆3.   Develop a Lyapunov technique which will work for discrete time systems. That is, to each state $\mathbf{x}$ of the system assign an "energy" $V(\mathbf{x})$. What condition(s) should $V(\mathbf{x})$ satisfy to justify the stability of a fixed point $\tilde{\mathbf{x}}$ of a discrete time system.

Use your theory to verify the stability of the fixed points of the following systems.

(a) $x(k+1) = \tan^{-1}[x(k)]$ at $x = 0$.

(b) $\begin{bmatrix} x(k+1) \\ y(k+1) \end{bmatrix} = \begin{bmatrix} -y(k)/2 \\ x(k) \end{bmatrix}$ at $\begin{bmatrix} 0 \\ 0 \end{bmatrix}$.

◆4.   For each of the following determine if the dynamical system is a gradient system. If so, find its fixed point(s) and assess its (their) stability.

(a) $\mathbf{x}' = f(\mathbf{x})$ where $f(\mathbf{x}) = \begin{bmatrix} -5x_1 + 7x_2 \\ 7x_1 - 10x_2 \end{bmatrix}$.

(b) $\mathbf{x}' = f(\mathbf{x})$ where $f(\mathbf{x}) = \begin{bmatrix} -x_1 + x_2 - x_1 x_2^2 e^{x_1^2} \\ x_1 - x_2 - x_2 e^{x_1^2} \end{bmatrix}$.

(c) $\mathbf{x}' = f(\mathbf{x})$ where $f(\mathbf{x}) = \begin{bmatrix} -x_1^2 + 2x_1 x_2 - x_2^2 \\ x_1^2 - 2x_1 x_2 + x_2^2 \end{bmatrix}$.

(d) $\mathbf{x}' = f(\mathbf{x})$ where $f(\mathbf{x}) = \begin{bmatrix} x_1^2 + x_2^2 \\ x_1^2 - x_2^2 \end{bmatrix}$.

(e) $\mathbf{x}' = f(\mathbf{x})$ where $f(\mathbf{x}) = \begin{bmatrix} 3x_1 + x_2^2 \\ -x_1 + 2x_2 \end{bmatrix}$.

◆5.   Show that if the linear system $\mathbf{x}' = A\mathbf{x}$ is a gradient system, then A must be a symmetric matrix, i.e., $A = A^T$.

◆6.   Are all one-dimensional continuous time systems gradient systems?

3.4   Examplification: Iterative methods for solving equations

Using iteration to solve equations.

We can harness the power of attractive fixed points to solve equations. Suppose we want to solve the equation $x = \cos x$. We can do this simply by iterating the function $\cos x$ starting from any guess, say $x_0 = 0.7$. After fewer than 20 iterations, we arrive at $\tilde{x} = 0.7391$.

This is a contrived example; we *know* from our earlier work that $\cos x$ has an attractive fixed point $\tilde{x}$, and if we iterate cosine starting near $\tilde{x}$ we gravitate to the solution to $\cos x = x$. Let's see if we can exploit this idea to solve other equations.

Consider the equation $e^{-x} = \sqrt{x}$. Solving this equation is equivalent to solving the equation $f(x) = 0$ where $f(x) = e^{-x} - \sqrt{x}$. Iterating f won't work; if the iterations $f^n(x)$ converge, they will converge to a solution to $f(x) = x$—not what we want. We need to iterate something else. The idea is to create a function $g(x)$ with the property that $g(x) = x$ exactly when $f(x) = 0$. Finding a fixed point of g is then the same as solving our equation. How can we pass from $f(x) = 0$ to $g(x) = x$? We simply add x to both sides! Let $g(x) = x + f(x)$. It is now clear that

Let $g(x) = f(x) + x$. Then $f(x) = 0$ exactly when $g(x) = x$.

$$f(x) = 0 \qquad \text{if and only if} \qquad g(x) = x.$$

In our example, $f(x) = e^{-x} - \sqrt{x}$, so $g(x) = x + e^{-x} - \sqrt{x}$. Let's iterate g and see what happens. Where should we begin? Let us plot graphs (see Figure 3.20) of both e^{-x} and $\sqrt{x}$ and notice that the curves cross around $x = 0.5$. Iterating g starting at $x = 0.5$, we attain the following values: $0.5 \mapsto 0.3994 \mapsto 0.4381 \mapsto 0.4215 \mapsto 0.4283 \mapsto 0.4254 \mapsto 0.4267 \mapsto 0.4262 \mapsto 0.4264 \mapsto 0.4263 \mapsto 0.4263 \mapsto \cdots$. Thus within 10 iterations, we have arrived at the value $\tilde{x} \approx 0.4263$, which is a fixed point of g and therefore a solution to $e^{-x} = \sqrt{x}$ (both sides are approximately 0.6529 at $x = 0.4263$).

If g's fixed point is stable, we can iterate g to solve $f(x) = 0$.

Why did this work? We succeeded in solving $g(x) = x$ because $\tilde{x} = 0.4263$ is an *attractive* fixed point of g. To check this, note that

$$g'(x) = 1 - e^{-x} - \frac{1}{2\sqrt{x}},$$

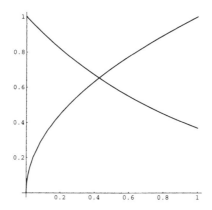

Figure 3.20. Graphs of the functions e^{-x} and $\sqrt{x}$.

so $g'(0.4263) \approx -0.4187$, which has absolute value less than 1.

Thus we have a possible method for solving equations: We write the equation in the form $f(x) = 0$, let $g(x) = x + f(x)$, iterate g starting at a reasonable guess, and hope that we converge to a fixed point of g. It would be nice to have a theorem which would guarantee that the fixed points of g are always attractive. Unfortunately, this is *not* the case.

Let's try another example. Consider the equation

$$\sin x = x^3 - 1.$$

Following our proposed method, we let

$$f(x) = \sin x - (x^3 - 1)$$

and let

$$g(x) = x + \sin x - (x^3 - 1).$$

In Figure 3.21 we plot the graphs of $\sin x$ and $x^3 - 1$ and observe that they cross at roughly $x = 1.2$. Indeed, the actual crossing is at $\tilde{x} \approx 1.24905$. To give our method a *really* good start, let us iterate $g(x)$ starting at $x = 1.25$. When we do, we obtain the following values: $1.25 \mapsto 1.24586 \mapsto 1.25975 \mapsto 1.21258 \mapsto 1.36618 \mapsto 0.795394 \mapsto 2.00633 \mapsto -4.16318 \mapsto 69.8464 \mapsto -340675. \mapsto 3.95385 \times 10^{16}$. Clearly, the method is blowing up. This unfortunate turn of events is due to

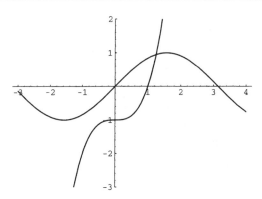

Figure 3.21. Graphs of the functions $\sin x$ and $x^3 - 1$.

the unstable nature of $\tilde{x} \approx 1.24905$. Since $g(x) = x + \sin x - (x^3 - 1)$, we have $g'(x) = 1 + \cos x - 3x^2$, and we compute $g'(1.24905) \approx -3.3642$, which has absolute value greater than 1.

The bad news is, the fixed point of $g(x) = x + f(x)$ might not be stable. The good news is, we can alter the definition of $g(x)$ so that (1) its fixed points still correspond to the roots of $f(x) = 0$, but (2) the fixed points are guaranteed to be stable.

A new g. Let $g(x) = x + af(x)$, where a is a constant.

Here is how we amend g. We chose $g(x) = x + f(x)$ because $g(x) = x$ exactly when $f(x) = 0$. However, g is *not* the only function with this property. Notice that if a is a nonzero constant, then $f(x) = 0$ exactly when $af(x) = 0$. We can now add x to both sides of $af(x) = 0$ and define our new g to be $g(x) = x + af(x)$. All we need to do now is to be clever in our choice of a to be sure that at $\tilde{x}$ (the root of $f(x) = 0$) we have $|g'(\tilde{x})| < 1$. Since $g(x) = x + af(x)$, we have $g'(x) = 1 + af'(x)$. Thus we want

$$-1 < g'(\tilde{x}) = 1 + af'(\tilde{x}) < 1.$$

We should now choose a so that $g'(\tilde{x})$ is safely between -1 and 1; let's choose a so that $g'(\tilde{x}) = 0$. The choice of a is now easy to derive: We want $1 + af'(\tilde{x}) = 0$, so we solve this for a and we find we want $a = -1/f'(\tilde{x})$.

Best choice for a is $-1/f'(\tilde{x})$. With this choice of a, we now have

$$g(x) = x - \frac{f(x)}{f'(\tilde{x})}$$

and so $g'(x) = 1 - f'(x)/f'(\tilde{x})$; therefore, $g'(\tilde{x}) = 0$. Thus if we iterate

g starting near $\tilde{x}$ we are guaranteed to converge to $\tilde{x}$.

This is encouraging, but there's a serious problem: If we don't know $\tilde{x}$, how can we compute $f'(\tilde{x})$? If our guess x is reasonably close to $\tilde{x}$, then (provided f' is continuous) it is reasonable to approximate $f'(\tilde{x})$ by $f'(x)$. Approximate $f(\tilde{x})$ by $f(x)$. As we improve the estimates for $\tilde{x}$, our approximation of $f'(\tilde{x})$ becomes more accurate. Thus we finally settle on the following definition for g:

$$g(x) = x - \frac{f(x)}{f'(x)}.$$

Aha! This is *exactly* Newton's method! (See §1.2.9 on page 26.) However, our "constant" a which we want equal to $-1/f'(\tilde{x})$ is no longer a constant: It varies with x. So our analysis—which depended on a being a constant—needs to be rechecked.

We notice first that $g(x) = x$ exactly when $f(x)/f'(x) = 0$, so finding a fixed point $\tilde{x}$ of g is the same as finding the solution $\tilde{x}$ of $f(x) = 0$ *unless* we also have $f'(\tilde{x}) = 0$. Next, we need to check that $|g'(\tilde{x})| < 1$ to be sure that $\tilde{x}$ is an attractive fixed point of g. We use the derivative of quotients rule to compute

$$g'(x) = 1 - \frac{f'(x)f'(x) - f(x)f''(x)}{[f'(x)]^2}$$

$$= \frac{f(x)f''(x)}{[f'(x)]^2}$$

$$\Longrightarrow g'(\tilde{x}) = \frac{f(\tilde{x})f''(\tilde{x})}{[f'(\tilde{x})]^2} = 0,$$

provided $f'(\tilde{x}) \neq 0$, and f'' is defined.

Higher dimensions

Newton's method is also applicable in higher dimensions. Suppose f is a function from $\mathbf{R}^n \to \mathbf{R}^n$, and we want to solve $f(\mathbf{x}) = \mathbf{0}$.

The idea is to iterate a function g whose fixed point $\tilde{\mathbf{x}}$ is a root of the equation $f(\mathbf{x}) = \mathbf{0}$. Let's try the same trick we did before: adding $\mathbf{x}$ to both sides of $f(\mathbf{x}) = \mathbf{0}$. We get

$$g(\mathbf{x}) = \mathbf{x} + f(\mathbf{x}).$$

Notice that $f(\mathbf{x}) = \mathbf{0}$ if and only if $g(\mathbf{x}) = \mathbf{x}$. Suppose $\tilde{\mathbf{x}}$ is a fixed point of g. We hope that the eigenvalues of $Dg(\mathbf{x})$ all have absolute value less than 1. Now, $Dg = I + Df$, and it is hard to know what the eigenvalues of $I + Df$ evaluated at $\tilde{\mathbf{x}}$ might be.

We can define g in a more complicated way that gives us more flexibility. We note that $f(\mathbf{x}) = \mathbf{0}$ if and only if $Af(\mathbf{x}) = \mathbf{0}$, where A is any $n \times n$ invertible matrix. Now we have

$$g(\mathbf{x}) = \mathbf{x} + Af(\mathbf{x}),$$

and therefore

$$Dg = I + ADf.$$

We want the eigenvalues of $Dg(\tilde{\mathbf{x}})$ to be near 0; the easiest way to assure this is to contrive g so that $Dg(\tilde{\mathbf{x}})$ is an all-0 matrix. Let's see if we can select A to make this happen:

$$\mathbf{0} = Dg(\tilde{\mathbf{x}}) = (I + ADf(\tilde{\mathbf{x}})) \quad \Longrightarrow \quad A = -[Df(\tilde{\mathbf{x}})]^{-1}.$$

Thus we want to set $g(\mathbf{x}) = \mathbf{x} + [Df(\tilde{\mathbf{x}})]^{-1} f(\mathbf{x})$; the problem is (just as before), we don't know $\tilde{\mathbf{x}}$. So we approximate $Df(\tilde{\mathbf{x}})$ by $Df(\mathbf{x})$, and we finally arrive at

$$g(\mathbf{x}) = \mathbf{x} - [Df(\mathbf{x})]^{-1} f(\mathbf{x}),$$

which is exactly Newton's method in higher dimensions.

Problems for §3.4

◆1. Use Newton's method to solve the following equations:

   (a) $x^2 - x - 1 = 0$.

   (b) $\cos x = \sin x$.

   (c) $x^3 - x^2 - 1 = 0$.

   (d) $1 - x^2 = \sin x$.

   (e) $e^x = x$.

◆2. Use the higher dimensional version of Newton's method to solve the following systems of equations:

   (a) $x^2 + y^2 = 9, x^2 - y^3 = 1$.

   (b) $x^2 - y^2 = 3, \sin x \sin y = \frac{1}{2}$.

   (c) $x + xy + y = 0, x^2 + y^2 = 1$.

   (d) $x + xy + xyz = 1, x^2 - y^2 + z^3 = 3, xy + yz = 1$.

◆3. Suppose computing $f'(x)$ is difficult; this might be the case when f' is not known analytically but is approximated using f. When we use Newton's method we compute

$$x(k+1) = x(k) - f[x(k)]/f'[x(k)].$$

Thus we compute f' every iteration. Suppose, instead, we compute f' only every *other* iteration. Will this modified Newton's method still converge to a root of $f(x) = 0$?

◆4. Our method for transforming the equation $f(x) = 0$ into $g(x) = x$ was to add x to both sides. Let's try another method. Suppose we *multiply* both sides by x. This can't work because $f(x) = 0 \iff xf(x) = 0$ (unless $x = 0$). So we need to transform 0 to 1. One idea is to add 1 to both sides and then multiply by x. This yields $g(x) = x(1 + f(x)) = x + f(x)$, and this is just what we had before. Another way to convert 0 to 1 is by exponentiation. Notice that $f(x) = 0 \iff e^{f(x)} = 1$.

Let's use this idea to come up with a variant of Newton's method. Let $g(x) = xe^{af(x)}$, where a is a nonzero number.

(a) With this newest definition of g, show that $g(x) = x$ if and only if $f(x) = 0$.

(b) Compute $g'(x)$.

(c) What value should we choose for a to ensure that $g'(\tilde{x}) = 0$ at a fixed point $\tilde{x}$ of g?

(d) In the previous part, the value you found for a depends on $\tilde{x}$. Since $\tilde{x}$ is unknown, this is a problem. Instead, use x to approximate $\tilde{x}$. What is your new formula for g?

(e) Unfortunately, your new a is no longer a constant, but you can still show that $g'(\tilde{x}) = 0$; please do so.

(f) Use this new iterative method to solve some equations.

(g) How does this new method compare with Newton's method? (Which converges more quickly to the answer? Which requires more computation?)

(h) Suppose the only root of $f(x) = 0$ is a negative number x, and suppose you use this alternative method, but your initial guess is positive. What will happen?

Chapter 4

Nonlinear Systems 2: Periodicity and Chaos

Dynamical systems do not live by fixed points alone.

Thus far we have seen three possible behaviors for dynamical systems: attraction to a fixed point, divergence to infinity, and (in continuous time) "cyclic" behavior (see the predator-prey example of §1.2.8 and the linear system with pure imaginary eigenvalues of Figure 2.22 on page 82).

In this chapter we see that periodic behavior can also occur in discrete time and that another type of behavior—chaos—is a possibility as well.

What is "periodic behavior"? A dynamical system exhibits periodic behavior when it returns to a previously visited state. We can write this as $\mathbf{x}(t_1) = \mathbf{x}(t_1 + T)$ for some $T > 0$. Notice that whatever trajectory the system took from time t_1 to time $t_1 + T$, the system is destined to repeat that same path again and again because the state at time $t_1 + T$ is *exactly the same* as the state at time t_1. Thus we realize that $\mathbf{x}(t_1) = \mathbf{x}(t_1 + T) = \mathbf{x}(t_1 + 2T) = \mathbf{x}(t_1 + 3T) = \cdots$. The system retakes the same steps over and over again, visiting the same states infinitely often. A fixed point is an extreme example of periodic behavior.

What is "chaos"? We discuss this concept later (see §4.1.4 and §4.2.5), but for now we want to point out that a system can behave in a nonperiodic and nonexplosive manner which, although completely determined, is utterly unpredictable!

As in the previous chapter we assume (unless we state otherwise) the following:

> Throughout this chapter, we assume f is differentiable with continuous derivative.

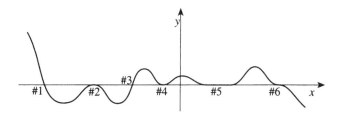

Figure 4.1. Graph of a function f for a one-dimensional dynamical system. Various fixed points are marked.

4.1 Continuous time

4.1.1 One dimension: no periodicity

One-dimensional continuous time systems either explode or tend to fixed points.

We begin by discussing the long-term fate of the simplest systems: continuous time dynamical systems in one variable, $x' = f(x)$.

Pick an x, any x. There are three possibilities: $f(x)$ is zero, positive, or negative. If $f(x)$ is zero, we know that x is a fixed point. If $f(x)$ is positive, then $x(t)$ must be increasing, and if $f(x)$ is negative, $x(t)$ is decreasing.

Periodic behavior is not possible for one-dimensional continuous time systems.

Our first observation is that periodic behavior is not possible (except for fixed points). Consider a state x_1 which we allegedly visit at times s and t, with $s < t$. This is possible if x_1 is a fixed point, but otherwise we have $f(x_1)$ either positive or negative. If $f(x_1)$ is positive, then, in the short run, the system moves to a state x_2 greater than x_1. Since f is continuous, we may assume that f is positive over the entire interval $[x_1, x_2]$. So we're at x_2 and still increasing. Now, how can we ever return to x_1? To get there, we must *decrease* through the interval $[x_1, x_2]$, but the equation $x' = f(x)$ says that x must *increase* throughout the same interval. Thus it's impossible to ever revisit the state x_1. By a similar analysis, we can never revisit a state with $f(x_1) < 0$.

Thus the only type of recurrent behavior one-dimensional continuous systems can exhibit is that of fixed points. Figure 4.1 shows the graph of a function f for a one-dimensional continuous time dynamical system $x' = f(x)$. Several fixed points are marked, with each somewhat different from the others.

The possible kinds of fixed points a one-dimensional continuous time system may have.

- Fixed point #1. This is a stable fixed point; to its left the system is increasing and to its right, decreasing.

- Fixed point #2. This is a "semistable" fixed point. To its left the system is decreasing, and so starting values less than $\tilde{x}$ move away from $\tilde{x}$. To the right the system is also decreasing, and so the fixed point behaves like an attractor on this side.

- Fixed point #3. This is an unstable fixed point. To its left, the system is decreasing and to its right, increasing.

- Fixed point #4. This is another semistable fixed point, but its action is opposite that of #2. This fixed point is an attractor on its left and a repellor on its right.

- Fixed points #5. This is an entire interval where $f(x) = 0$. These fixed points are marginally stable. Perturbing the system slightly away from one of these fixed points neither causes the system to return to the fixed points nor to fly away.

- Fixed point #6. This is another stable fixed point, but one where $f'(x) = 0$. Thus the linearization test of the previous chapter would fail at this fixed point.

In conclusion, the behaviors of one-dimensional continuous dynamical systems are rather limited. Ultimately, such a system must either gravitate toward a fixed point or explode to infinity.

4.1.2   Two dimensions: the Poincaré-Bendixson theorem

One-dimensional continuous systems either converge to a fixed point or diverge to infinity. These behaviors are exhibited by two-dimensional continuous systems as well. However, two-dimensional systems also exhibit another behavior: periodicity.

In two dimensions, continuous time systems may also be periodic.

Let $\mathbf{x}' = f(\mathbf{x})$ be a two-dimensional continuous time dynamical system. Each state of this system, $\mathbf{x}$, is a point in the plane (the phase space) of the system. If the system starts at state $\mathbf{x}_0$, we know that $\mathbf{x}(t)$ traces out a curve in the phase space; this curve is the trajectory (or orbit) of the system. If $\mathbf{x}_0$ is a fixed point of the system, then the trajectory starting at $\mathbf{x}_0$ is not very exciting: The system is "stuck" at $\mathbf{x}_0$ and remains there for all time. Otherwise ($\mathbf{x}_0$ is not a fixed point) the trajectory is a proper curve. In principle (and often in actuality) this curve can return to $\mathbf{x}_0$. Suppose the first return is at time T. Now, at time T it is as if we have started all over. Thus at time $t + T$ we are exactly in the same state as at time t. In other words, for any time t we have $\mathbf{x}(t + T) = \mathbf{x}(t)$. Such a curve is called

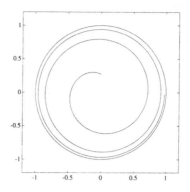

Figure 4.2. An orbit approaching a periodic orbit. The trajectory starts near the middle of the figure and spirals outward, becoming more and more like a circle.

periodic, and the smallest positive number T for which $\mathbf{x}(t + T) = \mathbf{x}(t)$ is called the *period* of the curve. For example, for the system

$$\begin{bmatrix} x_1' \\ x_2' \end{bmatrix} = \begin{bmatrix} 0 & 1 \\ -1 & 0 \end{bmatrix} \begin{bmatrix} x_1 \\ x_2 \end{bmatrix}$$

we find that (for any $\mathbf{x}_0$ other than $\mathbf{0}$) the trajectories are periodic with period 2π. [Please take a moment to work this out. You may wish to review equation (1.8) on page 9 and reread the material on page 68.]

If a dynamical system starts near, but not at, an stable fixed point $\tilde{\mathbf{x}}$, we expect the system to gravitate to $\tilde{\mathbf{x}}$. Note that we expect $\mathbf{x}(t)$ to *approach* $\tilde{\mathbf{x}}$; it need not be the case that $\mathbf{x}(t) = \tilde{\mathbf{x}}$ for any t.

Similarly, it is possible that a trajectory will *never* exhibit periodic behavior but will *approach* a periodic orbit; see Figure 4.2. A trajectory of this system begins at $\mathbf{x}_0 = \begin{bmatrix} 0 \\ 0.27 \end{bmatrix}$ and spirals outward approaching, but never quite reaching, the unit circle. Thus, as time progresses, the trajectory becomes more and more like the periodic orbit.

To be more specific, let $\mathbf{x}_1(t)$ and $\mathbf{x}_2(t)$ be two different trajectories of a system $\mathbf{x}' = f(\mathbf{x})$. We say that trajectory $\mathbf{x}_1$ *approaches* trajectory $\mathbf{x}_2$ provided $|\mathbf{x}_1(t) - \mathbf{x}_2(t + c)| \to 0$ (where c is a constant) as $t \to \infty$.

Trajectories cannot cross.

Two trajectories of a dynamical system, however, cannot cross; see Figure 4.3. Consider the point of intersection if two trajectories actually did intersect. The trajectory of the system starting at that point of intersection

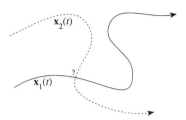

Figure 4.3. Two orbits of a dynamical system cannot cross.

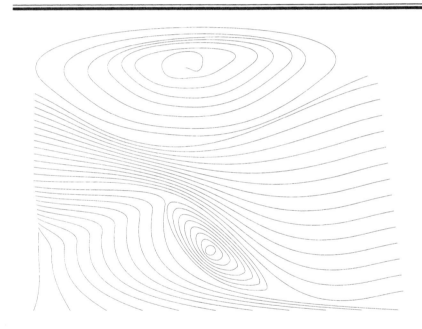

Figure 4.4. Many different orbits of a two-dimensional dynamical system.

is completely determined and therefore must proceed along a unique path. The situation in Figure 4.3 is therefore impossible.

Now imagine *all* the possible trajectories of a two-dimensional dynamical system drawn in a plane. You should see a situation akin to the one depicted in Figure 4.4. Since the curves cannot cross one another (or themselves), their behavior is greatly limited. Essentially they can (1) bunch together toward a point, (2) zoom off toward infinity, or (3) wrap more and more tightly around a simple closed curve.

The three behaviors open to continuous time two-dimensional systems.

These intuitive ideas are the heart of the Poincaré-Bendixson theorem, which states that a two-dimensional continuous time dynamical system $\mathbf{x}' = f(\mathbf{x})$ will have one of three possible behaviors as $t \to \infty$: It may (1) converge to a fixed point, (2) diverge to infinity, or (3) approach a periodic orbit.

Let's consider an example. Let

$$x_1' = x_1 + x_2 - x_1^3 \quad \text{and} \quad x_2' = -x_1. \tag{4.1}$$

The only fixed point (by solving $f(\mathbf{x}) = \mathbf{0}$) is $\tilde{\mathbf{x}} = \mathbf{0}$. Computing the Jacobian matrix, we have

$$Df = \begin{bmatrix} \partial f_1/\partial x_1 & \partial f_1/\partial x_2 \\ \partial f_2/\partial x_1 & \partial f_2/\partial x_2 \end{bmatrix} = \begin{bmatrix} -3x_1^2 + 1 & 1 \\ -1 & 0 \end{bmatrix},$$

so $Df(\mathbf{0}) = \begin{bmatrix} 1 & 1 \\ -1 & 0 \end{bmatrix}$, whose eigenvalues are $(1 \pm i\sqrt{3})/2$, which have positive real part. Thus $\mathbf{0}$ is an unstable fixed point of the system [equation (4.1)].

Trying to find a Lyapunov function for an unstable fixed point!?

Although $\mathbf{0}$ is an unstable fixed point, let's see if $V(\mathbf{x}) = x_1^2 + x_2^2$ is a Lyapunov function. This, of course, is crazy. It is impossible for V to be a Lyapunov function, since $\mathbf{0}$ is unstable. Let's do it anyway. We know that $V(\mathbf{x}) > 0$ for all $\mathbf{x} \neq \mathbf{0}$, and we now compute dV/dt:

$$\frac{dV}{dt} = \frac{\partial V}{\partial x_1}\frac{dx_1}{dt} + \frac{\partial V}{\partial x_2}\frac{dx_2}{dt}$$

$$= 2x_1\frac{dx_1}{dt} + 2x_2\frac{dx_2}{dt}$$

$$= 2x_1\left(x_1 + x_2 - x_1^3\right) + 2x_2(-x_1)$$

$$= -2x_1^4 + 2x_1^2$$

$$= 2x_1^2\left(1 - x_1^2\right).$$

We want $dV/dt < 0$ everywhere. What do we have? We see that when $|x_1| < 1$, then dV/dt is actually positive; this would be bad news if we really believed that V were a Lyapunov function. On the other hand, if $|x_1| > 1$, then $dV/dt < 0$. In other words, if x_1 is large, then the system is heading back toward $\mathbf{0}$. Now we can ask, Can this system explode? That

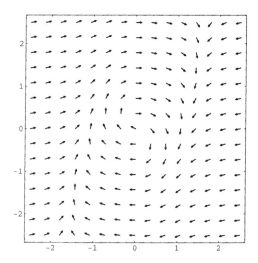

Figure 4.5. The dynamical system from equation (4.1) has **0** as an unstable fixed point, but the system does not blow up.

is, might $|\mathbf{x}(t)| \to \infty$ as $t \to \infty$? We claim the answer is no. Suppose $|\mathbf{x}(t)|$ were getting large. By our preceding analysis we cannot have $|x_1(t)|$ large, so it must be the case that $|x_2(t)|$ is large and $|x_1(t)|$ is bounded. But then since

$$x_1'(t) = x_2 - (\text{terms with } x_1 \text{ only}),$$

we see that dx_1/dt is unbounded, implying that $x_1(t)$ would be unbounded. In summary, we cannot have $\mathbf{x}(t)$ wandering too far from the origin.

We now know that the only fixed point, **0**, is unstable, but explosive behavior is impossible. What's left? The Poincaré-Bendixson theorem leaves us only one possible behavior: As $t \to \infty$ we must have $\mathbf{x}(t)$ tending to a periodic orbit. Figure 4.5 illustrates how this system behaves. Notice that near **0** all the arrows are pointing away from the origin. However, we know the system tends toward periodic behavior.

Figure 4.6 shows the trajectory of this system starting near **0** and rapidly settling into cyclic behavior. It is easy to see the periodic nature of this system in Figure 4.7, which shows how $x_1(t)$ varies with time.

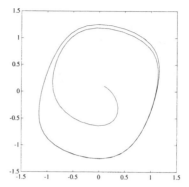

Figure 4.6. Starting near the unstable fixed point **0** of the system in equation (4.1) and approaching a stable cycle. (The trajectory starts near the origin and spirals outward.)

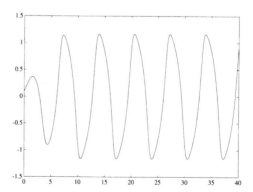

Figure 4.7. A plot of x_1 with respect to t for the system of equation (4.1). Observe that the system quickly becomes periodic.

4.1.3   The Hopf bifurcation*

We have considered the example $x' = f(x)$ where

$$f(x) = f \begin{bmatrix} x_1 \\ x_2 \end{bmatrix} = \begin{bmatrix} x_1 + x_2 - x_1^3 \\ -x_1 \end{bmatrix}.$$

We saw that 0 is an unstable fixed point, but the orbits don't escape to infinity. Rather, they approach a periodic orbit. We showed that the orbits cannot escape by considering $V(x) = x_1^2 + x_2^2$ and observing that dV/dt was negative once $|x_1|$ was large enough.

Let's see how far we can extend this idea. Let

A family of related dynamical systems.

$$f_a(x) = f_a \begin{bmatrix} x_1 \\ x_2 \end{bmatrix} = \begin{bmatrix} ax_1 + x_2 - x_1^3 \\ -x_1 \end{bmatrix}, \tag{4.2}$$

where a is a number we play with (i.e., a parameter). When $a = 1$, we have the system we had before [equation (4.1)].

First, please check that regardless of the value of a, the only fixed point of the system $x' = f_a(x)$ is 0. We ask, For which values a is 0 a stable fixed point of equation (4.2) and for which is 0 unstable? The Jacobian matrix is

$$Df_a = \begin{bmatrix} \partial f_1/\partial x_1 & \partial f_1/\partial x_2 \\ \partial f_2/\partial x_1 & \partial f_2/\partial x_2 \end{bmatrix} = \begin{bmatrix} a - 3x_1^2 & 1 \\ -1 & 0 \end{bmatrix},$$

which at $x = 0$ is $Df_a(0) = \begin{bmatrix} a & 1 \\ -1 & 0 \end{bmatrix}$. The characteristic polynomial of $Df_a(0)$ is $(\lambda - a)\lambda + 1 = \lambda^2 - a\lambda + 1$, which we set equal to zero to get

$$\lambda = \frac{a \pm \sqrt{a^2 - 4}}{2}.$$

Thus when $a < 0$, we know that 0 is a stable fixed point, and when $a > 0$, we know that 0 is an unstable fixed point.

A sudden change in the nature of the fixed point as the parameter a passes 0.

What becomes of our alleged Lyapunov function $V(x) = x_1^2 + x_2^2$? We compute, as before, that

$$\frac{dV}{dt} = 2x_1 \frac{dx_1}{dt} + 2x_2 \frac{dx_2}{dt}$$

$$= 2x_1 \left(ax_1 + x_2 - x_1^3 \right) + 2x_2(-x_1)$$

$$= 2ax_1^2 - 2x_1^4$$

$$= 2x_1^2 \left(a - x_1^2 \right).$$

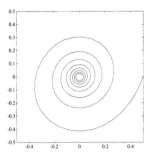 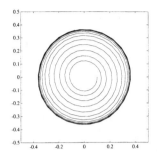

Figure 4.8. How the system $\mathbf{x}' = f_a(\mathbf{x})$ [from equation (4.2)] develops over time. On the left is the case where $a = -0.1$; the system spirals in toward the origin. On the right is the case where $a = 0.1$; here the origin is an unstable fixed point and the trajectory flies away from $\mathbf{0}$ and converges to a periodic orbit.

Notice that when $a < 0$, we really do have a Lyapunov function; this reconfirms the stability of $\mathbf{0}$ when $a < 0$. Otherwise $(a > 0)$ V is not a Lyapunov function, but it does tell us that the orbits of our system cannot go to infinity. Hence by the Poincaré-Bendixson theorem, $\mathbf{x}(t)$ must approach a periodic orbit.

Figure 4.8 illustrates how this system evolves in the cases where $a = -0.1$ and $a = 0.1$. When $a = -0.1$ (the left portion of the figure), we start our system at $\begin{bmatrix} 0.5 \\ 0.5 \end{bmatrix}$. Observe that as t increases, the orbit spirals inward toward the stable fixed point $\mathbf{0}$. When $a = 0.1$ (the right portion of the figure), we start the system at $\begin{bmatrix} 0.1 \\ 0.1 \end{bmatrix}$. Observe that in this case the system spirals outward, not to infinity but rather toward a periodic orbit.

Now imagine making a movie linking these two diagrams. The movie begins at $a = -0.1$ and ends at $a = 0.1$. In between, a gradually increases. What do we see? While a is negative we continue to observe the trajectory spiraling in to the origin. This is because $\mathbf{0}$ is stable and the linearized version of the system has complex eigenvalues with negative real parts. However, these real parts are creeping up toward zero, so the rate of descent into the origin is slowing. Suddenly, as we pass $a = 0$, there is a startling plot development. The fixed point $\mathbf{0}$ has lost its stability, and the orbits begin to spiral outward instead of inward. However, they don't go totally insane (to infinity) but are attracted toward periodic orbits.

This movie (transition from stable fixed point to stable cycle) is called

The Hopf bifurcation.

a *Hopf bifurcation*. A *bifurcation* is a change in the nature of a fixed point as we gradually change the function f. In this case we were varying our function f_a by adjusting a from negative to positive. The phenomenon we witnessed was the destabilization of the fixed point $\mathbf{0}$ and its ultimate demise into stable periodic orbits. We examine other examples of bifurcations in §4.2.3.

4.1.4 Higher dimensions: the Lorenz system and chaos

We have seen that one-dimensional continuous systems either gravitate toward a fixed point or diverge toward infinity. Two-dimensional systems can exhibit these behaviors, but they may also settle into periodic behavior. In three dimensions we can see these same behaviors and (here's the surprise) more!

Chaotic behavior in three dimensions.

We present the following celebrated dynamical system due to Lorenz, who was interested in modeling weather and the motion of air as it is heated. The physics behind the system is not critical to us; the nature of the system is.

The Lorenz system's state variable $\mathbf{x}(t)$ lives in $\mathbf{R}^3$. The system is

$$\frac{dx_1}{dt} = \sigma(x_2 - x_1),$$

$$\frac{dx_2}{dt} = rx_1 - x_2 - x_1x_3,$$

$$\frac{dx_3}{dt} = x_1x_2 - bx_3,$$

where σ, b, and r are constants. We take $\sigma = 10$, $b = \frac{8}{3}$, and $r = 28$. As an exercise (see problem 4 on page 167), find the three fixed points of this system and show (by linearization) that they are all unstable.

Let's begin the system in the state $\mathbf{x}_0 = \begin{bmatrix} 1 \\ 1 \\ 10 \end{bmatrix}$. Figures 4.9 through 4.11 show how the state variables $x_1(t)$, $x_2(t)$, and $x_3(t)$ fluctuate as time t progresses. The behavior of $\mathbf{x}$ is seen to be bounded (the values don't fly away to infinity) but aperiodic (they don't repeat). Thus this system stands in stark contrast to two-dimensional continuous systems, which must blow up, converge to a fixed point, or become periodic.

The full beauty of the Lorenz system is best appreciated by looking at a three-dimensional plot of the trajectory of the system. Figure 4.12 shows

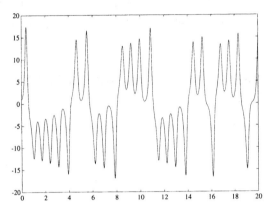

Figure 4.9. Plot of $x_1(t)$ for the Lorenz system.

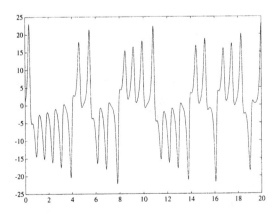

Figure 4.10. Plot of $x_2(t)$ for the Lorenz system.

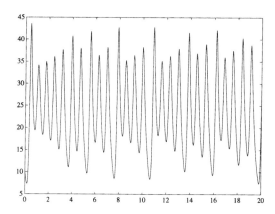

Figure 4.11. Plot of $x_3(t)$ for the Lorenz system.

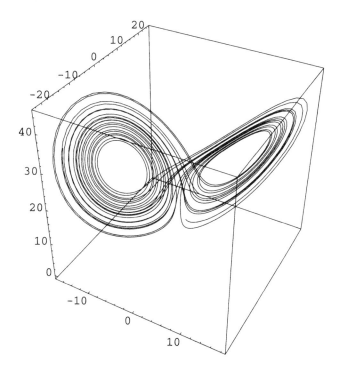

Figure 4.12. Three-dimensional plot of the trajectory of the Lorenz system.

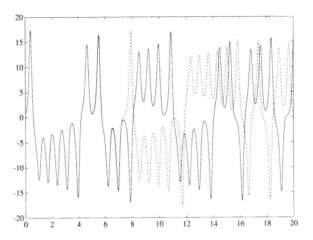

Figure 4.13. A plot of $x_1(t)$ for two different, but very similar starting vectors $\mathbf{x}_0$.

the trajectory of the system plotted in $\mathbf{R}^3$. It is hard to fully appreciate the intricacy of this three-dimensional trajectory from a two-dimensional picture. A movie of this figure would be much better. The point $\mathbf{x}(t)$ spirals around one loop of the diagram for a while and then suddenly jumps and spins around the other before returning to the first at seemingly random intervals. Of course, the intervals are not random: Each point on the trajectory $\mathbf{x}(t)$ is completely determined by the starting vector $\mathbf{x}_0$.

Numerical nonsense?

Now for the bad news. Figures 4.9 through 4.12, while correct in flavor, are probably inaccurate. We used good numerical methods (MATLAB's ode45 and *Mathematica*'s NDSolve), but we seriously doubt the numerical accuracy. The fault is not in the software but is inherent in the system itself.

Slight differences in the beginning yield enormous differences later.

Let us start the system at $\begin{bmatrix} 1 \\ 1 \\ 10 \end{bmatrix}$ and also at $\begin{bmatrix} 1 \\ 1 \\ 10.01 \end{bmatrix}$. There is a 0.1% difference between the two x_3 starting values. Surely this cannot make a big difference in the trajectory. Wrong! In Figure 4.13 we plot $x_1(t)$ for both starting vectors. We see that both curves are together initially, but suddenly (around $t = 7.5$) they fly apart and are very different thereafter.

Sensitive dependence on initial conditions.

Here is the problem. Any numerical method for finding solutions to differential equations works to only a fixed number of digits of precision. The slight errors in these computations result in enormous deviations between the value we compute and the real value of the system. The Lorenz system exhibits *sensitive dependence on initial conditions*, meaning that minute changes in the starting vector become huge differences in the state vector as time progresses. This has also been called the *butterfly effect* after the theory that a butterfly flapping its wings can cause significant changes in the weather because of the slight changes it makes in the atmosphere.

This is bad news for numerical methods and should make you very suspicious of any application of numerical methods for predicting the precise behavior of a nonlinear system beyond a short interval of time.

Have skepticism in numerical methods.

Problems for §4.1

◆1. Consider the double-spring system of problem 2 on page 144 (but suppose there is no friction). Suppose that the restoring force from the spring is *nonlinear*. For example, suppose that when we pull the spring a distance x, the restoring force is proportional to x^3.

Do computer simulations of such a system and observe its chaotic behavior.

◆2. Consider a *single* mass-and-spring system with a nonlinear spring. Do computer simulations and observe that the system is *not* chaotic. Explain.

◆3. The *van der Pol* equation is an example of mass-and-spring type system with nonlinear resistance. The equation is:

$$x'' + \mu(x^2 - 1)x' + x = 0.$$

(a) Use the ideas of Problem 13 on page 34 (from Chapter 1) to express the van der Pol equation as a dynamical system with two state variables. [Hint: Introduced a new variable y with $y = x'$.]

(b) Pick a value for μ (say $\mu = 1$), and draw a picture of the phase space of this system. Draw small arrows anchored at points (x, y) pointing in the direction the system is heading. Sketch several trajectories.

◆4. Find all fixed points of the Lorenz system (see §4.1.4 on page 163) with $\sigma = 10$, $b = \frac{8}{3}$, and $r = 28$. [Hint: There are three.]

Use linearization to show that all the fixed points are unstable.

◆5.* For each of the following families of dynamical systems $x' = f_a(x)$ the origin, 0, is a fixed point. As the parameter a changes, the system undergoes a Hopf bifurcation. Determine the value of a where this bifurcation occurs, and plot trajectories of the systems for values of a that are ± 0.1 of the bifurcation value.

(a) $f_a \begin{bmatrix} x_1 \\ x_2 \end{bmatrix} = \begin{bmatrix} -x_2 \\ \sin x_1 + ax_2 \end{bmatrix}$.

(b) $f_a \begin{bmatrix} x_1 \\ x_2 \end{bmatrix} = \begin{bmatrix} a(e^{x_1} - 1) + x_2 \\ -x_1 \end{bmatrix}$.

◆6.   Consider the four-dimensional system $\mathbf{x}' = A\mathbf{x}$ where

$$A = \begin{bmatrix} 0 & 1 & 0 & 0 \\ -1 & 0 & 0 & 0 \\ 0 & 0 & 0 & \sqrt{2} \\ 0 & 0 & -\sqrt{2} & 0 \end{bmatrix}, \quad \text{and} \quad \mathbf{x}_0 = \begin{bmatrix} 0 \\ 1 \\ 0 \\ 1 \end{bmatrix}.$$

(a) Find an exact formula for $\mathbf{x}(t)$.

(b) Sketch graphs of $x_1(t)$, $x_2(t)$, $x_3(t)$, and $x_4(t)$.

(c) Explain why

    (i) $\mathbf{x}(t)$ does not tend to a fixed point,

    (ii) $\mathbf{x}(t)$ does not tend to infinity, and

    (iii) $\mathbf{x}(t)$ does not tend to a periodic orbit.

    For (iii), you need to use the fact that $\sqrt{2}$ is an irrational number, i.e., $\sqrt{2}$ cannot be written as the quotient of two integers.

(d) Despite (iii) from the previous part of the problem, explain that this system can be decomposed into periodic subsystems.

(e) Explain why $\mathbf{x}(t)$ is *not* chaotic. In particular, explain why $\mathbf{x}(t)$ does not exhibit sensitive dependence on initial conditions.

4.2   Discrete time

We now turn to studying the behavior of discrete time dynamical systems $\mathbf{x}(k + 1) = f(\mathbf{x}(k))$. Recall that a fixed point of such a system is a state vector $\tilde{\mathbf{x}}$ for which $f(\tilde{\mathbf{x}}) = \tilde{\mathbf{x}}$. The fixed point $\tilde{\mathbf{x}}$ is stable if the eigenvalues of $Df(\tilde{\mathbf{x}})$ have absolute value less than 1.

We limit our discussion to one-dimensional systems. In this case we have a simpler condition to check for stability: The fixed point $\tilde{x}$ is stable if $|f'(\tilde{x})| < 1$ and unstable if $|f'(\tilde{x})| > 1$.

In linear discrete time systems there are essentially only two behaviors: convergence to a fixed point or divergence to infinity. There is one notable exception: If $f(x) = b - x$, then as we iterate f, we achieve the values

$$x_0 \mapsto b - x_0 \mapsto x_0 \mapsto b - x_0 \mapsto x_0 \mapsto \cdots.$$

The system oscillates between two values.

We now explore periodic behavior of discrete systems.

4.2.1 Periodicity

Let $x(k + 1) = f(x(k))$ be a one-dimensional discrete time dynamical system. We can write $x(k) = f^k(x)$.

A *fixed point* of this system is a value $\tilde{x}$ for which $f(\tilde{x}) = \tilde{x}$. More generally, a *periodic point* of this system is a value $\tilde{x}$ for which $f^k(\tilde{x}) = \tilde{x}$. We call the number k a *period* of x. Now if x is a periodic point with period k we know that $x = f^k(x)$, but it then follows that

$$f^{2k}(x) = f^k[f^k(x)] = f^k(x) = x,$$

so x is also periodic with period $2k$. The same reasoning shows that x is periodic with periods $3k$, $4k$, etc. These are not the fundamental period of x. We call the least positive integer k for which $x = f^k(x)$ the *prime period* of x.

The term *prime period* can cause some linguistic confusion because prime periods need not be prime numbers. It is possible for a function to have a periodic point x of prime period 4. This simply means that $f(x) \neq x$ and $f^2(x) \neq x$ and $f^3(x) \neq x$, but $f^4(x) = x$. In this context *prime* means *indecomposable*. (See problems 7 and 8 on page 218.)

Let us consider an example. Suppose f is the function $f(x) = 1 - x^2$. What are the fixed points of f? They are the solutions to the equation $f(x) = x$, i.e, we solve

$$1 - x^2 = x \quad \Rightarrow \quad x^2 + x - 1 = 0 \quad \Rightarrow \quad x = \frac{-1 \pm \sqrt{5}}{2}.$$

In Figure 4.14 we see the graph of the function $y = f(x)$ and the line $y = x$; the points where these graphs cross determine the fixed points of f.

To check the stability of these fixed points we note that $f'(x) = -2x$, so f' evaluated at these fixed points give $1 + \sqrt{5} \approx 3.236$ and $1 - \sqrt{5} \approx -1.236$. Since $|f'(x)| > 1$ at both of these fixed points, they are unstable.[1] Thus when we iterate f, we do *not* expect the iterates to tend toward either of $(-1 \pm \sqrt{5})/2$.

What do we see? Let's experiment and see what happens when we iterate f starting at, say, $x = \frac{1}{2}$. We compute

$$0.50 \mapsto 0.75 \mapsto 0.44 \mapsto 0.81 \mapsto 0.35 \mapsto 0.88 \mapsto 0.23 \mapsto \cdots.$$

From these few values, it is hard to see what's going on. Let's plot several iterations to try to understand what's happening. Figure 4.15 plots the first

Recall that $f^k(x)$ does *not* mean the kth power of $f(x)$ but rather the kth iteration of f starting at x.

Primality of periods is not the same as primality of numbers.

---

[1] That $f'[(-1 - \sqrt{5})/2] > 1$ is visible in Figure 4.14. Less clear, but also true, is that $f'[(-1 + \sqrt{5})/2] < -1$.

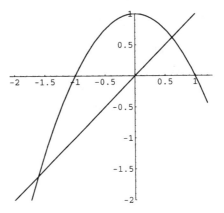

Figure 4.14. The graph of the function $f(x) = 1 - x^2$. The fixed points of f are the points of intersection with the line $y = x$.

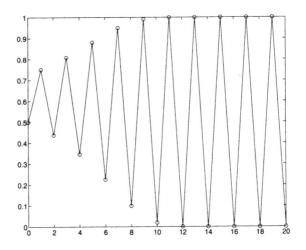

Figure 4.15. The first several iterations of $f(x) = 1 - x^2$.

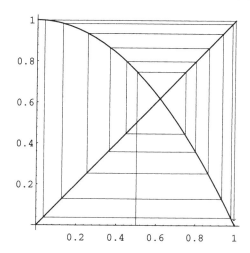

Figure 4.16. Graphical analysis of iterating $f(x) = 1 - x^2$ starting at $x = \frac{1}{2}$.

several values of $f^k(\frac{1}{2})$. It is now clear that as $k \to \infty$ the values $f^k(\frac{1}{2})$ oscillate between 0 and 1.

We also show the iterations graphically in Figure 4.16. Notice that the iterations spiral out from the unstable fixed point $\frac{1}{2}(-1 + \sqrt{5})$ and start to alternate between values approaching 0 and 1.

Notice that $f(0) = 1$ and $f(1) = 0$, hence 0 and 1 are periodic points of prime period 2. We might wonder if there are other points of prime period 2. Such points must satisfy the equation $f^2(x) = f(f(x)) = x$. To solve this equation, we first work out a formula for $f(f(x))$:

Finding all points of prime period 2.

$$f(x) = 1 - x^2, \quad \text{so}$$
$$f^2(x) = f[f(x)]$$
$$= f\left(1 - x^2\right)$$
$$= 1 - \left(1 - x^2\right)^2$$
$$= 1 - \left(1 - 2x^2 + x^4\right)$$
$$= 2x^2 - x^4.$$

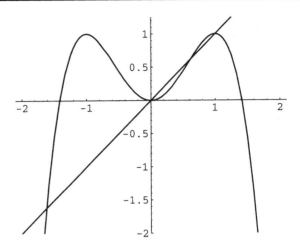

Figure 4.17. Graph of the function $y = f^2(x)$ where $f(x) = 1 - x^2$. The intersections with the line $y = x$ give the four points of period 2.

Now, we need to solve $f^2(x) = x$, i.e., we solve

$$x = 2x^2 - x^4 \quad \Rightarrow \quad x^4 - 2x^2 + x = 0.$$

We can factor $x^4 - 2x^2 + x$ as $x(x - 1)(x^2 + x - 1)$, so the points of period 2 of f are 0, 1, and $(-1 \pm \sqrt{5})/2$. Can this be? Aren't those last two values the *fixed points* of f? Yes, they are and everything is OK. You see, the fixed points of f are also left unchanged when we compute $f(f(x))$. Fixed points are periodic with period 2, but their *prime* period is 1. Thus the only points of prime period 2 are 0 and 1.

We can also use graphical methods to verify that 0, 1, and $(-1 \pm \sqrt{5})/2$ are the only points of period 2 of f. Figure 4.17 shows the graph of the function $f^2(x)$. The four points where the curve crosses the line $y = x$ correspond to f's four points of period 2.

Finding points of period 3.

Now we can ask, Does f have points of prime period 3? If so, they satisfy $f^3(x) = x$. We can compute that

$$\begin{aligned}
f^3(x) &= f(f^2(x)) \\
&= f(2x^2 - x^4) \\
&= 1 - [2x^2 - x^4]^2 \\
&= 1 - 4x^4 + 4x^6 - x^8.
\end{aligned}$$

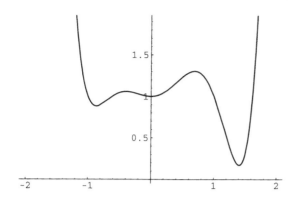

Figure 4.18. The graph of $y = x^6 - x^5 - 2x^4 + x^3 + x^2 + 1$.

To solve $f^3(x) = x$ we solve $x = 1 - 4x^4 + 4x^6 - x^8$, or equivalently, $1 - x - 4x^4 + 4x^6 - x^8 = 0$. This equation factors (a little) to give

$$\left(1 + x^2 + x^3 - 2x^4 - x^5 + x^6\right)\left(1 - x - x^2\right) = 0.$$

The two roots of the quadratic factor are $(-1 \pm \sqrt{5})/2$, the fixed points of f. This is no surprise, since if x is a fixed point, then certainly $f^3(x) = x$ as well. Any *other* points of period 3 are the roots of the 6th degree polynomial $x^6 - x^5 - 2x^4 + x^3 + x^2 + 1$. We plot the graph of this polynomial in Figure 4.18. From the graph it is clear that $x^6 - x^5 - 2x^4 + x^3 + x^2 + 1$ has no real roots. Indeed, we can find the six roots of this polynomial by numerical methods, and they are $0.0871062 \pm 0.655455i$, $-1.00914 \pm 0.324759i$, and $1.42203 \pm 0.114188i$. Thus f has no periodic points of prime period 3.

For any function f it is simple *in principle* to find the points of period k. All one has to do is solve the equation $f^k(x) = x$. *In practice* this can be extremely difficult. If $f(x)$ is a quadratic polynomial, then the equation $f^k(x) = x$ is a polynomial of degree 2^k. (When $k = 10$, this means finding the roots of a polynomial of degree over 1000.)

How to find periodic points.

4.2.2 Stability of periodic points

We have considered the behavior of the function $f(x) = 1 - x^2$. We saw that its two fixed points, $(-1 \pm \sqrt{5})/2$, are unstable. Using numerical

methods, we observed that starting at $x = \frac{1}{2}$ the iterates approach the period
2 points 0 and 1. Why does this happen?

To understand why fixed points of a function f are stable, we can
examine the graph $y = f(x)$. To understand f's points of period 2, it
is best to look at the graph $y = f^2(x)$. The four fixed points of f^2 are
the four points of period 2 of f. Look at Figure 4.17. The fixed points
$(-1 \pm \sqrt{5})/2$ (at about -1.6 and 0.6) are clearly unstable—it is easy to
see that the slope of the curve $y = f^2(x)$ through those points is steeper
than the slope of the line $y = x$. How about the other two fixed points of
f^2, namely 0 and 1? From the graph, it appears that the slopes at these
points are nearly flat. Thus we need to compute $(f^2)'(0)$ and $(f^2)'(1)$. To
compute these derivatives, we could work out $f^2(x)$, take the derivative,
and then substitute 0 and 1 for x. That method is fine, but here is another
way using the chain rule:

$$\frac{d}{dx} f^2(x) = \frac{d}{dx} f[f(x)] = f'[f(x)] \cdot f'(x).$$

Now, $f(x) = 1 - x^2$, so $f'(x) = -2x$. We compute

$$(f^2)'(0) = f'[f(0)]f'(0) = f'(1)f'(0) = 0, \quad \text{and}$$
$$(f^2)'(1) = f'[f(1)]f'(1) = f'(0)f'(1) = 0,$$

which verifies our impression that the curve $y = f^2(x)$ is flat at $x = 0$ and
$x = 1$. Thus 0 and 1 are attractive fixed points of f^2; see Figure 4.19. It is
clear that when we start at $x_0 = \frac{1}{2}$ and proceed two steps at a time (i.e., iterate
f^2), we approach 0. Had we started just to the right of $(-1+\sqrt{5})/2 \approx 0.62$,
we would have approached 1. In either case, we see that the iterates of f
tend to the alternating sequence 0, 1, 0, 1, etc.

We know that $\{0, 1\}$ is an *attractive* orbit of period 2 because $|(f^2)'(0)| < 1$ (and also $|(f^2)'(1)| < 1$).

Let us summarize what we have learned.

- To find the points of period k, solve the equation $f^k(x) = x$. Let p
  be a point of period k.

- If $|f'(p)| < 1$, then if the system starts near p, it gravitates to the
  orbit $\{p, f(p), f^2(p), \ldots, f^{k-1}(p)\}$. This is a stable periodic orbit.

- Otherwise, if $|f'(p)| > 1$, then $\{p, f(p), f^2(p), \ldots, f^{k-1}(p)\}$ is
  an unstable orbit. If the system is started near (but not at) one of

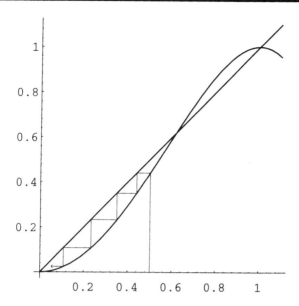

Figure 4.19. Iterating $f^2(x)$ (where $f(x) = 1 - x^2$) starting at $x_0 = \frac{1}{2}$.

these points, subsequent iterations move farther away from the orbit.

4.2.3 Bifurcation

We have studied how to find fixed and periodic points of discrete time
dynamical systems $x(k + 1) = f(x(k))$. We are now interested in gently
changing f and observing what happens to the fixed and periodic points.
We assume that we have a *family* of functions f_a where a is a parameter—a
number we can adjust. We assume that the function f_a changes gradually
as we change a. In particular, we can think of $f_a(x)$ as a function of
two numbers: a and x. As such, we require f to be differentiable with
continuous derivatives.

How a slight change in f can
dramatically alter the nature
of its periodic points.

   A *bifurcation* is a sudden change in the number or nature of the fixed
and periodic points of the system. Fixed points may appear or disappear,
change their stability, or even break apart into periodic points!

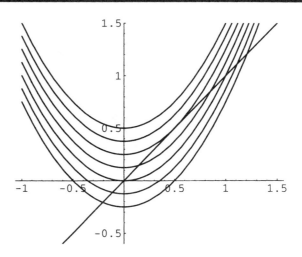

Figure 4.20. Graphs of the functions $f_a(x) = x^2 + a$ for various values of a near $\frac{1}{4}$.

Tangent (saddle node) bifurcations

Consider the functions

$$f_a(x) = x^2 + a.$$

We ask, What are the fixed points of f_a? We solve the equation $f_a(x) = x$, i.e.

$$x^2 + a = x \quad \Rightarrow \quad x^2 - x + a = 0 \quad \Rightarrow \quad x = \frac{1 \pm \sqrt{1 - 4a}}{2}.$$

Notice that if $a > \frac{1}{4}$, then f_a has no fixed points (because $f_a(x) = x$ has no real roots). For $a = \frac{1}{4}$ there is a unique fixed point, and for $a < \frac{1}{4}$ there are two fixed points. This can be seen most clearly in Figure 4.20. When $a > \frac{1}{4}$, the curve $y = f_a(x)$ doesn't intersect the line $y = x$, so there are no fixed points. Then, just when $a = \frac{1}{4}$, there is a unique fixed point at $\tilde{x} = \frac{1}{2}$. Graphical analysis shows that this point is semistable; it attracts on the left and repels on the right. Now, just as we decrease a below $\frac{1}{4}$, the fixed point $\frac{1}{2}$ splits in two—it bifurcates. When a is just below $\frac{1}{4}$, the two fixed points are $(1 \pm \sqrt{1 - 4a})/2$. Glancing at the graph in Figure 4.20, we see that the larger fixed point is unstable (the curve is steep), while the smaller fixed point is stable (the curve is relatively flat). Let's verify this analytically.

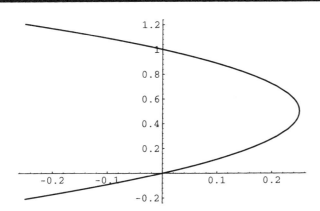

Figure 4.21. Bifurcation diagram for $f_a(x) = x^2 + a$ over the range $-\frac{1}{4} \leq a \leq \frac{1}{4}$.

The derivative of $f_a(x) = x^2 + a$ is $f_a'(x) = 2x$. At the larger fixed point, $\tilde{x}_1 = (1 + \sqrt{1 - 4a})/2$, we have $f_a'(\tilde{x}_1) = 1 + \sqrt{1 - 4a} > 1$, confirming that $\tilde{x}_1$ is unstable.

At the smaller fixed point, $\tilde{x}_2 = (1 - \sqrt{1 - 4a})/2$, we have $f_a'(\tilde{x}_1) = 1 - \sqrt{1 - 4a}$; this value is clearly less than 1, and if a is not too much below $\frac{1}{4}$, it is also greater than -1. In particular, if $-\frac{3}{4} < a < \frac{1}{4}$, then $\sqrt{1 - 4a}$ is real and less than 2. In this range, the smaller fixed point is stable. (We discuss what happens near $a = -\frac{3}{4}$ in just a moment.)

It is interesting to plot both fixed points of f_a as a function of a; Figure 4.21 does this. The horizontal axis represents a, and the vertical axis is x. For each value of a we plot the fixed points of f_a. Notice that to the right of $a = \frac{1}{4}$ there are no fixed points, then as a decreases, we suddenly have a unique semistable fixed point at $a = \frac{1}{4}$ which splits in two below $\frac{1}{4}$.

This sudden change in fixed point behavior is called a *bifurcation*. This particular example (with the sudden appearance and then splitting of a fixed point) is called a *tangent* (or *saddle node*) bifurcation. It is called a *tangent* bifurcation because the curves $y = f_a(x)$ become tangent to the line $y = x$ at the bifurcation value (in this example $\frac{1}{4}$.)

Notice that at a tangent bifurcation we have $f_a'(\tilde{x}) = 1$, since the curve $y = f_a(x)$ is just touching the line $y = x$.

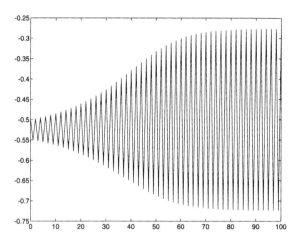

Figure 4.22. Several iterations of $f_a(x) = x^2 + a$ with $a = -0.8 < -\frac{3}{4}$, starting at $x_0 = -0.5$, which is near $\tilde{x}_2$.

Period-doubling (pitchfork) bifurcations

Now let's see what happens near $a = -\frac{3}{4}$. The larger fixed point, $\tilde{x}_1 = (1 + \sqrt{1 - 4a})/2$, has $f'(x_1) = 1 + \sqrt{1 - 4a} > 1$ and so is unstable. The other fixed point, $\tilde{x}_2 = (1 + \sqrt{1 - 4a})/2$, has $f'(\tilde{x}_2) = 1 - \sqrt{1 - 4a}$. When $a > -\frac{3}{4}$, we have $-1 < f'(\tilde{x}_2) < 1$, so x_2 is stable. However, when $a < -\frac{3}{4}$, we have $f'(\tilde{x}_2) < -1$ and therefore $\tilde{x}_2$ becomes unstable. As a drops through $-\frac{3}{4}$ we see a sudden change in the nature of the fixed point $\tilde{x}_2$ from stable to unstable.

Let's see what happens if we iterate $f_a(x)$ for a just less than $-\frac{3}{4}$. Let us take $a = -0.8 < -\frac{3}{4}$. When $a = -0.8$, we have $\tilde{x}_2 \approx -0.5246951$, so let us take $x_0 = -0.5$. Figure 4.22 plots the first 100 iterations. Notice that the values appear to be periodic with period 2. Why does this happen?

To understand this phenomenon, it helps to first find the points of period 2 for f_a. In other words, we need to solve the equation

$$f_a(f_a(x)) = f_a^2(x) = x.$$

Now,

$$f_a^2(x) = f_a(x^2 + a) = (x^2 + a)^2 + a = x^4 + 2ax^2 + a^2 + a.$$

Thus we must solve

$$x^4 + 2ax^2 + a^2 + a = x \quad \Rightarrow \quad x^4 + 2ax^2 - x + a^2 + a = 0.$$

Fortunately, this polynomial factors (a bit) to give

$$x^4 + 2ax^2 - x + a^2 + a = (x^2 - x + a)(x^2 + x + a + 1) = 0.$$

Thus we have four roots to the equation $f_a^2(x) = x$, namely,

$$\frac{1 \pm \sqrt{1 - 4a}}{2} \quad \text{and} \quad \frac{-1 \pm \sqrt{-3 - 4a}}{2}.$$

The first two roots, $(1 \pm \sqrt{1 - 4a})/2$, are the fixed points of f_a. The other two roots, which we call p_1 and p_2, are therefore points of prime period 2. Notice, however, that we need the term under the square-root sign to be positive in order for these points to be real. This happens provided $-3 - 4a \geq 0 \Rightarrow a \leq -\frac{3}{4}$.

Take a moment to verify that $f_a(p_1) = p_2$ and that $f_a(p_2) = p_1$.

Aha! This can't be a coincidence! Furthermore, when $a = -\frac{3}{4}$, we have $p_1 = p_2 = \tilde{x}_2 = -\frac{1}{2}$. Just as the point $\tilde{x}_2$ goes from stable to unstable it gives birth to a pair of points of period 2. We'll see how this happens in just a moment. Figure 4.23 illustrates this birth. In this graph we plotted for each value of a on the horizontal axis the fixed and periodic points of f_a on the vertical axis. For example, at $a = -1$ we plotted three points: 0 (a point of period 2), -0.6 (a fixed point), and -1 (a point of period 2).

Notice that as a drops past $-\frac{3}{4}$ we see two new curves in the bifurcation diagram. The pitchfork has three branches: the middle is the fixed point $\tilde{x}_2$, and the other two tines are p_1 and p_2.

Let's check the stability of these points of period 2. We want to know $|(f_a^2)'(x)|$ for $x = p_1$ and $x = p_2$. Now, $(f_a^2)'(x) = f_a'[f_a(x)] \cdot f_a'(x)$. For these points, we have

$$
\begin{aligned}
(f_a^2)'(p_1) &= f_a'[f_a(p_1)] \cdot f_a'(p_1) \\
&= f_a'(p_2) \cdot f_a'(p_1) \\
&= 2p_2 \cdot 2p_1 \\
&= \left(-1 - \sqrt{-3 - 4a}\right)\left(-1 + \sqrt{-3 - 4a}\right) \\
&= 1 - (-3 - 4a) = 4 + 4a.
\end{aligned}
$$

Likewise, $(f_a^2)'(p_2) = 4 + 4a$. When $a < -\frac{3}{4}$ we know that $4 + 4a < 1$. So long as $a > -\frac{5}{4}$ we have $4 + 4a > -1$ and we know that these periodic points are stable.

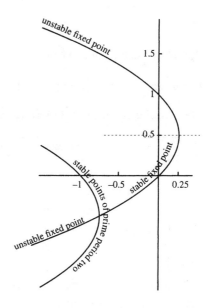

Figure 4.23. Bifurcation diagram for $f_a(x) = x^2 + a$ showing the pitchfork bifurcation at $a = -\frac{3}{4}$.

This bifurcation—where a stable fixed point becomes unstable and casts off two stable points of period 2—is called a *pitchfork* or *period-doubling* bifurcation.

Let's see how f_a and f_a^2 behave during this bifurcation. Figures 4.24 through 4.26 are graphs of the functions f_a and f_a^2 for values of a above, equal to, and below $-\frac{3}{4}$.

Prior to bifurcation.

In the first graph (Figure 4.24) we have $a = -0.6 > -\frac{3}{4}$. For this value of a the fixed point $\tilde{x}_2$ is stable and there are no points of prime period 2. The curve $y = f_a(x)$ crosses the line $y = x$ at $\tilde{x}_2$ with a slope $f_a'(\tilde{x}_2) \approx -0.84$, hence it is stable. Notice that the curve $y = f_a^2(x)$ crosses the line $y = x$ just at the point $\tilde{x}_2$; the slope at the crossing is positive but less than 1.

At the bifurcation.

In the second graph (Figure 4.25) we have $a = -\frac{3}{4}$, the critical value. The curve $y = f_a(x)$ crosses the line $y = x$ at a slope exactly equal to -1. Thus linearization does not tell us the stability of this fixed point. Notice that $y = f_a^2(x)$ crosses the line $y = x$ at exactly the same place but with slope $+1$. Again, we cannot tell from linearization the stability of $\tilde{x}_2$. Graphical analysis (or numerical experimentation) reveals, however, that

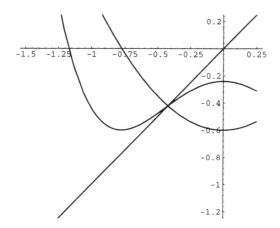

Figure 4.24. Plot of $y = f_a(x)$ (parabola) and $y = f_a^2(x)$ (S-shaped) where $f_a(x) = x^2 + a$. In this plot $a = -0.6 > -\frac{3}{4}$.

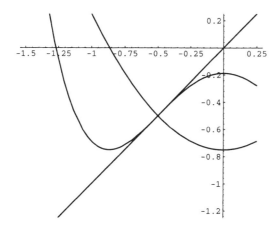

Figure 4.25. Plot of $y = f_a(x)$ and $y = f_a^2(x)$ with $a = -0.75 = -\frac{3}{4}$.

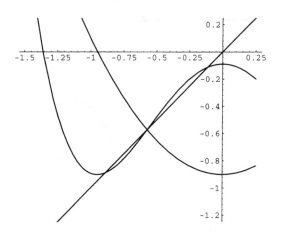

Figure 4.26. Plot of $y = f_a(x)$ and $y = f_a^2(x)$ with $a = -0.9 < -\frac{3}{4}$.

$\tilde{x}_2$ is stable (but barely!).

After the bifurcation.

Finally, the third graph (Figure 4.26) shows the case with $a = -0.9 < -\frac{3}{4}$. Notice that now the curve $y = f_a(x)$ crosses $y = x$ with a steeper slope (less than -1), so the fixed point $\tilde{x}_2$ is now unstable. However, the curve $y = f_a^2(x)$ has twisted even further and now intersects the line $y = x$ three times: once at $\tilde{x}_2$ (with slope greater than 1) and also at p_1 and p_2 (with positive slope less than 1). This confirms the calculation that p_1 and p_2 form a stable periodic orbit of period 2.

Imagine Figures 4.24 through 4.26 linked by a movie. As a drops through $-\frac{3}{4}$ we see the curve $y = f_a(x)$ twisting as it crosses the line $y = x$ with steeper and steeper slope, destabilizing the fixed point $\tilde{x}_2$ as a passes $-\frac{3}{4}$. Likewise, the curve $y = f_a^2(x)$ twists more and more (becoming more S-like) until (at $a = -\frac{3}{4}$) it crosses $y = x$ three times: once at the now unstable fixed point and twice at the stable points of period 2.

We checked that the points p_1 and p_2 form a stable orbit of period 2 once $a < -\frac{3}{4}$, but we also required $a > -\frac{5}{4}$. What happens as we drop past $a = -\frac{5}{4}$? First, let's do a numerical experiment. In Figure 4.27 we take $a = -1.3 < -\frac{5}{4}$ and start iterating. Observe that the values tend to repeat after *four* iterations. They eventually settle down into an orbit whose

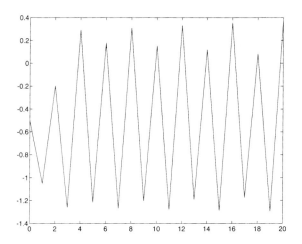

Figure 4.27. Iterations of $f_a(x) = x^2 + a$ with $a = -1.3$ and $x_0 = 0.5$.

values are approximately

$$-1.149 \mapsto 0.0194 \mapsto -1.2996 \mapsto 0.389 \mapsto -1.149.$$

What has happened? The points p_1 and p_2 of period 2 have destabilized and undergone another period-doubling bifurcation, each splitting in two, giving four points of period 4. The same type of twisting action can be seen in Figure 4.28. In this figure we have plotted the graphs of the function $y = f_a^2(x)$ and $y = f_a^4(x)$ for $a = -\frac{5}{4}$. (The graph of f_a^2 is roughly S-shaped, and the graph of f_a^4 is more wiggly.) You should note that the curve $y = f_a^2(x)$ crosses the line $y = x$ three times in the figure. The middle crossing is an original fixed point $\tilde{x}_2$. The other two crossings are the nearly destabilized points of period 2, p_1 and p_2; at these latter two crossings the curve $y = f_a^2(x)$ crosses with slope -1. The curve $y = f_a^4(x)$ crosses $y = x$ at the same points but with slope $+1$. As a drops below $-\frac{5}{4}$ the $y = f_a^4(x)$ curve will twist even more and give birth to four points, forming a stable orbit of prime period 4.

The period-doubling (or pitchfork) bifurcation consists of a stable fixed point $\tilde{x}$ becoming unstable and giving rise to a stable orbit of period 2. At the bifurcation value we have $f'(\tilde{x}) = -1$, and $(f^2)'(\tilde{x}) = 1$.

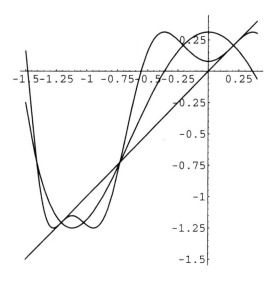

Figure 4.28. Graphs of the functions $y = f_a^2(x)$ and $y = f_a^4(x)$ for $a = -\frac{5}{4}$.

Bifurcation diagrams

In our example ($f_a(x) = x^2 + a$) we noticed that at $a = \frac{1}{4}$ we have a sudden appearance of two fixed points: the unstable $\tilde{x}_1$ and the stable (for the moment) $\tilde{x}_2$. As a drops past $-\frac{3}{4}$ the fixed point $\tilde{x}_2$ becomes unstable and breaks apart into two stable points of period 2: p_1 and p_2; then as a drops through $-\frac{5}{4}$ these points become unstable and give rise to four points of period 4.

What happens next? Not surprisingly, as a drops a bit farther these four points destabilize and give rise to a stable orbit of period 8. As a drops a tiny bit more the eight points of period 8 bifurcate again to give 16, then 32, etc. These bifurcations become increasingly hard to compute exactly, so we switch to numerical methods.

If we take $a = -1.3$ (as we did in Figure 4.27), we can find the points of period 4 by two means. One is to work out the polynomial $f_a^4(x)$ and get its roots. This involves solving a 16th order polynomial equation—yuck! A simpler method is to start with some arbitrary value, say $x_0 = -1$, and iterate f_a many times (100 should do). The next several values will tell us the behavior of f_a for this value of a. What we see is a repeating pattern every four stages. Try it! It is not hard to write a simple computer program

Finding periodic points through iteration.

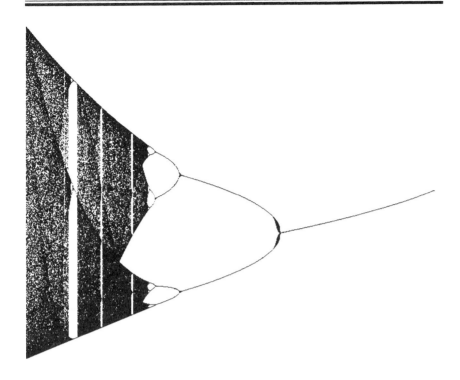

Figure 4.29. Computer-generated bifurcation diagram for $f_a(x) = x^2 + a$. The horizontal axis is a (and runs from -2 to 0). The vertical axis runs from -2 to 2.

or to use spreadsheet software to do the computations. What you *do* see are the *stable* periodic points of the system for your chosen value of a. What you *don't* see are the unstable ones.

We can do this for several values of a and then plot a graph. On the horizontal axis we record the values of a, and on the vertical axis we plot the periodic points the computer finds; see Figure 4.29. Look at the diagram from right to left. We start at $a = 0$ (although we could start at $a = \frac{1}{4}$, where f_a first has a fixed point). From $a = 0$ down to $a = -\frac{3}{4}$ we see a single curve representing the attractive fixed point. At $a = -\frac{3}{4}$ the curve splits in two (bifurcates). The two branches represent the points of period 2. The fixed point $\tilde{x}_2$ (which forms the middle tine of the fork in Figure 4.22) has disappeared; this corresponds to the fact that it is an unstable fixed point, so when we iterate f_a, the iterates stay away from $\tilde{x}_2$. At $-\frac{5}{4}$ the diagram bifurcates again into four branches: the attractive orbit of period

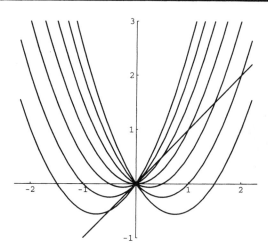

Figure 4.30. Plots of the function $g_a(x) = x^2 + ax$ for various values of a.

4. A bit farther (just below $a = -1.3$) the split into an orbit of period 8 is visible. The branches of subsequent splits into orbits of period 16, 32, 64, etc., are too tightly clustered to see in the low-resolution picture we have presented. After all the period-doubling has happened (somewhere around $a = -1.4$), we enter a chaotic region. For some values between $a = -1.5$ and $a = -2$ we have periodic behavior. For example, just below $a = -1.75$ it looks like we have an attractive orbit of period 3 (this is, in fact, true, as we discuss later). For other values of a it is unclear whether we have a periodic orbit (with some high period) or *chaotic* behavior (we discuss the idea of chaos in §4.2.5).

Transcritical bifurcations

Before we leave this section, we discuss one more type of bifurcation: the *transcritical* bifurcation.

To illustrate the transcritical bifurcation, we use a different family of functions: Let $g_a(x) = x^2 + ax$. Figure 4.30 shows plots of the function $y = g_a(x)$ (and the line $y = x$) for various values of a. It is clear from the figure that $g_a(x)$ has two fixed points: 0 and something else. Let's work this out analytically. To find the fixed points, we solve the equation

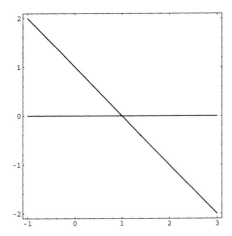

Figure 4.31. Bifurcation diagram for the family of functions $g_a(x) = x^2 + ax$. This is a transcritical bifurcation.

$g_a(x) = x$, i.e.,

$$x^2 + ax = x \quad \Rightarrow \quad x^2 + (a-1)x = x[x + (a-1)] = 0;$$

hence the fixed points of g_a are 0 and $1 - a$. Notice that at the special value $a = 1$ these two fixed points become one. Something interesting is happening there!

Let's check the stability of these fixed points. We note that $g_a'(x) = 2x + a$, hence

$$g_a'(0) = a, \quad \text{and}$$
$$g_a'(1 - a) = 2(1 - a) + a = 2 - a.$$

When $-1 < a < 1$, we see that 0 is a stable fixed point and $1 - a$ is unstable, but in the interval $1 < a < 3$ it's $1 - a$ that's stable and 0 is unstable. At the value $a = 1$ they swap roles.

Figure 4.31 shows the bifurcation diagram for the functions g_a. Notice that at the bifurcation value $a = 1$ the two fixed points merge, and when they split apart, they have swapped stability.

This is a *transcritical bifurcation*: two fixed points that merge and then split apart.

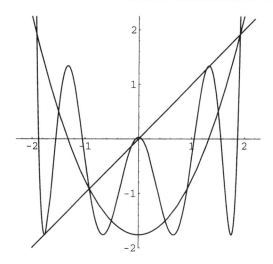

Figure 4.32. Graphs of the functions $f_a(x) = x^2 + a$ and $f_a^3(x)$ for $a = -1.76$.

4.2.4 Sarkovskii's theorem*

A function with points of prime period 3

Having points of one prime period can imply having points of another. In particular, this example has points of prime period 3. In this section we show that this implies having points of all prime periods.

Let us return to the main example of the preceding section: the family of functions $f_a(x) = x^2 + a$. At $a = -1.76$, f_a has points of period 3. If we perform a numerical experiment, it is plausible that the iterates of f_a settle into an orbit of period 3. Let's show this analytically. First, let us compute the points of period 3 for f_a. To do this we must solve the equation $f_a^3(x) = x$. Now, $f_a(x) = x^2 + a$, and $a = -1.76$ and so (after some messy computations for which a package such as *Mathematica* is very helpful) we have

$$f_a^3(x) = 0.0291738 - 9.4167\,x^2 + 15.0656\,x^4 - 7.04\,x^6 + x^8.$$

We can use the computer to find the eight roots of the equation $f_a^3(x) = x$, namely, -1.75943, -1.74226, -0.917745, -0.133203, 0.0238308, 1.27546, 1.3356, and 1.91774. Of these, -0.917745 and 1.91774 are actually fixed points of f_a. Figure 4.32 shows the graph of the function $y = f_a^3(x)$ as well as the function $y = f_a(x)$ and the line $y = x$. The two fixed points of f_a are clearly visible. The other six roots of $f_a^3(x) = x$ are not so clear. In Figure 4.33 we zoom in on the three places where the curve $y = f_a^3(x)$ nicks the line $y = x$. These graphs are useful in distinguishing

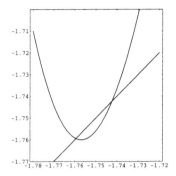

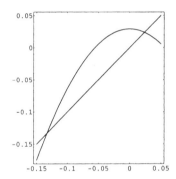

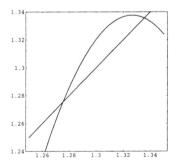

Figure 4.33. Graphs of the function $y = f_a^3(x)$ where $f_a(x) = x^2 + a$ (with $a = -1.76$) and the line $y = x$ near $x = -1.75$ (top), $x = 0$ (middle), and $x = 1.3$ (bottom).

between the root $x = -1.75943$ of $f_a^3(x) = x$, which crosses the line $y = x$ with a gentle slope, and $x = -1.74226$, where $y = f_a^3(x)$ crosses $y = x$ with a steep slope. Of course, we can compute the derivatives at each of the roots analytically. We find

$$(f_a^3)'(-1.75943) = -0.448,$$

$$(f_a^3)'(-1.74226) = 2.368,$$

$$(f_a^3)'(-0.133203) = 2.368,$$

$$(f_a^3)'(0.0238308) = -0.448,$$

$$(f_a^3)'(1.27546) = 2.368, \quad \text{and}$$

$$(f_a^3)'(1.3356) = -0.448.$$

Thus $\{-1.75943, 0.0238308, 1.3356\}$ form a stable orbit of period 3 (the other three values form an unstable orbit of period 3). Indeed,

$$f_a(-1.75943) = 1.3356,$$

$$f_a(1.3356) = 0.0238308, \quad \text{and}$$

$$f_a(0.0238308) = -1.75943.$$

(You should also check that the other three values also form an orbit of prime period 3.)

Consequences of period 3

We have expended much effort to show that $f_a(x)$ for $a = -1.76$ has orbits of prime period 3. So what? The "so what" is nothing short of miraculous. We will soon know that this function has points of prime periods 13,098 and 93 and 2 and 83 and 100,000,000.

How, you might ask, can we know this? It turns out that if a continuous function f (of one variable) has points of prime period 3, then it has points of prime period k for any k. This is a special case of Sarkovskii's theorem (which we explain in full detail later). To begin, let's consider a very simple version of the Sarkovskii result:

Prime period 2 implies a fixed point.

Let $f: \mathbf{R} \to \mathbf{R}$ be a continuous function. If f has points of prime period 2, then f must have a fixed point.

To see why this works, suppose the points of prime period 2 are a and b. This means that $a \neq b$, $f(a) = b$, and $f(b) = a$. Now let's look at a

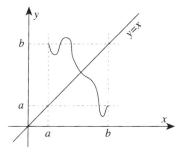

Figure 4.34. Graph of a continuous function f with points $\{a, b\}$ of prime period 2.

graph of the function f. It must look something like Figure 4.34. The curve $y = f(x)$ must connect the points (a, b) and (b, a). These points lie on opposite sides of the line $y = x$. Since f is continuous, the curve $y = f(x)$ must cross the line $y = x$; indeed, it must cross[2] somewhere between a and b. The point where $y = f(x)$ crosses $y = x$ must correspond to a fixed point of f.

Thus if f has points of prime period 2, then it must have a point of prime period 1 (i.e., a fixed point). We note, however, that having prime period 2 does not *imply* having fixed points of any other prime period (except period 1). For example, consider the simple function $f(x) = -x$. Consider any number x. If $x \neq 0$, then x is a point of prime period 2, since $x \mapsto -x \mapsto x$. Thus f has (many) points of prime period 2. From our preceding argument we know that f must have a fixed point; indeed, $f(0) = 0$. This discussion exhausts *all* possible values of x, so we know there are no points of prime period 3 or higher.

We are now ready to work on the following:

Let $f : \mathbf{R} \to \mathbf{R}$ be a continuous function. If f has points of prime period 3, *then f has points of prime period k for all positive integers k.*

This special case of Sarkovski's theorem is due to Li and Yorke. We assume only that f is continuous; it need not be differentiable.

To say that f has points of prime period 3 means there are three distinct numbers a, b, and c for which

$$f(a) = b, \quad f(b) = c, \quad \text{and} \quad f(c) = a.$$

---

[2]See problem 9 on page 218.

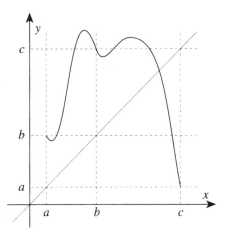

Figure 4.35. Graph of f with $f(a) = b$, $f(b) = c$, and $f(c) = a$.

By (perhaps) renaming[3] the points, we can assume that b is between a and c, i.e., either $a < b < c$ or $a > b > c$. For the rest of our discussion, we will consider only the case $a < b < c$. The discussion in the other case is essentially the same.[4]

We now work to show that f has points of prime period 1, 2, 3, 4, etc.

Step 1: Why f has a fixed point (a point of prime period 1).

We know that $a < b < c$ and $f(a) = b$, $f(b) = c$, and $f(c) = a$. Thus if we draw the graph of f, it must go through the points (a, b), (b, c), and (c, a). This is shown in Figure 4.35. Notice that the point (a, b) is above the line $y = x$, and the point (c, a) is below. Since f is continuous, the curve $y = f(x)$ must cross the line $y = x$ somewhere between b and c. Where it crosses, we have a fixed point.

Step 2: Why f has points of prime period 2.

A point of period 2 satisfies $f^2(x) = x$, so we sketch a graph of

---

[3]If b isn't between the other two, but say c is, we can rename the points a', b', c', where $b' = c$, $c' = a$, and $a' = b$. We still have $f(a') = b'$, $f(b') = c'$, and $f(c') = a'$. Notice that now b' (the 'old' c) is between a' and c'.

[4]Indeed, it would be a good exercise for you to reproduce everything we are about to do for the case $a > b > c$ to check your understanding. Do this the *second* time you read this material.

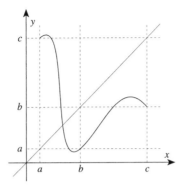

Figure 4.36. Plot of the graph $y = f^2(x)$ where $f(a) = b$, $f(b) = c$, and $f(c) = a$. The curve $y = f^2(x)$ must go through the points (a, c), (b, a), and (c, b).

$y = f^2(x)$. Now,

$$f^2(a) = f[f(a)] = f(b) = c,$$

$$f^2(b) = f[f(b)] = f(c) = a, \quad \text{and}$$

$$f^3(c) = f[f(c)] = f(a) = b.$$

Thus the curve $y = f^2(x)$ must go through the points (a, c), (b, a), and (c, b). This is illustrated in Figure 4.36. Notice that (a, c) and (b, a) are on opposite sides of the line $y = x$, so the curve $y = f^2(x)$ must cross $y = x$ between a and b. Where they cross we must have a point of period 2. But wait! Is this a point of *prime* period 2? It is conceivable that this crossing represents a fixed point of f (and would also satisfy $f^2(x) = x$) and is not a point of *prime* period 2. We need to be more careful.

It is possible that f has fixed points in the interval $[a, b]$. Let d be the *last* fixed point of f in the interval $[a, b]$. Thus $f(d) = d$, but for all x with $d < x \leq b$ we have $f(x) \neq x$ [note that $f(b) \neq b$, since $f(b) = c$]; see Figure 4.37. So far we know $a < d < b < c$, $f(a) = b$, $f(b) = c$, $f(d) = d$, $f(c) = a$, and for no x between d and b do we have $f(x) = x$. So the curve $y = f(x)$ must go from the point (d, d) to the point (b, c) *without crossing the line $y = x$*.

Now, since $d < b < c$ and the function f is continuous, we can't get from (d, d) to (b, c) without crossing the line $y = b$ (the middle, horizontal

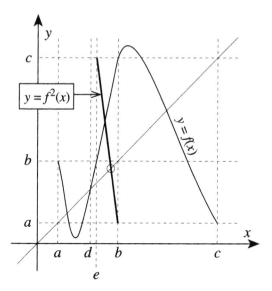

Figure 4.37. A detailed look at $y = f(x)$ and $y = f^2(x)$ where $f(a) = b$, $f(b) = c$, and $f(c) = a$. The point d is the last fixed point of f in $[a, b]$. The point e satisfies $d < e < b$ and $f(e) = b$.

dotted line in Figure 4.37). Let e be a point where the curve $y = f(x)$ crosses $y = b$ with $d < e < b$.

Let's summarize what we know thus far:

- $a < d < e < b < c$,

- $f(a) = b$, $f(b) = c$, and $f(c) = a$,

- $f(d) = d$, and for no x between d and b does $f(x) = x$, and

- $f(e) = b$.

Now it's time to consider the curve $y = f^2(x)$. Let's consider $f^2(e)$ and $f^2(b)$. We have

$$f^2(e) = f(f(e)) = f(b) = c, \quad \text{and}$$
$$f^2(b) = f(f(b)) = f(c) = a.$$

Hence the curve $y = f^2(x)$ must go through the points (e, c) and (b, a), which *are on opposite sides of the line* $y = x$. Thus somewhere between

e and b we have a point where $f^2(x) = x$ (circled in Figure 4.37). This point x must be a point of period 2, but (and this is what all the fuss was about) is it of *prime* period 2? Is it possible that x is just a fixed point of f? The answer, happily, is that x cannot possibly be a fixed point of f, since $d < x < b$, and d was the *last* fixed point of f between a and b. We therefore are delighted to declare that x is indeed a point of prime period 2. Whew!

Step 3: Why f has points of prime period 3.

This is a freebie. Our assumption is that f has points (a, b, and c) of prime period 3.

Step 4: Why f has points of prime period 4.

Before we begin, we need to develop a few ideas. First, let I be a closed interval, i.e., $I = [p, q] = \{x : p \leq x \leq q\}$. We know that f is a function which, for every number x, returns a number $f(x)$. We use the notation $f(I)$ to stand for the set of all values $f(x)$ where $x \in I$. In other words,

$$f(I) = \{f(x) : x \in I\}.$$

For example, if $f(x) = x^2$ and $I = [-1, 3]$, then $f(I) = [0, 9]$, because if we square all the numbers from -1 to 3, the answers we get are the numbers from 0 to 9.

Now we present two claims we use in showing that f has points of prime period 4.

Claim 1: If I is a closed interval and $f(I) \supseteq I$, then $f(x) = x$ for some $x \in I$.

The hypothesis that $f(I) \supseteq I$ means that for any number y in the interval I there must be an $x \in I$ for which $f(x) = y$. In other words, every number in I must be hit by $f(x)$ as x ranges over the interval I.

This is best illustrated geometrically. Consider Figure 4.38. We focus on the section of the curve $y = f(x)$ for $x \in I$, i.e., the portion of the curve between the vertical lines $x = p$ and $x = q$. The condition that $f(I) \supseteq I$ means that the curve $y = f(x)$ must cover every y-value between p and q. This means that for some x we have $f(x)$ down at p, and for another x we have $f(x)$ up at q. This means that the curve $y = f(x)$ must wander from points below (or on) the line $y = x$ to points above it as x ranges between p and q. Therefore, there must be a fixed point $x = f(x)$ for some x in the interval $I = [p, q]$. This justifies Claim 1.

Lest panic ensue in the reader, rest assured that this is the last step in showing that f has points of prime period k for all k. Although what we are about to do is tailored to showing that f has points of prime period 4, the methods we now discuss can be used to show that f has points of prime periods 5, 6, and so on.

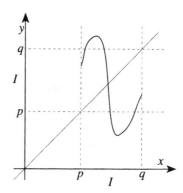

Figure 4.38. If $f(I) \supseteq I$, then f has a fixed point in the interval I.

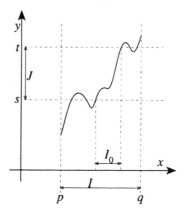

Figure 4.39. If $f(I) \supseteq J$, then for some $I_0 \subseteq I$ we have $f(I_0) = J$.

Claim 2: If I and J are closed intervals and $f(I) \supseteq J$, then I contains a closed subinterval $I_0 \subseteq I$ for which $f(I_0) = J$.

The hypothesis is that for every number y in J, there must be some number x in I for which $f(x) = y$. Of course, there may be some numbers $x \in I$ for which $f(x)$ is outside J. (We only assumed "$\supseteq$".) This is depicted graphically in Figure 4.39. Suppose $I = [p, q]$ and $J = [s, t]$. The condition that $f(I) \supseteq J$ means that the portion of the curve $y = f(x)$

between the vertical lines $x = p$ and $x = q$ must include all y-values between $y = s$ and $y = t$. As the curve wanders around, it may start across the gap from $y = s$ to $y = t$, only to wander outside that region. But at some stage, we must cross from $y = s$ to $y = t$ in an uninterrupted path. The interval I_0 represents the x-values of this uninterrupted path from $y = s$ to $y = t$. Thus we have $f(I_0) = J$ as required.

Armed with Claims 1 and 2, we are ready to hunt for points of prime period 4. Here we go:

(a) Let $I_0 = [b, c]$.

Since $f(b) = c$ and $f(c) = a$, we note that $f(I_0) \supseteq [a, c] \supseteq [b, c] = I_0$. So using Claim 2, ...

(b) ... choose $I_1 \subseteq I_0$ so that $f(I_1) = I_0$.

Now, $f(I_1) = I_0 \supseteq I_1$. So using Claim 2, ...

(c) ... choose $I_2 \subseteq I_1$ so that $f(I_2) = I_1$.

Now,

$$f^3(I_2) = f^2[f(I_2)]$$
$$= f^2(I_1) \qquad \text{by (c)}$$
$$= f[f(I_1)]$$
$$= f(I_0) \qquad \text{by (b)}$$
$$\supseteq [a, c] \supseteq [a, b].$$

So using Claim 2, ...

(d) ... choose $I_3 \subseteq I_2$ so that $f^3(I_3) = [a, b]$.

Now,

$$f^4(I_3) = f[f^3(I_3)]$$
$$= f([a, b]) \qquad \text{by (d)}$$
$$\supseteq [b, c] \qquad \text{because } f(a) = b \text{ and } f(b) = c$$
$$= I_0 \qquad \text{by (a)}$$
$$\supseteq I_1 \qquad \text{by (b)}$$
$$\supseteq I_2 \qquad \text{by (c)}$$
$$\supseteq I_3 \qquad \text{by (d)}.$$

In short, $f^4(I_3) \supseteq I_3$, so by Claim 1, ...

(e) f^4 has a fixed point p in I_3, i.e., $p \in I_3$ and $f^4(p) = p$.

We have found a point, p, of period 4, but we worry if p is of *prime* period 4. It is and let's see why.

Let's recap what we know so far:

$$I_0 = [b, c] \supseteq I_1 \supseteq I_2 \supseteq I_3,$$

$$f(I_1) = I_0,$$

$$f(I_2) = I_1,$$

$$f^3(I_3) = [a, b], \quad \text{and}$$

$$f^4(p) = p \quad \text{for some } p \in I_3.$$

Since $p \in I_3$, we know that $f^3(p) \in [a, b]$ (because (d) $f^3(I_3) = [a, b]$).

Since $p \in I_3$ and $I_3 \subseteq I_2$, we have

$$p \in I_2 \quad \Rightarrow \quad f(p) \in I_1 \quad \Rightarrow \quad f^2(p) = f[f(p)] \in I_0 = [b, c].$$

(The first $\Rightarrow$ is by (c), $f(I_2) = I_1$, and the second $\Rightarrow$ is by (b), $f(I_1) = I_0$.)

Now, since $I_2 \subseteq I_1 \subseteq I_0 = [b, c]$, we know that p, $f(p)$, and $f^2(p)$ are all in $[b, c]$, but $f^3(p) \in [a, b]$.

We know that p is periodic of period 4. We are now ready to explain why p is of *prime* period 4. Suppose, for sake of contradiction, that p is periodic with some period less than 4. This means that $p = f(p)$ or $p = f^2(p)$ or $p = f^3(p)$. Since p, $f(p)$, $f^2(p) \in [b, c]$ it follows that *all* iterates $f^k(p)$ are in $[b, c]$. However, we also know that $f^3(p) \in [a, b]$. Since b is the only number $[a, b]$ and $[b, c]$ have in common, we must have $f^3(p) = b$. But then, $p = f^4(p) = f(b) = c$, and therefore $f(p) = f(c) = a$. The trouble is, $f(p) \in [b, c]$, but $f(p) = a \notin [b, c]$. This is a contradiction. Therefore our supposition that p is periodic with period less than 4 is absurd, and we conclude that p is periodic with prime period 4.

We can use the preceding technique to show that f must have points of prime period 5, 6, and so on. For example, to show that f has a point of prime period 5, we argue (as we did in (a), (b), etc., previously) that there are intervals $[b, c] = I_0 \supseteq I_1 \supseteq I_2 \supseteq I_3 \supseteq I_4$ for which $f(I_1) = I_0$, $f(I_2) = I_1$, $f(I_3) = I_2$, and $f^4(I_4) = [a, b]$. We find a point $q \in I_4$ so that $f^5(q) = q$ and check that q, $f(q)$, $f^2(q)$, $f^3(q) \in [b, c]$, but $f^4(q) \in [a, b]$. We then conclude that q has prime period 5.

We have shown, then, that any continuous function f with points of prime period 3 (such as $f(x) = x^2 - 1.76$) must have points of prime period k for any k.

The Sarkovskii order

We have learned the following about a continuous function $f: \mathbf{R} \to \mathbf{R}$:

- If f has points of prime period 2, then f has points of prime period 1 but not necessarily points of any other prime period.

- If f has points of prime period 3, then f has points of prime period k for any positive integer k.

Further, functions such as $f(x) = x^3$ have points of prime period 1 (i.e., fixed points) but *no* points of primes period k for any $k > 1$.

We introduce a convenient shorthand for expressing such results. We write $k \triangleright j$ to stand for the following sentence: "If a continuous function $f: \mathbf{R} \to \mathbf{R}$ has points of prime period k, then it must have points of prime period j."

We can utter the sentence "$k \triangleright j$" tersely as "k forces j."

For example, $2 \triangleright 1$ means "continuous functions with points of prime period 2 must have fixed points." The sentence $2 \triangleright 1$ is true. However, it is *not* true that $1 \triangleright 2$; the function $f(x) = x^3$ has points of prime period 1 (fixed points) but *no* points of prime period 2. Of course, *some* functions with fixed points do have points of prime period 2; however, not all do, so the sentence $1 \triangleright 2$ is false.

The sentences $1 \triangleright 1$, $2 \triangleright 2$, $3 \triangleright 3$, etc., are all trivially true. We also know that all the following sentences are true:

$$3 \triangleright 1, \quad 3 \triangleright 2, \quad 3 \triangleright 3, \quad 3 \triangleright 4, \quad 3 \triangleright 5, \dots.$$

We can ask questions with this notation: Does $15 \triangleright 16$? (The answer is yes, as explained next.) Does $80 \triangleright 90$? (The answer is no.)

Sarkovskii's theorem enables us to answer all questions of the form: Does $k \triangleright j$? It is an amazing fact that the relation $\triangleright$ is a *total order* on the positive integers; this means:

$\triangleright$ is a total order on the positive integers.

- For any two numbers j and k, either $j \triangleright k$ or $k \triangleright j$.

- If $j \triangleright k$ and $k \triangleright j$, then it must be the case that $j = k$.

- If $j \triangleright k$ and $k \triangleright \ell$, then $j \triangleright \ell$.

You are familiar with ordering the positive integers with the $\geq$ relation; note that $\geq$ satisfies the same three properties that $\triangleright$ satisfies.

We know that $3 \triangleright k$ for all positive integers k. This says that 3 is first in the Sarkovskii ordering. What comes next? The answer is 5. There is even a last number in the Sarkovskii ordering (which you already know!)—it's 1. We know that $1 \triangleright j$ is false for any $j \neq 1$.

Here is the full Sarkovskii ordering (which we will explain fully):

$$3 \triangleright 5 \triangleright 7 \triangleright 9 \triangleright 11 \triangleright \cdots$$
$$6 \triangleright 10 \triangleright 14 \triangleright 18 \triangleright 22 \triangleright \cdots$$
$$12 \triangleright 20 \triangleright 28 \triangleright 36 \triangleright 44 \triangleright \cdots$$
$$24 \triangleright 40 \triangleright 56 \triangleright 72 \triangleright 88 \triangleright \cdots$$
$$\triangleright \cdots \triangleright$$
$$\cdots \triangleright 32 \triangleright 16 \triangleright 8 \triangleright 4 \triangleright 2 \triangleright 1.$$

Explaining the Sarkovskii ordering.

The first row contains the odd numbers (other than 1) in their natural ascending order. The second row is obtained by doubling the numbers in the first row. The third row is double the second, and the fourth is double the third. This pattern continues for *infinitely* many rows: each is double the previous. (There is more to come; please stand by.) In which row would we find the number 100? Since $100 = 4 \times 25 = 2^2 \times 25$ we find 100 in row number 3, just between $4 \times 23 = 92$ and $4 \times 27 = 104$. Numbers of the form $m2^n$, where m is an odd number (other than 1), can be found in row $n + 1$. What numbers have we missed? The powers of 2. Numbers such as 16 can't be written in the form $m2^n$, where m is odd and greater than 1. The last row in our chart (occurring after all the infinitely many other rows) contains the powers of 2 in descending numerical order, with 2 and then 1 at the end of the list.

Sarkovskii's theorem is the following: If j appears before k in the preceding list, then $j \triangleright k$ is true. Otherwise (if j is after k in the ordering), $j \triangleright k$ is false.

For example, 5 is the second number on the list. Sarkovskii's Theorem tells us that for any integer k *other than* 3, we have $5 \triangleright k$. The proof of this fact is similar to (but more involved than) our argument for $3 \triangleright k$.

Why $5 \triangleright 3$ is false.

The following example shows that the sentence $5 \triangleright 3$ is false. Let

$$f(x) = \frac{2x^3 - 21x^2 + 61x - 24}{6}.$$

Please compute that $f(1) = 3$, $f(2) = 5$, $f(3) = 4$, $f(4) = 2$, and $f(5) = 1$, so $1 \mapsto 3 \mapsto 4 \mapsto 2 \mapsto 5 \mapsto 1$. In lieu of computing these values by hand, examine the graph of f in Figure 4.40. Thus f has

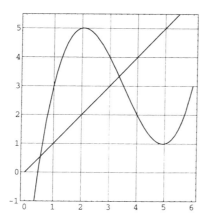

Figure 4.40. Graph of the function $f(x) = \frac{1}{6}\left(2x^3 - 21x^2 + 61x - 24\right)$. Notice that $f(1) = 3$, $f(2) = 5$, $f(3) = 4$, $f(4) = 2$, and $f(5) = 1$.

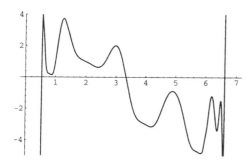

Figure 4.41. Graph of $f^3(x) - x$ where $f(x) = \frac{1}{6}\left(2x^3 - 21x^2 + 61x - 24\right)$. There are only three zeros, and these correspond to the fixed points of f; hence f has no points of prime period 3.

$\{1, 2, 3, 4, 5\}$ as points of prime period 5. Now, f has three fixed points (no surprise, since $5 \triangleright 1$) at 0.543195, 3.33737, and 6.61943. (The first two you can see in the graph; the third is just beyond the range we plotted.) The question is, Does f have points of prime period 3? To answer this, we plot the graph of $f^3(x) - x$ in Figure 4.41. We see that the curve crosses the x-axis only at the values corresponding to the fixed points of f. The plot

The only solutions to $f^3(x) = x$ are the three fixed points of f; there are no other real roots.

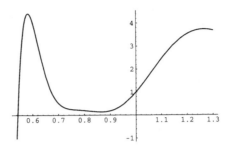

Figure 4.42. Graph of $f^3(x) - x$ where $f(x) = \frac{1}{6}\left(2x^3 - 21x^2 + 61x - 24\right)$, near $x = 0.9$.

is a bit worrisome: Does the graph dip below the x-axis around $x = 0.9$? In Figure 4.42 we enlarge that region to assure ourselves there are no other crossings. We can also solve the 27th order polynomial equation $f^3(x) = x$ and find there are only 3 real roots; the other 24 roots are complex.

This example shows that $5 \rhd 3$ is false.

4.2.5   Chaos and symbolic dynamics

We have been exploring the family of functions $f_a(x) = x^2 + a$. In this section we consider two particular values of a, the case $a = -1.95$, and the case $a = -2.64$. The particular values are not especially important. What is important is that -1.95 is just slightly greater than -2 and that -2.64 is less than -2.

Suppose $a > -2$ and we iterate f_a. If the initial value is in the interval $[-2, 2]$, then the iterations stay within $[-2, 2]$ as well (see the bifurcation diagram Figure 4.29 on page 185). When $a < -2$, we will see that for most values x, the iterates $f^k(x)$ tend to infinity. The set of values x for which $f^k(x)$ stays bounded is quite interesting, and the behavior of f on that set can be worked out precisely.

The case $a = -1.95$: chaos observed

Let's perform numerical experiments with the function $f(x) = x^2 - 1.95$. We compute the first 1000 iterations of f, starting with $x = -0.5$, i.e., we compute $f(-0.5)$, $f^2(-0.5)$, ..., $f^{1000}(-0.5)$. The best way to see these values is to look at a graph. Figure 4.43 shows the first 100 values.

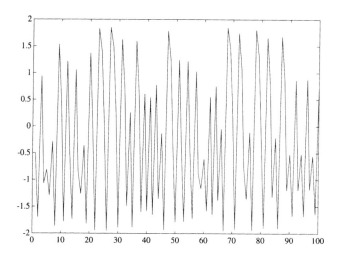

Figure 4.43. One hundred iterations of $f(x) = x^2 - 1.95$ starting with $x = -\frac{1}{2}$.

The iterations do not seem to be settling down into a periodic behavior. Perhaps we have not done enough iterations? You should experiment on your computer, and compute more iterations to see if you find a pattern. You won't. The iterations continue in what appears to be a random pattern. Of course, the pattern isn't random at all! The numbers are generated by a simple *deterministic* rule, $f(x) = x^2 - 1.95$.

Now let's throw a monkey wrench into the works. Let's repeat the experiment, only this time starting with $x = -0.50001$. When we do the iterations $f(x)$, $f^2(x)$, $f^3(x)$, etc., the first dozen or so are numerically very close to the iterations we computed starting with $x = -0.5$. After that, however, the iterates move apart. You should do the computations yourself. You can see the results in Figure 4.44. Observe that for $k \geq 20$ the values $f^k(-0.5)$ and $f^k(-0.50001)$ are wildly different. Subtle differences in x lead to enormous differences in $f^k(x)$. We are witnessing *sensitive dependence on initial conditions*. [This is similar to what we saw in the case of the Lorenz attractor in §4.1.4.]

A slight change in the initial value yields huge changes in subsequent iterations.

The news is even worse. How did we compute the values $f^k(-0.5)$? We used a computer, of course. The computer doesn't compute each value *exactly* but only to an accuracy of several digits (depending on the machine and the software it is running). So my machine *claims* that

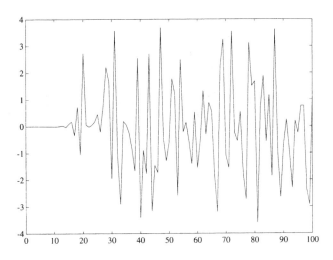

Figure 4.44. The difference $f^k(-0.5) - f^k(-0.50001)$ for $k = 1$ to 100.

$f^{22}(-0.5) = 1.82968746165504$, but the last decimal place or so is a round-off from the previous calculation. Now we know that in several iterations this round-off error will become enormous. Although my computer says that $f^{1000}(-0.5) = -1.9106$, the truth is, I haven't the vaguest idea if this value is even remotely correct! The preceding value was computed using MATLAB on my Macintosh computer. Switching to *Mathematica* and redoing the computation (on the *same* computer!), I get $f^{1000}(-0.5) = 0.93856$. Make a game out of this! See how many different answers you can get for $f^{1000}(-0.5)$.

Different answers to the same problem.

In short, we haven't the vaguest idea what $f^{1000}(-0.5)$ is. The iterations of this function, although deterministic, are unpredictable. They are chaotic.

Can we "beat the system" by using greater accuracy? Yes and no. Using *Mathematica*, we can coerce the computer to use 1000 digits worth of accuracy to the right of the decimal point. This slows the computer down considerably and makes each computation more accurate, but is 1000-digit accuracy enough to overcome the mistakes introduced in each iteration? Yes—let's see why.

We begin our analysis by comparing two different evaluations of $f(x)$ using slightly different values of x. Let ε be a very small number and let's compare $f(x + \varepsilon)$ and $f(x)$

$$f(x) = x^2 - 1.95,$$

$$f(x + \varepsilon) = (x + \varepsilon)^2 - 1.95$$

$$= x^2 + 2\varepsilon x + \varepsilon^2 - 1.95,$$

so

$$|f(x + \varepsilon) - f(x)| = |2\varepsilon x + \varepsilon^2|.$$

As we compute the iterations $f(-0.5)$, $f^2(-0.5)$, $f^3(-0.5)$, ... please observe that we always have $|f''(-0.5)| < 2$. (You can convince yourself of this either algebraically or geometrically. If you like, look ahead in §6.2 on page 327 and at Figure 6.13.) Therefore, if the error ε is small (say $|\varepsilon| < 0.01$), we have

$$|f(x + \varepsilon) - f(x)| = |2\varepsilon x + \varepsilon^2|$$

$$\leq |2\varepsilon x| + |\varepsilon^2|$$

$$< 4.01|\varepsilon|$$

$$< 10|\varepsilon|.$$

We can interpret this result as follows. If x and y are the same to 1000 decimal places, then $f(x)$ and $f(y)$ are the same to at least 999 decimal places. Therefore, if our computer's calculations are correct to more than 1000 digits of accuracy, then our calculation of $f^{1000}(-0.5)$ should be correct to at least a few decimal places.

Using 1000-digit accuracy, my computer reports that $f^{1000}(-0.5) = 1.468974$. However, if we wanted to calculate $f^{1000000}(-0.5)$, then we should use 1 million digits of accuracy, and my computer isn't powerful enough to do that!

Instead of using 100-digit or 1000-digit precision, a computer can perform "infinite" precision arithmetic. Instead of storing a number such as $\frac{1}{3}$ to a finite number of decimal places, the computer can work with this number as a pair of integers, 1 and 3. Will using infinite precision arithmetic enable us to compute $f^{1000000}(-0.5)$? Let's try and see what happens. We can write $f(x)$ as $x^2 - \frac{39}{20}$ (since $\frac{39}{20} = 1.95$) and -0.5 as $-\frac{1}{2}$. Using *Mathematica* we find that $f(-\frac{1}{2}) = -\frac{17}{10}$ and $f^2(-\frac{1}{2}) = \frac{47}{50}$. So far so good. It looks as if we ought to be able to evaluate the iterates of f exactly. Now for the bad news. Let's look at $f^7(-\frac{1}{2})$. How bad could that be? We get (and I promise that this is the actual computer output):

$$f^7\left(\frac{-1}{2}\right) = -\frac{20245637286145626631359577772304769861101 99733}{10842021724855044340074528008699417 11425781250}.$$

I'm afraid to even *think* about how many digits would be in the numerator and denominator of $f^{1000000}(-1/2)$.

The iterations of $f(x) = x^2 - 1.95$ are chaotic. No matter how accurately we try to perform our computations, eventually small round-off errors overwhelm the true values of the iterations and our computer begins to spew nonsense. Greater accuracy mearly delays the deterioration.

The case $a = -2.64$: chaos again

Now we consider the case $a = -66/25 = -2.64$. Why this value? The only thing special (from our perspective) about this number is that it is less than -2. When a drops past -2, there is a fundamental change in the behavior of f_a. When $a \geq -2$, and if x_0 is between -2 and $+2$, then the iterates $f^k(x_0)$ always remain between ± 2. However, when $a < -2$ it can be proved that the iterates typically explode. We illustrate this in the case $a = -2.64$. The other reason we chose $a = -2.64$ is that the fixed points of f are simple numbers. We solve $f(x) = x$ to find

$$x^2 - \frac{66}{25} = x \quad \Rightarrow \quad x = \frac{11}{5}, \ -\frac{6}{5}$$

or in decimal notation, the fixed points are (exactly) 2.2 and -1.2.

What do we hope to learn about this function? We will note that for most values x_0, the iterates $f^k(x_0)$ tend to infinity. Then we will focus on the values x_0 for which $f^k(x_0)$ stays bounded. We will then explore the structure of the set of these bounded values and how f moves them around.

If $x \notin [-2.2, +2.2]$, then $f^k(x) \to \infty$.

The first thing we need to notice is that if $|x_0| > 2.2$, we have $f^k(x_0) \to \infty$ as $k \to \infty$. We can see this graphically in Figure 4.45, which shows what happens as we iterate f for any x_0 just less than -2.2. Note that if $x_0 < -2.2$, then $f(x_0) = x_0^2 - 2.64 > 4.84 - 2.64 = 2.2$, so in one step we are to the right of the fixed point 2.2. Now, $f'(2.2) = 2 \times 2.2 = 4.4$, so 2.2 is an unstable fixed point, and the subsequent iterations are sent speeding off to $+\infty$. What's more, even if $|x_0| \leq 2.2$, if we ever have $f^k(x_0) > 2.2$ for some number k, then we know that the iterates must explode.

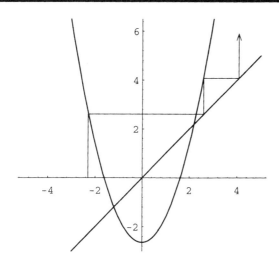

Figure 4.45. The iterates of $f(x) = x^2 - 2.64$ with $|x_0| > 2.2$ tend to infinity.

We are interested in studying the set B of all values x_0 for which $f^k(x_0)$ remains bounded (in the interval $[-2.2, 2.2]$) for all k. What we know so far is that $B \subseteq [-2.2, 2.2]$.

Definition of the set B: all values that remain bounded under iteration.

We know that $f(2.2) = 2.2$ (it's a fixed point), so $f^k(2.2) = 2.2$ for all k, thus $2.2 \in B$. Similarly, $f(-2.2) = 2.2^2 - 2.64 = 2.2$, so $f^k(-2.2)$ remains bounded, and we conclude that $-2.2 \in B$. Perhaps B is the entire interval $[-2.2, 2.2]$? No! Consider $x_0 = 0$. We have $f(0) = 0^2 - 2.64$, so $|f(0)| > 2.2$. We now know that subsequent iterations must go to infinity, so $0 \notin B$. Indeed, there's a whole chunk of numbers (an open interval) around 0 which get blown away by f. We can see why by examining Figure 4.46. Let A_1 be the set of all numbers x for which $f(x) < -2.2$, i.e.,

$$x^2 - 2.64 < -2.2 \quad \Rightarrow \quad x^2 < 0.44 = \frac{11}{25} \quad \Rightarrow \quad |x| < \frac{\sqrt{11}}{5} \approx 0.6633$$

so $A_1 = \left(-\sqrt{11}/5, +\sqrt{11}/5\right) \approx (-0.6633, +0.6633)$. Thus for any number x in A_1 we have $|f(x)| > 2.2$, so we know that $f^k(x) \to \infty$. Thus none of the numbers in A_1 are in B, so $B \subseteq [-2.2, 2.2] - A_1$. Incidentally, the endpoints of A_1, namely, $\pm\sqrt{11}/5$ *are* in B. Observe that

The notation $X - Y$, where X and Y are sets stands for the set of all elements which are in X but not in Y.

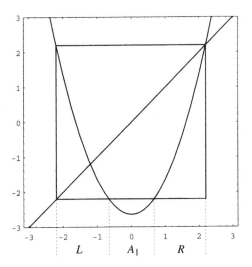

Figure 4.46. A graph of the function $y = f(x) = x^2 - 2.64$. For all numbers x in the open interval A_1, the iterates $f^k(x)$ must go to infinity. Closed intervals L and R are the left and right parts of $[-2.2, 2.2] - A_1$.

$$f\left(\pm\frac{\sqrt{11}}{5}\right) = \frac{11}{25} - 2.64 = 0.44 - 2.64 = -2.2,$$

so $f^2(\pm\sqrt{11}/5) = +2.2$, and for all future iterations, $f^k(\pm\sqrt{11}/5) = 2.2$, so the numbers $\pm\sqrt{11}/5$ are in B.

Now, the set A_1 breaks the interval $[-2.2, 2.2]$ into two pieces: the closed intervals

$$L = \left[-2.2, -\sqrt{11}/5\right] \quad \text{on the left, and}$$

$$R = \left[\sqrt{11}/5, 2.2\right] \quad \text{on the right,}$$

so $B \subseteq L \cup R$.

f stretches L to [−2.2, +2.2] but reverses its direction.

It is important to understand what happens to the intervals L and R when we compute f. Observe that $f(L) = f(R) = [-2.2, 2.2]$; this is visible in Figure 4.46. For L, we see that the left endpoint of L (namely, -2.2) maps to the right endpoint of $[-2.2, 2.2]$ (because $f(-2.2) = 2.2$) and the right endpoint of L (namely, $-\sqrt{11}/5$) maps to the left endpoint of $[-2.2, 2.2]$ (because $f(-\sqrt{11}/5) = -2.2$). The function f maps the

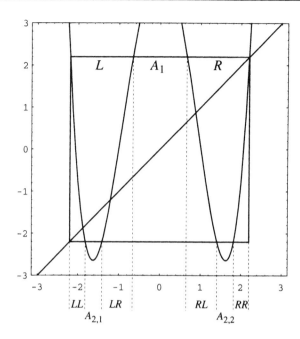

Figure 4.47. Graph of $y = f^2(x)$ where $f(x) = x^2 - 2.64$. Points x in the open intervals $A_{2,1}$ and $A_{2,2}$ are sent to infinity when we iterate f. The intervals L and R are broken into two pieces each.

points in L in a one-to-one[5] manner onto the interval $[-2.2, 2.2]$. Further, the *order* of the points in L is *reversed*: The smaller numbers in L become the larger numbers in $[-2.2, 2.2]$.

Similarly, $f(R) = [-2.2, 2.2]$. The function f maps the interval R in a one-to-one fashion onto the interval $[-2.2, 2.2]$, but in this case, the order of the points in R is *preserved*.

f stretches R to $[-2.2, +2.2]$ but preserves its orientation.

In both cases the effect of doing one iteration of the function f to the numbers in L (or in R) is to stretch the short interval out and have it look like $[-2.2, 2.2]$. In essence L and R behave just like the interval $[-2.2, 2.2]$. To see this, consider Figure 4.47. This figure shows the graph of $y = f^2(x)$. Within each of the intervals L and R we see a section where $f^2(x) < -2.2$; these sections are the open intervals $A_{2,1}$ and $A_{2,2}$. We know that points in these intervals are sent to infinity when we iterate f. What we see is that L and R each behaves like the full interval $[-2.2, 2.2]$.

---

[5]This means that if x_1 and x_2 are different numbers in L, then $f(x_1) \neq f(x_2)$. Distinct points in L retain their distinctiveness, i.e., are sent by f to distinct points in $[-2.2, 2.2]$

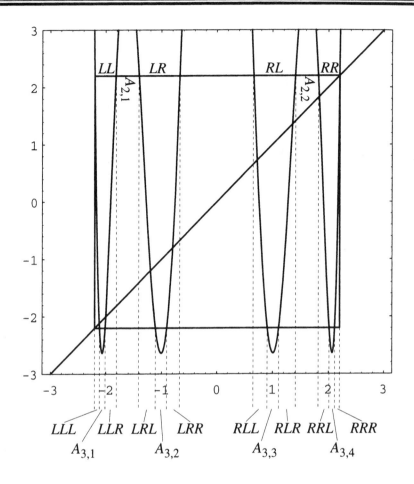

Figure 4.48. Plot of $y = f^3(x)$.

One can work out the exact values of the endpoints of the intervals $A_{2,1}$ and $A_{2,2}$, but it's a burdensome chore. Numerically, we have $A_{2,1} = (-1.81751, -1.40594)$ and $A_{2,2} = (1.40594, 1.81751)$. The four endpoints of $A_{2,1}$ and $A_{2,2}$ are the four solutions to the equation $f^2(x) = -2.2$. Thus the four endpoints are in the set B.

The open intervals $A_{2,1}$ and $A_{2,2}$ subdivide the intervals L and R into four new closed intervals we call LL, LR, RL, and RR. We now know that $B \subseteq LL \cup LR \cup RL \cup RR$.

We now look at what happens in f^3; see Figure 4.48. Within each of the intervals LL, LR, RL, and RR (which behave just like $[-2.2, 2.2]$ after two iterations of f) there is an open interval where $f^3(x) < -2.2$; we call these open intervals $A_{3,1}$, $A_{3,2}$, $A_{3,3}$, and $A_{3,4}$. Points inside any of these A's are destined to be iterated to infinity. The endpoints, however, belong to B, since they satisfy $f^3(x) = 2.2$, and will be stuck at 2.2 thereafter. The new A's break the intervals LL through RR into eight pieces we designate LLL through RRR. We now know that

$$B \subseteq LLL \cup LLR \cup \cdots \cup RRR.$$

Of course, we can now go to f^4 and find 16 closed intervals named LLLL through $RRRR$, and then on to f^5, and so on. With each successive iteration, we bite an open interval $A_{k,j}$ from the midst of each of the $LRRLLRLLRL$-type closed intervals.

What happens in the long run? The points *not* in B are exactly those that end up in an $A_{k,j}$. What remains is the set B, i.e.,

$$B = [-2.2, 2.2] - \bigcup_{k,j} A_{k,j}.$$

The question remains, What are the points in B? We can describe a few in conventional notation: for example, -2.2 and $\sqrt{11}/5$ are in B. The others are not readily described this way. We develop a better notation that enables us to understand how points in B are affected by f.

Consider an *infinite* sequence of the letters L and R. For example: $LLRRLRLL\cdots$. From this sequence, we construct a nested collection of closed intervals by taking the first letter (L), then the first and second letters (LL), then the first three (LLR), etc.:

$$L \supset LL \supset LLR \supset LLRR \supset LLRRL \supset \cdots$$

Each successive interval is smaller and smaller. Are there numbers that are in all of these intervals? The answer is yes. To see that

$$L \cap LL \cap LLR \cap LLRR \cap LLRRL \cap \cdots \neq \emptyset$$

think about the left endpoints of the successive intervals. They form a bounded, nondecreasing sequence of numbers which converges to some number ℓ. Also, the successive right endpoints form a bounded, nonincreasing sequence which converges to some number r. The numbers ℓ and r must be in *all* the intervals in our list and therefore in the intersection. Also, the numbers ℓ and r, since they are in *none* of the A's must be in B. So $f^k(\ell)$ and $f^k(r)$ are always in $[-2.2, 2.2]$.

Symbolic representation of points in B based on their L,R address.

The next question is, Are r and ℓ different? We claim no. Suppose $\ell \neq r$. Note that since r and ℓ belong to $L = [-2.2, -0.6633]$ and since $f'(x) = 2x$, we know that $|f'(x)| > 2 \times 0.6633 > 1.2$ for all numbers in L (this is also true in R). Now, $(f^2)'(x) = f'[f(x)]f'(x)$. Since $x \in B$, we know that $|f'[f(x)]| > 1.2$ and $|f'(x)| > 1.2$, so $|(f^2)'(x)| > (1.2)^2$. By similar reasoning we work out that $|(f^k)'(x)| > (1.2)^k$. Now let's apply what we've been discussing to ℓ and r. We apply the mean value theorem (see §A.3.1) to the function f^k. We learn that for some number c between ℓ and r,

$$\left| \frac{(f^k)(r) - (f^k)(\ell)}{r - \ell} \right| = |(f^k)'(c)| > 1.2^k.$$

So for k large enough, we have the distance between $f^k(\ell)$ and $f^k(r)$ is bigger than 4.4. But this is impossible, since $f^k(\ell)$ and $f^k(r)$ are always in $[-2.2, 2.2]$. The only possibility is that $\ell = r$.

Let us recap what we have learned: Suppose we have an infinite sequence of symbols L and R such as $LLRRLRLL \cdots$. Then the intersection of the intervals

$$L \supset LL \supset LLR \supset LLRR \supset LLRRL \supset \cdots$$

must contain a *unique* point of B; further, every point of B is described by such a sequence.

For example, the point $LLLLL \cdots$ must be the number -2.2. The point $RLLLLL \cdots$ is the left endpoint of the interval R, i.e., the sequence $RLLLLL \cdots$ identifies the point $\sqrt{11}/5$. Every point of B can be described by a "code" word consisting of an infinite sequence of L's and R's.

Although we do not have a method for specifying all the points in B by conventional notation (such as $\sqrt{11}/5$ or -2.2), we can name the points using the LR notation. We also want to understand how f treats the points in B. The beauty of the LR notation is we can understand how f works just by looking at the LR symbols. When we work with these symbols in place of conventional numbers we are dealing with *symbolic dynamics*.

Computing f using the LR notation.

Recall what f does to the intervals L and R: f stretches them out to the entire interval $[-2.2, 2.2]$. Further, f preserves the order of R but reverses L. Now ,what does f do to the intervals LL, LR, RL, and RR? Since f stretches and reverses L, we know that $f(LL) = R$ and $f(LR) = L$, and since f simply stretches R, we have $f(RL) = L$ and $f(RR) = R$. Now what does f do to the intervals LLL through RRR? Draw pictures to understand that

$$f(LLL) = RR, \qquad f(RLL) = LL,$$
$$f(LLR) = RL, \qquad f(RLR) = LR,$$
$$f(LRL) = LR, \qquad f(RRL) = RL,$$
$$f(LRR) = LL, \qquad f(RRR) = RR.$$

Do you see the pattern? Since f simply stretches out R, then we compute $f(Rxxxx\cdots)$ by just dropping the leading R. For example,

$$f(RLLRLRRLRL\cdots) = LLRLRRLRL\cdots.$$

To compute $f(Lxxxx\cdots)$ is almost as easy. We drop the initial L, but then we swap R's and L's in the remaining portion (the $xxxx\cdots$). For example,

$$f(LRRLRLRRLLR\cdots) = LLRLRLLRRL\cdots.$$

Let's see how this works with some of the values we know. For example, we know that $RRRRR\cdots$ corresponds to the values 2.2. Now, in conventional notation we know $f(2.2) = 2.2^2 - 2.64 = 2.2$, and again in symbolic notation $f(RRRRR\cdots) = RRRRR\cdots$; either way, we see that $2.2 = RRRRR\cdots$ is a fixed point of f. What about the other fixed point, -1.2? Since $-1.2 < 0$, we know that -1.2 must be in L, so the symbolic form of -1.2 must begin with an L. What is the rest of the code for -1.2? Since -1.2 is a fixed point of f [note that $f(-1.2) = (-1.2)^2 - 2.64 = 1.44 - 2.64 = -1.2$], then we know

$$f(Lx_2x_3x_4x_5\cdots) = \overline{x}_2\overline{x}_3\overline{x}_4\overline{x}_5 = Lx_2x_3x_4x_5\cdots,$$

where $\overline{L} = R$ and $\overline{R} = L$. In order to make the preceding equation work out, we need

$$\overline{x}_2 = L, \quad \overline{x}_3 = x_2, \quad \overline{x}_4 = x_3, \quad \overline{x}_5 = x_4, \quad \ldots,$$

from which we know that

$$x_2 = R, \quad x_3 = L, \quad x_4 = R, \quad x_5 = L, \quad \ldots,$$

and therefore the code for -1.2 is $LRLRLRLR\cdots$.

We can use the symbolic notation to find points of period 2. We know that $RRRRR\cdots$ and $LRLRLR\cdots$ are fixed points and therefore have period 2, so what we really seek are points of prime period 2. Now we know that these other points *cannot* start $RRx_3x_4x_5\cdots$ because

$$f^2(RRx_3x_4x_5\cdots) = x_3x_4x_5\cdots = RRx_3x_4x_5\cdots,$$

Finding periodic points using LR codes.

so we have $R = x_3 = x_4 = x_5 = \cdots$. Also, we can learn that a point of prime period 2 *cannot* begin $LR \cdots$. To see why, we compute

$$f^2(LRx_3x_4x_5 \cdots) = f(\overline{R}\overline{x}_3\overline{x}_4\overline{x}_5 \cdots)$$

$$= f(L\overline{x}_3\overline{x}_4\overline{x}_5 \cdots)$$

$$= \overline{x}_3\overline{x}_4\overline{x}_5 \cdots = LRx_3x_4x_5 \cdots,$$

from which we see that $x_3 = L$, $x_4 = R$, $x_5 = L$, and so on, giving the code word $LRLRLRLR \cdots$, the other fixed point.

So, if p is a point of prime period 2, then p must begin with either LL or RL. If $p = LL \cdots$, then we have

$$f^2(p) = f^2(LLx_3x_4x_5 \cdots)$$

$$= f(R\overline{x}_3\overline{x}_4\overline{x}_5 \cdots)$$

$$= \overline{x}_3\overline{x}_4\overline{x}_5 \cdots$$

$$= LLx_3x_4x_5 \cdots = p.$$

So we must have $L = \overline{x}_3$, $L = \overline{x}_4$, $x_3 = \overline{x}_5$, $x_4 = \overline{x}_6$, and so on. Thus we have $p = LLRRLLRRLLRR \cdots$. Double checking, we see that

$$p = LLRRLLRRLL \cdots \mapsto f(p) = RLLRRLLRRL \cdots$$

$$\mapsto f^2(p) = LLRRLLRRLL \cdots,$$

and we also learn that $f(p) = RLLRRLLRRLLR \cdots$ is another point of prime period 2 (one that begins RL as we noted before). You can work out that $RLLRRLLR \cdots$ is the only point of prime period 2 which begins RL, so p and $f(p)$ are the only points of prime period 2.

Now we can work out explicitly the points of prime period 2 by solving the quartic equation $f(f(x)) = x$. The four roots of this equation are

$$\frac{11}{5}, \quad -\frac{6}{5}, \quad \frac{-25 \pm 5\sqrt{189}}{50}.$$

Since $11/5$ and $-6/5$ are the fixed points of f, we know that

$$LLRRLLRRLL \cdots = \frac{-25 - 5\sqrt{189}}{50} \approx -1.8748, \quad \text{and}$$

$$RLLRRLLRRL \cdots = \frac{-25 + 5\sqrt{189}}{50} \approx 0.8748.$$

Note, however, that it is *much* easier to work with the symbol version.

We next consider points of period 3. There are eight possible ways the points' codes might begin: from LLL to RRR. For each of these possibilities, we see there is only one way to complete the sequence to make a point of period 3. Here are the ways we can do this (the commas are for clarity):

Points of period 3 and higher via LR codes.

$$LLL, RRR, LLL, RRR, \cdots \quad RLL, LRR, RLL, LRR, \cdots$$
$$LLR, LLR, LLR, LLR, \cdots \quad RLR, RLR, RLR, RLR, \cdots$$
$$LRL, RLR, LRL, RLR, \cdots \quad RRL, LLR, RRL, LLR, \cdots$$
$$LRR, LRR, LRR, LRR, \cdots \quad RRR, RRR, RRR, RRR, \cdots.$$

Of these, $LRLRLRL\cdots$ and $RRRRRR\cdots$ are fixed points of f, but the other six form two orbits of prime period 3:

$$
\begin{array}{cc}
LLL, RRR, \cdots & LLR, LLR, \cdots \\
\downarrow & \downarrow \\
RRL, LLR, \cdots & RLR, RLR, \cdots \\
\downarrow & \downarrow \\
RLL, LRR, \cdots & LRR, LRR, \cdots \\
\downarrow & \downarrow \\
LLL, RRR, \cdots & LLR, LLR, \cdots.
\end{array}
$$

We can continue in this way to learn about the 16 points of period 4, of which 2 are fixed points and 2 are points of prime period 2, leaving 12 points of prime period 4 (arranged in three orbits). And there are 32 points of period 5, of which 2 are fixed points, leaving 30 points of prime period 5 (arranged in six orbits). Indeed, we know there are points of prime period k for any k (of course, this follows from Sarkovskii's theorem and the existence of points of prime period 3), and we can work out their symbolic forms.

Chaos in B

We know that in B (the set of points which remain bounded when we iterate $f(x) = x^2 - 2.64$) we can find points of prime period k for any k, and we can work out how many we have of each. Our symbolic approach, however, tells us that there is much more to the story. The periodic points in B all have repeating code names, that is, after a certain number of letters, the pattern of L's and R's repeats verbatim over and over again. These repeating code words are, in a sense, atypical. A more typical sequence is generated by the following method: Take a fair coin and mark one side

'L' and the other side 'R'. Flip the coin over and over again to generate an infinite sequence of L's and R's. *That* is what a typical element of B looks like. What happens when we iterate f on a typical point in B? We observe that the trajectory of the point moves erratically around, coming near all points in the set B as it wanders around.

What happens if we iterate f for two points that are very close together in B, but different? This means their code words might match for dozens and dozens of places, but eventually one will have an L where the other has an R. From that point forth, the trajectory of the two nearly identical starting points will be wildly different—f exhibits sensitive dependence on initial conditions. The behavior of f on B is chaotic.

Problems for §4.2

◆1. For each of the following functions find points of period 1, 2, and 3. Which are prime periodic points? When reasonable, find exact answers; otherwise, use numerical methods. For each periodic point you find, classify it as stable or unstable.

   (a) $f(x) = 3.1x(1 - x)$.
   (b) $f(x) = (-3x^2 + 11x - 4)/2$.
   (c) $f(x) = \frac{53}{40}(1 - x^2)$. (Note: $\frac{53}{40} = 1.325$.)
   (d) $f(x) = \cos x$
   (e) $f(x) = 3\cos x$.
   (f) $f(x) = \cos 3x$.
   (g) $f(x) = e^x - 1$.
   (h) $f(x) = e^x - 2$.
   (i) $f(x) = \frac{1}{3}e^x$.
   (j) $f(x) = -2\tan^{-1} x$.
   (k) $f(x) = -3/(1 + x^2)$.
   (l) $f(x) = 2(1 - |x|)$.

◆2. For each of the following families of functions f_a find values a at which the family undergoes bifurcations. Categorize the bifurcations you find (as saddle node, etc.).

   (a) $f_a(x) = ae^x$.
   (b) $f_a(x) = a\sin x$.
   (c) $f_a(x) = \sin(ax)$.
   (d) $f_a(x) = a + 2\cos x$.
   (e) $f_a(x) = e^{a-x^2}$.

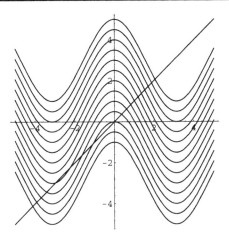

Figure 4.49. Graphs of the functions $f_a(x) = a + 2\cos x$ for various values of a between -3 and 3. The line $y = x$ is shown as well.

[Note: It is helpful to plot several members of family of functions $f_a(x)$ on the same set of axes. For example, in Figure 4.49 we display the family $f_a(x) = a + 2\cos x$ by plotting the curves $y = f_a(x)$ for various values of a between -3 and 3.]

◆3. The function $f_a(x) = ax(1 - x)$ is called the *logistic map*. Prepare a computer generated bifurcation diagram for the family $f_a(x)$. You will need to do some exploring to figure out the right range of values for the parameter a.

Compare your diagram with that in Figure 4.29 on page 185.

Find an exact value of a at which there is a tangent bifurcation.

Find two exact values of a at which there is a period doubling bifurcation.

◆4.* Given that $2 \rhd 1$, prove that $4 \rhd 1$. [Hint: Think about f^2.]

◆5.* Explain why $\rhd$ is transitive. In other words, prove that for any positive integers a, b, c if $a \rhd b$ and $b \rhd c$, then $a \rhd c$.

◆6. Consider the following function f which acts on infinite sequences ("words") made up of the letters A, B, and C according to the following rules:

(a) If the word begins with an A, drop the initial A and swap B's for C's in the remainder.

(b) If the word begins with a B, drop the initial B and swap A's for C's in the remainder.

(c) If the word begins with a C, drop the initial C and swap A's for B's in the remainder.

Some examples:

$$f(ABACCABAC\cdots) = CABBACAB\cdots,$$

$$f(CBBCACCAC\cdots) = AACBCCBC\cdots,$$

$$f(BBAACCBBA\cdots) = CCAABBC\cdots.$$

Find the fixed points and the points of prime periods 2 and 3 for this function.

◆7.   Give an example of a function f which has points of *prime* period 4. Note that 4 is not prime in the arithmetic sense. Primality of numbers and primality of periods are not the same. However, see the next problem.

◆8.   Let f be a function with a point x of period p where p is a prime number. Explain why x is either a fixed point or a point of prime period p.

◆9.   Let $f\colon \mathbf{R} \to \mathbf{R}$ be continuous. Suppose $a < b$, $f(a) = b$, and $f(b) = a$. Let $g(x) = f(x) - x$. Use the intermediate value theorem (see §A.3.1 on page 349) to show that $g(c) = 0$ for some c between a and b. Conclude that f has a fixed point.

4.3   Examplification: Riffle shuffles and the shift map

4.3.1   Riffle shuffles

A mathematical model of shuffling cards.

Before playing a card game, the dealer shuffles the deck. A common way to do this is known as a *riffle* (or *dovetail*) shuffle. First the deck is cut in half. The dealer then merges the two halves by alternately dropping the bottom card of the half-decks into a new pile. Of course, the dealer can't do this perfectly. At the first stage, the deck is split in half, but perhaps not perfectly.[6] At the second stage, the cards drop from the half-decks more or less alternately, but, again, not perfectly.

Let us develop a mathematical model for how people shuffle cards. This description of shuffling is called the GSR model after the inventors of this formulation: Gilbert, Shannon, and Reeds.

Cutting the deck.

Consider the first step. We want to cut the deck in half. One way to do this is to flip a coin 52 times and record how many heads we see; call this

---

[6]Well, most dealers can't. Some expert magicians *can* do a perfect riffle shuffle. If one does eight consecutive perfect shuffles the deck will be restored to its original state. This fact forms the basis for some amazing card tricks.

number h. We split the deck so the top half has h cards and the bottom half
has $52 - h$ cards. We expect this split to be close to 26-26.

Next, consider the riffle step. From which half-deck do we drop the *Merging the two halves.*
first card? This is hard to say; you might like to run some experiments to
see if people have a tendency one way or the other. However, if one of the
two half-decks has more cards than the other, it seems reasonable that the
heavier "half" is more likely to drop a card than the lighter. Let's make this
more precise. As the cards are dropped from the two half-decks, we arrive
at a situation where there are a cards in one half and b cards in the other.
The side with more cards is more likely to be the next to drop; let us set
the probability that we drop a card from the size a subdeck to be $a/(a+b)$
and the probability we drop from the other subdeck to be $b/(a+b)$. Thus
the probability we choose one side or the other is proportional to its size.

In summary, we split the deck by flipping a fair coin 52 times and we
record the number of heads h; we split the deck into two subdecks with h
and $52 - h$ cards each. Call the two subdecks *left* and *right*. We now drop
the bottom cards of left or right one at a time onto the combining pile. When
there are a cards in the left subdeck and b cards in the right, we choose left
with probability $a/(a+b)$ [and right with probability $b/(a+b)$].

Is the GSR shuffle a good mathematical model of shuffling cards? Yes, *Is this a good model?*
in the mathematical sense that it is mathematically precise and we can
use this model to solve problems such as, How many times should a person shuffle the deck to make sure it is thoroughly mixed up? It is also a
reasonable model of how humans actually shuffle cards. However, many
people tend to shuffle cards in a more clumped fashion than the GSR model
predicts. (See [7] pages 77ff.)

In this section our aim is to understand riffle shuffles using dynamical
systems methods.

4.3.2 The shift map

We now investigate the function

$$\sigma(x) = 2x \bmod 1.$$

The function σ takes a real number, multiplies it by 2, and returns the
fractional part of the result, i.e., it rounds the result *down* to an integer,
dropping the digits to the left of the decimal point. For example, let us
compute $\sigma(0.8)$. We know that $2 \times 0.8 = 1.6$ and the fractional part of 1.6
is 0.6, so $\sigma(0.8) = 0.6$. Let's do another example: Let $x = 2.4$ and let us

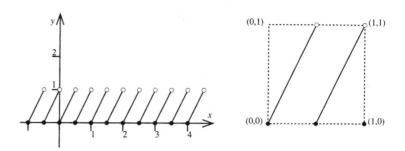

Figure 4.50. (Left) the graph of the function $\sigma(x) = 2x \bmod 1$; (right) a detailed look at the graph inside the unit square.

compute $\sigma(x)$. Note that $2x = 4.8$, and since $4.8 \bmod 1 = 0.8$, we have $\sigma(2.4) = 0.8$. The graph $y = \sigma(x)$ is plotted in Figure 4.50.

We are interested in exploring the effect of iterating the function σ and, ultimately, in relating σ to GSR shuffles.

Notice first that for any number x we know that $\sigma(x)$ must be between 0 and 1; indeed, $0 \le \sigma(x) < 1$. We focus our attention on the effect of $\sigma(x)$ just for x in the unit interval $[0, 1]$.

Notice that 0 is the only fixed point of σ. What are the points of period 2? We want to solve $\sigma(\sigma(x)) = x$. It is awkward to write down a formula for $\sigma^2(x)$; it is

$$\sigma^2(x) = [2(2x \bmod 1)] \bmod 1.$$

Setting that equal to x and solving is messy.[7] However, if you plot the graphs $y = \sigma^2(x)$ and $y = x$ on the same set of axes, you quickly discover that $\sigma(\frac{1}{3}) = \frac{2}{3}$ and $\sigma(\frac{2}{3}) = \frac{1}{3}$ and that $\frac{1}{3}, \frac{2}{3}$ are the only points of prime period 2.

The formula $2x \bmod 1$ is a bit cumbersome. There is an alternate way to understand the function σ which justifies calling it the *shift map*.

Let x be a number between 0 and 1. If $x < \frac{1}{2}$, then $\sigma(x) = 2x$; otherwise $(x \ge \frac{1}{2})$, $\sigma(x) = 2x - 1$. If x is written as a decimal number, we can judge whether x is less than $\frac{1}{2}$ or at least $\frac{1}{2}$ by looking at the first digit after the decimal point. If that digit is 0, 1, 2, 3, or 4 we know that $x < \frac{1}{2}$, but if that digit is 5 or larger, we know that $x \ge \frac{1}{2}$. There's a slight ambiguity if $x = \frac{1}{2}$. In this case we can write $x = 0.5$ or $x = 0.49999 \cdots$.

---

[7] Actually, its not too bad. See problems 1 and 2 on page 229.

It makes sense to write $\frac{1}{2} = 0.5$ in order for our "check the first digit" test to work properly.

How do the digits of $2x$ relate to the digits of x? This is a difficult problem (because of carries and the like) when we are working in base *ten*. However, multiplying by 2 is easy when we work in *binary* (base 2). Let us restart our digit discussion, but this time working in binary.

Let x be a number between 0 and 1. If $x < \frac{1}{2}$, then $\sigma(x) = 2x$;   *Working in binary.*
otherwise $\sigma(x) = 2x - 1$. If x is written in *binary*, we can judge whether x is less than $\frac{1}{2}$ or at least $\frac{1}{2}$ just by looking at the first digit after the decimal point.[8] If the first digit is a 0, then we know that $x < \frac{1}{2}$, and if the first digit is a 1, then we know that $x \geq \frac{1}{2}$. Again, there is an ambiguity when $x = \frac{1}{2}$. In this case we can write (in binary) $x = 0.1$ or $x = 0.01111 \cdots$. We opt for the first notation not only because it is simpler but because it makes our "check the first digit" test work properly. Furthermore, some other real numbers can be written with a finite binary expansion. For example, $\frac{3}{8}$ in binary is 0.011. It can also be written as $0.01011111 \cdots$. In such cases, let us agree to use the simpler notation.

To summarize, we can check whether or not $x < \frac{1}{2}$ by examining x's first digit (in binary).

Next, let us understand how $\sigma(x)$ works when we use binary notation. Suppose we write x in binary as

$$x = 0.d_1 d_2 d_3 d_4 \cdots,$$

where each d_i is either 0 or 1. We may assume that this sequence of digits does not end in an infinite string of 1's. Now let us calculate $\sigma(x)$.

Notice that $2x = d_1.d_2 d_3 d_4 \cdots$; multiplying by 2 simply shifts all the digits one place to the left.

Now let us compute $\sigma(x)$. When $x < \frac{1}{2}$ (so $d_1 = 0$), we know that $\sigma(x) = 2x$, so $\sigma(x) = 2x = d_1.d_2 d_3 d_4 \cdots = 0.d_2 d_3 d_4 \cdots$ (because $d_1 = 0$). On the other hand, suppose $x \geq \frac{1}{2}$ (so $d_1 = 1$). Then $\sigma(x) = 2x - 1 = (d_1.d_2 d_3 d_4 \cdots) - 1 = 0.d_2 d_3 d_4 \cdots$ (because $d_1 = 1$). Something wonderful has happened! We do *not* have to check the first digit (d_1) of x. In both cases we have

$$\sigma(x) = \sigma(0.d_1 d_2 d_3 d_4 \cdots) = 0.d_2 d_3 d_4 \cdots.$$

In words, to compute σ of $0.d_1 d_2 d_3 d_4 \cdots$ we drop the first digit d_1 and *shift* all the digits one place to the left. The binary notation gives us a *symbolic* way to work with the shift map. (This is similar to—and even simpler

---

[8]For the purists: ... after the *binary point*.

than—the symbolic LR dynamics of the function $f(x) = x^2 - 2.64$ from §4.2.5.)

Using the symbolic representation to find periodic points of σ.   Let us work with the symbolic dynamics for σ to learn about its fixed points and other properties. To solve the equation $\sigma(x) = x$, we have

$$\sigma(0.d_1 d_2 d_3 \cdots) = 0.d_2 d_3 d_4 \cdots = 0.d_1 d_2 d_3 \cdots,$$

so we know that $d_1 = d_2 = d_3 = d_4 = \cdots$. There are only two choices: all the d_i's equal 0, or they all equal 1. If they are all 0, then we have $x = 0$ (which we know is a fixed point). If they all equal 1, then we have $x = 0.1111 \cdots = 1$; but we don't allow sequences which become an endless list of 1's (indeed, $\sigma(1) \neq 1$). So 0 is the only fixed point of σ.

    To solve $\sigma^2(x) = x$ we write

$$\sigma^2(0.d_1 d_2 d_3 d_4 \cdots) = 0.d_3 d_4 d_5 \cdots = 0.d_1 d_2 d_3 \cdots,$$

from which we learn

$$d_1 = d_3 = d_5 = \cdots \qquad \text{and} \qquad d_2 = d_4 = d_6 = \cdots.$$

We now know that the d_{odd}'s are all 0 or all 1, and the d_{even}'s are also all 0 or all 1 (but we don't have both $d_{\text{odd}} = d_{\text{even}} = 1$). This gives three solutions to the equation $\sigma^2(x) = x$:

$$0.000000 \cdots = 0, \quad 0.010101 \cdots = \frac{1}{3}, \quad \text{and} \quad 0.101010 \cdots = \frac{2}{3}.$$

    By a similar analysis we can solve the equation $\sigma^3(x) = x$. The solutions are of the form $x = 0.abcabcabc \cdots$, where a, b, and c are either 0 or 1, but they are not all 1. There are two orbits of period 3:

$$.001001 \cdots \mapsto .010010 \cdots \mapsto .100100 \cdots \mapsto .001001 \cdots, \quad \text{and}$$

$$.011011 \cdots \mapsto .110110 \cdots \mapsto .101101 \cdots \mapsto .011011 \cdots,$$

or in more conventional notation, $\frac{1}{7} \mapsto \frac{2}{7} \mapsto \frac{4}{7} \mapsto \frac{1}{7}$ and $\frac{3}{7} \mapsto \frac{6}{7} \mapsto \frac{5}{7} \mapsto \frac{3}{7}$.

    In a similar manner, one can find the points of higher periods. It is worthwhile to note, however, that *none* of the periodic points of σ are stable. Let's see why. The unique fixed point of σ is 0, and this is clearly an unstable fixed point. If x is a little bit bigger than 0, then $\sigma(x), \sigma^2(x), \sigma^3(x), \ldots$ are equal to $2x, 4x, 8x, \ldots$, so the iterates move away from 0. (If x is a

All periodic points of σ are unstable.   little bit less than 0, we have a huge jump to $\sigma(x)$, which is nearly 1.)

    Let's now consider the other periodic points of σ. We need to look only inside the unit interval. First, σ is neither continuous nor differentiable; however, it is continuous and differentiable for all values of x strictly between 0 and 1 except for $x = \frac{1}{2}$ (see Figure 4.50 on page 220). Now, $x = \frac{1}{2}$ is not a periodic point of σ because $\sigma(\frac{1}{2}) = 0$ and then $\sigma(0) = 0$. Thus if p is a periodic point of σ (other than 0) we know that $\sigma'(p)$ is defined; indeed, $\sigma'(p) = 2$ for *any* p between 0 and 1 (other than $\frac{1}{2}$). Thus all periodic points of σ are unstable.

What happens when we iterate σ starting from a typical value x between 0 and 1? A typical x, when written in binary, consists of a random string of 0's and 1's; the kth digit of x is either 0 or 1 with probability 50%, and each digit is independent of the others. Thus, as we iterate $\sigma^k(x)$ we see the iterations jump wildly about the unit interval coming close to every point between 0 and 1. For a typical starting value of x the iterations $\sigma^k(x)$ are chaotic!

For typical x the iterations $\sigma^k(x)$ are chaotic.

4.3.3 Shifting and shuffling

At first glance the GSR model for riffle shuffling and the shift map σ have no connection. We are now ready to explore how we can use the shift map to understand riffle shuffling.

Introducing the shift shuffle.

To forge the connection, we consider another method for mixing up the cards of a deck. Let us call this method, which uses the function σ, a *shift shuffle*.

To begin, we choose 52 random numbers between 0 and 1, and we sort these numbers numerically from smallest to largest; let's call these numbers $x_1 < x_2 < \cdots < x_{52}$.

We write these numbers (in blue ink), in order, on the cards in the deck. The top card is labeled with the number x_1, the next card is labeled x_2, etc., and the last card gets x_{52}.

Next, we compute σ of each card's number and write the result on the card also (in red ink). So the top card is marked with the numbers x_1 (in blue) and $\sigma(x_1)$ in red, etc.

Now we reorder the deck so the cards appear in correct numerical sequence according to their red numbers (the $\sigma(x_i)$'s).

This procedure is illustrated in Table 4.1 and Figure 4.51. For simplicity, we use a 13-card deck (ace through king in one of the suits) in place of the full 52-card deck.

Look at the upper half of Table 4.1. The first column lists the names of the cards (A through K) in order. The second column gives the values x_1 through x_{13}—13 random numbers. The third column gives the same

Card name	x decimal	x binary	$\sigma(x)$ decimal	$\sigma(x)$ binary
A	0.1484	0.001001011111	0.2968	0.01001011111
2	0.2321	0.001110110110	0.4642	0.01110110110
3	0.3116	0.010011111100	0.6232	0.10011111100
4	0.3296	0.010101000101	0.6592	0.10101000101
5	0.4036	0.011001110101	0.8072	0.11001110101
6	0.4306	0.011011100011	0.8612	0.11011100011
7	0.5365	0.100010010101	0.0731	0.00010010101
8	0.6395	0.101000111011	0.2790	0.01000111011
9	0.8505	0.110110011011	0.7011	0.10110011011
T	0.8675	0.110111100001	0.7350	0.10111100001
J	0.9669	0.111101111000	0.9338	0.11101111000
Q	0.9873	0.111111001100	0.9746	0.11111001100
K	0.9976	0.111111110110	0.9952	0.11111110110

$\sigma(x)$ decimal	$\sigma(x)$ binary	New order
0.0731	0.00010010101	7
0.2790	0.01000111011	8
0.2968	0.01001011111	A
0.4642	0.01110110110	2
0.6232	0.10011111100	3
0.6592	0.10101000101	4
0.7011	0.10110011011	9
0.7350	0.10111100001	T
0.8072	0.11001110101	5
0.8612	0.11011100011	6
0.9338	0.11101111000	J
0.9746	0.11111001100	Q
0.9952	0.11111110110	K

Table 4.1. Relating the shift map σ to riffle shuffling.

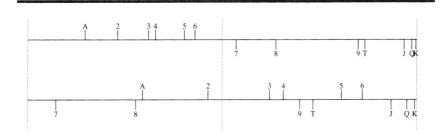

Figure 4.51. Understanding how the shift map induces a shuffle.

numbers, but in binary notation. In the fourth and fifth columns of the upper table are the computed values $\sigma(x_1)$ through $\sigma(x_{13})$.

The first two columns of the lower chart in Table 4.1 repeat the last two columns of the upper chart, but now the values are rearranged in ascending numerical order. The rightmost column of the lower chart shows the order of the cards after this procedure is finished.

It is instructive to have a geometric view of the shift shuffle. Look at Figure 4.51. The long horizontal lines represent the unit interval. The dotted vertical lines mark the points 0, $\frac{1}{2}$, and 1. In the upper portion we show the locations of the numbers x_1 through x_{13}. In the lower portion, we show the locations of the numbers $\sigma(x_1)$ through $\sigma(x_{13})$.

A geometric view of the shift shuffle.

The geometric effect of σ on the unit interval is to stretch the intervals $[0, \frac{1}{2}]$ and $[\frac{1}{2}, 1]$ to twice their length and then overlay the results.

In our example, the first six cards (A through 6) have labels which are less than $\frac{1}{2}$, and the remaining seven cards (7 through K) have labels larger than $\frac{1}{2}$. We know that after we shift shuffle these cards together, cards A–6 and cards 7–K are still in their natural order, but the two lists are interspersed.

Does this look similar to a GSR riffle shuffle? The punch line, of course, is that the shift shuffle *is* the GSR shuffle, only in a disguised form. Let's understand why.

The two models are the same!

The first step of the GSR shuffle is to cut the deck by coin flips. We flip a coin 52 times and record the number of heads, h. We cut the deck after h cards. In lieu of flipping a coin, we may choose random numbers uniformly between 0 and 1, count the number of numbers that are less than $\frac{1}{2}$, and then cut that many cards. This is the same as the number of card labels which begin with the binary digit 0 (see the horizontal line in the middle of Table 4.1). Thus the cut step of both shuffles is the same.

Now that the deck is cut, we need to riffle the two parts together. In the GSR shuffle, if the two decks have a and b cards, we drop the bottom card from the first part with probability $a/(a+b)$ and from the second part with probability $b/(a+b)$. In the shift shuffle, the bottom card depends on which new label $\sigma(x)$ is largest. Notice that the cut step in the shift shuffle uses only the first binary digit in the label, while the relative order of the $\sigma(x)$'s is determined by the remaining digits.

Consider Figure 4.51. Those cards whose first digit is 0 (A–6) are uniformly distributed over the interval $[0, 1]$, as are those whose first digit is 1 (7–K). We can see these two subdecks merging by reading the lower portion of the figure from right to left. The cards drop in the following order: first the king, queen, and jack from the "1" subdeck, followed by the 6 and 5 from the "0" subdeck, then the 10 and 9 from "1", etc., until the last card to drop onto the combined pile is the 7 from the "1" subdeck.

The question is, If we have two collections of points uniformly spread over an interval (say a points of type "0" and b points of type "1") what is the probability that a type "0" point is the furthest to the right? The answer is $a/(a+b)$ (hurray!). Let's see why. There are $(a+b)!$ ways to arrange the points in the interval; each of these is equally likely to occur (because we are spreading the points uniformly). The number of orders which end with a type "0" point is $a(a+b-1)!$ (there are a ways to choose which a type "0" point to be last and $(a+b-1)!$ ways to arrange the remaining points). Thus the probability that a type "0" point is last is just

$$\frac{a(a+b-1)!}{(a+b)!} = \frac{a}{a+b},$$

as promised. By a similar analysis, the probability that a type "1" point is last is $b/(a+b)$. Thus as we scan the values of $\sigma(x)$ from right to left we encounter cards whose label begins with 0 and cards whose label begins with 1. When we reach a stage where there are a and b such cards left to scan, the probability that the next card we see will be of type "0" is $a/(a+b)$, and the probability that the next card will be of type "1" is $b/(a+b)$. Thus the merging step of the shift shuffle and the merging step of the GSR shuffle are equivalent!

4.3.4   Shuffling again and again

One shuffle isn't enough.

Shuffling a deck of cards only once before playing is a bad idea. Although we have mixed up the order of some of the cards, many cards are still in their same relative order. To thoroughly mix up a deck, we need to do several shuffles.

Let's look carefully at what information we use in shuffling the cards by means of the shift shuffle. We use the first digit in each card's label x to determine where to cut the deck. Then we need the *order* of the $\sigma(x)$'s to riffle the parts together. We don't need the exact values of the $\sigma(x)$'s—only their order.

We are ready to do another shuffle. We could assign a new label to each vertex and repeat the whole procedure; however, there is another choice. We can use the random labels $\sigma(x)$ already on the cards! *Recycle your numbers!*

When we perform a second shuffle on our deck, we use $\sigma(x)$ as the labels and reorder the deck by the $\sigma^2(x)$'s. Please refer again to Table 4.1. A second shuffle on this deck will cut the deck between the 2 and the 3 (see the horizontal line in the lower portion of the table) and then reorder the deck according to $\sigma^2(x)$.

For a full deck, the order of the cards after k shuffles is determined by looking at the order of $\sigma^k(x_1)$ through $\sigma^k(x_{52})$.

We can give a geometric interpretation of doing k consecutive shuffles. *The geometric view, again.* As before, we choose 52 numbers uniformly at random in the unit interval $[0, 1]$ and assign the smallest to the first card, etc. We divide the unit interval into 2^k subintervals of equal size and stretch each of these subintervals to the full length of the interval $[0, 1]$.

Notice that two labels lie in the same subinterval exactly when the first k (binary) digits of their labels agree. In this case, their relative order during the first k shuffles is unchanged.

On the other hand, if the 2^k subintervals contain at most one label each, then when we stretch the subintervals to full length there will be no telling what order the cards are in. The deck will be thoroughly mixed.

So the question is, How many shuffles do we need to do? Stated another way, how large do we need to make k so that no two labels have the same first k binary digits?

Since there are 52 labels, we claim that $k = 5$ is not sufficient. Here is why: There are only $2^5 = 32$ strings of five 0's and 1's. So there must be two (or more) labels with the same first five digits. When $k = 6$, there are $2^6 = 64$ different strings of 0's and 1's (corresponding to the 64 subintervals of $[0, 1]$) so there is a chance that the labels are all in distinct subintervals. How good a chance?

More specifically, if we select 52 points uniformly at random from $[0, 1]$ what is the probability that no 2 of them occupy a common subinterval of length $1/2^k$? We drop the points into $[0, 1]$ one at a time. The first point *What is the probability that 52 randomly selected points will each be in its own subinterval?*

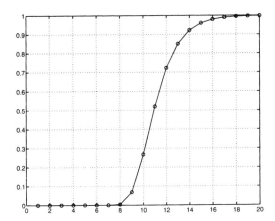

Figure 4.52. How many times do we need to shuffle a deck of 52 cards? The horizontal axis shows the number k of riffle shuffles, and the vertical axis shows the probability that 52 random points selected uniformly in the unit interval will each be in its own subinterval of length 2^{-k}.

dropped has a 100% chance of falling into an empty subinterval. The second point dropped might end up in the same subinterval as the first, but the probability of that happening is $1/2^k$. The probability that the second point is in its own compartment is therefore $1 - 1/2^k$. Given that the first 2 points are each in their own subinterval, the third point falls into an empty subinterval with probability $1 - 2/2^k$. In general, if the first t points are each in their own subinterval, the probability that the next point (number $t + 1$) goes into an empty compartment is $1 - t/2^k$. Thus the probability that all 52 points end up each in their own subinterval is precisely

$$\left(1 - \frac{0}{2^k}\right) \times \left(1 - \frac{1}{2^k}\right) \times \left(1 - \frac{2}{2^k}\right) \times \left(1 - \frac{3}{2^k}\right) \times \cdots \times \left(1 - \frac{51}{2^k}\right). \quad (4.3)$$

If we take $k = 6$, equation (4.3) evaluates to approximately 3.17×10^{-14}; it is *extremely* unlikely that all 52 cards are each in their own compartment when we partition $[0, 1]$ into just 64 subintervals.

In Figure 4.52 we plot equation (4.3) for various values of k. Notice that we need to take $k = 14$ to be at least 90% sure that no two labels have the same first k digits. If we take $k = 20$, we can be *very* sure we have thoroughly randomized the deck.

Twenty shuffles, in fact, is overkill. More sophisticated techniques show that for a 52-card deck eight or nine GSR shuffles provide a good level of randomization.

Problems for §4.3

◆1. Show that $\sigma^k(x) = 2^k x \mod 1$. [Hint: Use the symbolic/binary representation for σ, not the formula $\sigma(x) = 2x \mod 1$.]

◆2. Use the formula $\sigma^k(x) = 2^k x \mod 1$ to find all the roots of the equation $\sigma^k(x) = x$.

◆3. Write a computer program to compute iterates of the shift map σ. Run your program starting with $x_0 = 1/\pi$. Compute 100 iterations (i.e., compute $\sigma(1/\pi), \sigma^2(1/\pi), \ldots, \sigma^{100}(1/\pi)$.

Notice that most of the iterations you compute are zero. This is incorrect. Explain why your computer made this mistake.

◆4.* How many points of *prime* period n does the shift map have? Call this number $f(n)$. Prove that

$$2^n - 1 = \sum_{d|n} f(d),$$

where the sum is over all divisors d of n.

Prove that

$$f(n) = \sum_{d|n} \left(2^d - 1\right) \mu(n/d),$$

where

$$\mu(n) = \begin{cases} (-1)^t & \text{if } n \text{ is the product of } t \text{ distinct primes, and} \\ 0 & \text{otherwise.} \end{cases}$$

◆5. Show that performing eight "perfect" riffle shuffles of a standard 52-card deck restores the deck to its original order.

◆6. Consider the function $f(x) = 3x \mod 1$. Develop a symbolic representation of this function and find its points of prime period k for $k = 1, 2, 3, 4$.

◆7. The inhabitants of the planet Zorkan have three arms (each with only one hand). When a Zorkanite shuffles a deck of cards, it divides the deck into three roughly equal piles and then riffles the three piles together. Develop random models akin to the GSR shuffle and the shift shuffle to describe this shuffle.

How many times should a Zorkanite shuffle to be sure its deck of 216 cards is thoroughly mixed? (Zorkanologists believe the deck has 216 cards because Zorkanites have 6 fingers on each of their hands.)

◆8.   Here is another way to shuffle cards. First cut the deck into roughly four equal-size subdecks; call these decks (in order from the top) A, B, C, and D. Now riffle decks A and C together, then riffle decks B and D together, and finally riffle the two combined packs together.

Show that the effect of this procedure is the same as doing two ordinary riffle shuffles in a row.

What happens if the first pair of riffles is A with B, and then C with D?

What happens if the first pair of riffles is A with D, and then B with C?

Chapter 5

Fractals

Euclidean geometry is wonderful. Not only is it a beautiful mathematical theory, but it is also tremendously useful for modeling the world around us. If you are building a house, you use geometry every time you measure. Toss a ball and it flies along a parabolic arc. And, as its name implies, geometry is invaluable in surveying.

But there are shapes that classical (Euclidean) geometry isn't designed to discuss. Look at a tree, or the coast of the Chesapeake Bay. There are many shapes in nature (from the surface of your brain to the surface of the moon) for which Euclidean geometry falls short as a model. Rough shapes such as these are difficult to describe using lines, circles, ellipses, and the like. However, a different category of geometric objects—fractals—does provide an excellent model for these shapes.

It is difficult to describe a tree with words such as line, circle, and parabola.

In this chapter we explore the notion of fractals, how they are fixed "points" of dynamical systems, and (as their name suggests) how to see them as objects of fractional dimension.

5.1 Cantor's set

The first fractal we explore in depth is known as *Cantor's set*. If as we discuss Cantor's set it seems very familiar to you, that's good. The set B from the previous chapter (the set of values which stay bounded on iteration of the function $f(x) = x^2 - 2.64$ discussed in §4.2.5) has a structure that's nearly identical with that of Cantor's set.

We use the letter C to denote Cantor's set, and here is how we build it. Beginning with the unit interval $[0, 1]$, we delete from $[0, 1]$ the open interval covering its middle third, i.e., we delete $\left(\frac{1}{3}, \frac{2}{3}\right)$. So far we are left

0   1

Figure 5.1. Construction of Cantor's set.

with two intervals: $\left[0, \frac{1}{3}\right] \cup \left[\frac{2}{3}, 1\right]$.

Now we repeat the process: we delete the middle third of each of the intervals $\left[0, \frac{1}{3}\right]$ and $\left[\frac{2}{3}, 1\right]$, and we are left with

$$\left[0, \frac{1}{9}\right] \cup \left[\frac{2}{9}, \frac{1}{3}\right] \cup \left[\frac{2}{3}, \frac{7}{9}\right] \cup \left[\frac{8}{9}, 1\right].$$

This might be getting complicated to hold in your mind, so it's best to look at Figure 5.1. The first line of the figure is the unit interval [0, 1]. The second line represents [0, 1] with its middle third removed. In the third line we have removed the middle third of each of the closed intervals in the previous line. This pattern continues *ad infinitum*. At each stage we remove the open middle third of all the closed intervals in the previous stage.

How much have we taken away? The length of the unit interval is 1. At the first step, we deleted one-third of its length. At the second step, we deleted one third of the length of the remainder. In general, at the kth step the total length of the 2^k closed intervals is $(\frac{2}{3})^k$, which tends to zero as $k \to \infty$. Thus the total length of C is 0. You might wonder, Is anything left after we have done this process? The answer is, Quite a lot. Certainly, the numbers 0 and 1, as well as $\frac{1}{9}$, $\frac{2}{9}$, $\frac{1}{3}$, $\frac{2}{3}$, $\frac{7}{9}$, and $\frac{8}{9}$ are all in C.

We now describe *all* the points in C in two manners: first, in a symbolic (LR) notation, and then in more conventional notation.

5.1.1   Symbolic representation of Cantor's set

An LR representation of Cantor's set.

When we take our first bite out of the unit interval (when we delete $\left(\frac{1}{3}, \frac{2}{3}\right)$), we are left with two pieces. We call the left piece L and the right piece R, i.e.,

$$L = \left[0, \frac{1}{3}\right] \qquad \text{and} \qquad R = \left[\frac{2}{3}, 1\right].$$

Each of the pieces L and R is, in turn, broken in two. We can name the two pieces on the left as $LL = \left[0, \frac{1}{9}\right]$ and $LR = \left[\frac{2}{9}, \frac{1}{3}\right]$, and the two pieces on the right as $RL = \left[\frac{2}{3}, \frac{7}{9}\right]$ and $RR = \left[\frac{8}{9}, 1\right]$. We can name the eight closed intervals from the third stage as LLL through RRR, and so on.

An infinite sequence of L's and R's, such as $LLRRLRL \cdots$, gives rise to a nested sequence of closed intervals:

$$L \supset LL \supset LLR \supset LLRR \supset LLRRL \supset \cdots,$$

and as we discussed in the previous chapter, the intersection of such a sequence is nonempty. So

$$L \cap LL \cap LLR \cap LLRR \cap LLRRL \cap \cdots$$

is nonempty and contains points of the Cantor set C. The intersection cannot contain more than a single point, for otherwise it would contain an entire interval, contradicting the fact that the total length of the Cantor set is 0. So each point of the Cantor set is uniquely described by a sequence of L's and R's, just like the set B from §4.2.5 of the previous chapter.

5.1.2 Cantor's set in conventional notation

The LR notation shows us that the structure of Cantor's set C and that of the set B are the same. Now, it would be nice to be able to decide if numbers such as $\frac{3}{4}$ are or are not in C.

To understand which numbers are in C it is helpful to work with numbers in ternary (base 3) notation rather than decimal (base 10). This should not be too shocking: Base 3 is especially convenient if you plan to do a lot of dividing by 3.

Working in ternary, base 3.

Let's do a quick review of base 3. Integers in base 3 look like 1012_3. Reading this number from right to left, we have the 2 is in the ones column, a 1 in the threes column, then a 0 in the nines column, and lastly (leftmost) a 1 in the twenty-sevens column. Thus

$$1012_3 = \underline{1} \times 27 + \underline{0} \times 9 + \underline{1} \times 3 + \underline{2} \times 1 = 32.$$

Just as we can express any real number in decimal, so too can we express real numbers in ternary. The places to the right of the decimal point[1] are the thirds, ninths, etc., places, so

$$1.012_3 = \underline{1} \times 1 + \underline{0} \times \frac{1}{3} + \underline{1} \times \frac{1}{9} + \underline{2} \times \frac{1}{27} = \frac{32}{27}.$$

---

[1] Actually, in base 3 it's a *ternary* point, but that sounds funny.

How do we express the number $\frac{1}{2}$ in base 3? We have $\frac{1}{2} = 0.111\cdots_3$. We check that this is correct as follows:

$$\text{Let } x = 0.1111111\cdots_3.$$

$$\text{So } 3x = 1.111111\cdots_3.$$

$$\text{Subtracting } 3x - x \Rightarrow 2x = 1.$$

$$\therefore x = \frac{1}{2}.$$

As an exercise, you should check that $\frac{3}{4} = 0.20202020\cdots_3$.

Now armed with base 3 notation, let us return to the problem of determining which numbers are in Cantor's set C. Let's begin with the ones we *know*: the endpoints of the closed intervals of the form LLR, etc. Consider the eight intervals LLL through RRR. You should check that they are

$$LLL = \left[\frac{0}{27}, \frac{1}{27}\right] \qquad RLL = \left[\frac{18}{27}, \frac{19}{27}\right]$$

$$LLR = \left[\frac{2}{27}, \frac{3}{27}\right] \qquad RLR = \left[\frac{20}{27}, \frac{21}{27}\right]$$

$$LRL = \left[\frac{6}{27}, \frac{7}{27}\right] \qquad RRL = \left[\frac{24}{27}, \frac{25}{27}\right]$$

$$LRR = \left[\frac{8}{27}, \frac{9}{27}\right] \qquad RRR = \left[\frac{26}{27}, \frac{27}{27}\right]$$

Next we write the 16 endpoints of these eight intervals in ternary:

$0/27$	$=$	0.000_3	$18/27$	$=$	0.200_3
$1/27$	$=$	0.001_3	$19/27$	$=$	0.201_3
$2/27$	$=$	0.002_3	$20/27$	$=$	0.202_3
$3/27$	$=$	0.010_3	$21/27$	$=$	0.210_3
$6/27$	$=$	0.020_3	$24/27$	$=$	0.220_3
$7/27$	$=$	0.021_3	$25/27$	$=$	0.221_3
$8/27$	$=$	0.022_3	$26/27$	$=$	0.222_3
$9/27$	$=$	0.100_3	$27/27$	$=$	1.000_3

Ambiguity in place value notation.

Do you see a pattern? If so, bravo! If not, that's OK. It's not obvious yet. The difficulty is that the way we write numbers (in either base 10 or base 3) is slightly imperfect in the sense that there can be two ways to write the same number. For example, in base 10, the number 81 can also be written $80.9999\cdots = 80.\overline{9}$. Similarly, in base 3 the number $\frac{1}{3}$ can be written 0.1_3, but it can also be written $0.022222\cdots_3 = 0.0\overline{2}_3$. This ambiguity arises only with numbers whose notation is finite (doesn't run on forever) and whose last digit is 1. We can always convert that last 1 into $0\overline{9}$ (in decimal) or $0\overline{2}$ (in ternary). For understanding Cantor's set it is better to those convert numbers which end with 1 into their alternate, lengthy notation.[2] Let's do that for the 16 endpoints we listed above. We get

$0/27$	$=$	0.000_3	$18/27$	$=$	0.200_3
$1/27$	$=$	$0.000\overline{2}_3$	$19/27$	$=$	$0.200\overline{2}_3$
$2/27$	$=$	0.002_3	$20/27$	$=$	0.202_3
$3/27$	$=$	$0.00\overline{2}_3$	$21/27$	$=$	$0.20\overline{2}_3$
$6/27$	$=$	0.020_3	$24/27$	$=$	$0.20\overline{2}_3$
$7/27$	$=$	$0.020\overline{2}_3$	$25/27$	$=$	$0.220\overline{2}_3$
$8/27$	$=$	0.022_3	$26/27$	$=$	0.222_3
$9/27$	$=$	$0.02\overline{2}_3$	$27/27$	$=$	$0.\overline{2}_3$

Notice that all the numbers listed are written exclusively with 0's and 2's; the digit 1 never appears. We claim this is true for *all* numbers in Cantor's set C:

A number $x \in [0, 1]$ is in Cantor's set if and only if it can be written in base 3 using only 0's and 2's.

Cantor's set in ternary.

In fancy notation, $x \in C$ if and only if

$$x = \sum_{j=1}^{\infty} a_j 3^{-j} \qquad \text{with all } a_j \in \{0, 2\}.$$

Let's see why this is true. What do we know about numbers whose first digit (after the decimal point) is 1 (in base 3, of course)? We know they are at least 0.1_3 and less than 0.2_3, i.e., they are between $\frac{1}{3}$ and $\frac{2}{3}$. Furthermore, the number 0.1_3 doesn't count as having a 1 in its first position because it can also be written as $0.0\overline{2}_3$. So the numbers with an indelible 1 in their third's column are the numbers in the open interval $\left(\frac{1}{3}, \frac{2}{3}\right)$, and these are the numbers we eliminated in the first step of constructing Cantor's set!

Now, which numbers have a 1 in their second digit (ninths place), i.e., have the form $0.\square 1 \square \square \square_3$? We've already handled the numbers with a 1 in the first digit, so we are left to consider those numbers of the form $0.01\square\square\square_3$ and $0.21\square\square\square_3$. Of course, the numbers 0.01_3 and 0.21_3 don't worry us; they can be rewritten as $0.00\overline{2}_3$ and $0.20\overline{2}_3$. The numbers which *really* have their first 1 in the second digit are of the form

Numbers with a 1 in their ternary notation lie in intervals that get deleted.

$$0.01_3 < x < 0.02_3 \qquad \text{or} \qquad 0.21_3 < x < 0.22_3$$

---

[2]This is in contrast with how we denoted binary numbers in §4.3.2, where we preferred the terse notation.

or in more customary notation,

$$\frac{1}{9} < x < \frac{2}{9} \qquad \text{or} \qquad \frac{7}{9} < x < \frac{8}{9}.$$

Aha! These are the numbers in the open intervals we deleted at stage 2. I hope the situation is clear now. The numbers whose first *real* 1 is in the kth base 3 digit are exactly those numbers we eliminate in the kth stage of constructing C. Those numbers which remain (those numbers with only 0's and 2's) form Cantor's set.

Now we can answer the question, Is $\frac{3}{4}$ in C? Since $\frac{3}{4} = 0.202020\cdots_3 = 0.\overline{20}_3$, the answer is yes. On the other hand, $\frac{3}{8} = 0.1\overline{0}_3$ is not in C.

5.1.3   The link between the two representations

L is 0 and R is 2!

We have two ways to describe points in C: the symbolic (LR) notation and the ternary notation. Is there a nice way to see what the point, say, $LRLRLRLR\cdots$ might be? The answer is a beautiful yes.

What is the interval L? It is the interval $\left[0, \frac{1}{3}\right]$, but let's think about this in another way. The interval L contains all those numbers whose first (base 3) digit is 0 (from $0 = 0.0_3$ to $\frac{1}{3} = 0.0\overline{2}_3$). The interval R contains those numbers whose first digit is 2.

What is the interval LR? We know that $LR = \left[\frac{2}{9}, \frac{1}{3}\right] = \left[0.020_3, 0.02\overline{2}_3\right]$. Thus the numbers in LR are exactly those numbers whose ternary notation begins with 0.02.

What are the numbers in $LLRLR$? The answer is: those whose ternary expansion begins 0.00202. Thus the point $LRLRLRLR\cdots$ must be $0.02020202\cdots_3 = 0.\overline{02}_3 = \frac{1}{4}$.

Is the pattern clear? The symbol L becomes a 0, and the symbol R becomes a 2 (and we stick a decimal point in front). Thus we can easily switch between the symbolic and ternary notations.

5.1.4   Topological properties of the Cantor set

Cantor's set is a *set* of real numbers. We can also view it geometrically. Because C has no length, the usual sort of geometric questions we might ask about it don't seem to be relevant. We consider more basic properties of C: *topological* properties. Topology is (roughly) the study of properties of objects that are unchanged by stretching (but not tearing). If Cantor's set were drawn on an rubber band, then its shape might change as we stretch the rubber band in various places, but its topological properties would not change. It is well beyond the scope of this text to do justice to the ideas of topology. Nonetheless, we describe C's more important topological properties (it is bounded, closed, compact, totally disconnected, and perfect), as we need some of these ideas (especially compact sets) later.

Cantor's set is bounded

When we say a set is *bounded*, we mean (roughly) that we can draw a circle around it. More precisely, a set S is bounded means that there is a number b such that all points in the set are within distance b of one another. For example, a line segment in the plane is bounded. The unit interval $[0, 1]$ of the real line is bounded. Circles, triangles, and spheres are all bounded. However, lines and parabolas are unbounded.

It is clear that Cantor's set is bounded; all its points are within distance 1 of one another, since the entire set lies inside the unit interval $[0, 1]$.

Cantor's set is closed

The idea of a *closed* set is more technical. Roughly, it means that points *not* in the set are *not* extremely close to the set. Let's make this precise. Let S be a set and let p be a point *not* in S. We say that p is *separated* from S Separation. if there is some number d so that *all* points within distance d of p are also not in S. (The number d might depend on the point p in question.)

For example, let's see why the point 1.1 is separated from the set $[0, 1]$. Note that all numbers within distance $d = 0.05$ of 1.1 are also *not* in $[0, 1]$. Thus 1.1 is separated from $[0, 1]$. On the other hand, 1 is *not* separated from the open interval $(0, 1)$. Although $1 \notin (0, 1)$, we see that for any distance d, there are points a distance d away from 1 which are in $(0, 1)$. For similar reasons, the point $(1, 0)$ (in the plane) is not separated from the set $\{(x, y) : x^2 + y^2 < 1\}$ (the interior of the unit circle).

Now we are ready to define *closed* sets.[3]   Closed.

A set S is *closed* means that every point not in S is separated from S. For example, the unit interval $[0, 1]$ is closed. So are the x-axis of the coordinate plane, and the unit circle $\{(x, y) : x^2 + y^2 = 1\}$. A set consisting of a single point (such as the origin) is a closed set. However, the following sets are not closed: the open interval $(0, 1)$, the interior of a triangle, and the points in the plane (x, y) for which $x < y$.

---

[3]There are other ways to define closed sets which are equivalent to the one we have given. One definition says that a set is closed if boundary points of the set must be members of the set. Another definition says that if a sequence of points from the set converges, the limit must also be in the set.

Cantor's set is a closed set. To see why, think about a point p *not* in Cantor's set. If $p > 1$ or $p < 0$, we see that p is separated from C (choose d equal to half the distance to 1 or 0, respectively). Otherwise (p is between 0 and 1) we know that p is deleted at some stage in the construction of C. For example, if p is in $\left(\frac{1}{3}, \frac{2}{3}\right)$, then p must be separated from C: We choose d to be half the distance to either $\frac{1}{3}$ or $\frac{2}{3}$, whichever is closer. More generally, p is in some open interval (s, t) which was deleted at some stage of the construction of C. We choose d to be half the smaller of the distances from p to s or from p to t. Thus every point not in C is separated from C; therefore, C is closed.

A restatement of this definition is that every point *not* in a closed set S is a positive distance away from S. More precisely, the distance between a point p and a closed set S is the minimum distance between p and any point in S. For example, if S is the set $\{(x, y) : x^2 + y^2 \leq 1\}$ (the unit disk) and $p = (3, 0)$, then the distance from p to S is 2: the point of S closest to p is $(1, 0)$, and it is at distance 2 from p.

For a closed set S, the distance from a point to p to S is zero if and only if $p \in S$.

Cantor's set is compact

A *compact* set is a set which is closed and bounded.[4] Thus a circle is a compact set, but neither a line (it's unbounded) nor the interior of a square (it's not closed) is compact.

Clearly, Cantor's set is compact. We have just seen that it is bounded and closed.

All the fractals we consider in this book are compact sets. We will have more to say about compact sets later in this chapter.

Cantor's set is totally disconnected*

The words *totally disconnected* are virtually self-defining. Intervals, such as $[0, 1]$ are connected sets.[5] A set such as $[0, 1] \cup [2, 3]$, while not connected, does contain two connected pieces.

---

[4]This is not 100% correct. The actual definition of compact is a bit more technical. However, for the sets we are considering (i.e., for subsets of $\mathbf{R}^n$ for some n) the "closed and bounded" definition is equivalent to the full definition.

[5]There is a broad topological definition of *connected*, but we don't need it. For subsets of the real line $\mathbf{R}$, it suffices to say that a set S is *connected* if whenever p, q are in S, then the entire interval $[p, q]$ must be a subset of S.

By contrast, Cantor's set is *totally disconnected*. [We define what this means only for subsets of the real line **R**.] A set S is *totally disconnected* means that whenever p and q are points of S, then there is some point between p and q which is *not* in S.

To see why Cantor's set is totally disconnected, think about two numbers p and q in Cantor's set. We have to find a number between them which is not in C. To do this, we write p and q in base 3 notation. At some digit they must disagree (since they are unequal), say

$$p = 0.22020\underline{0}?????\cdots_3 ,$$

$$q = 0.22020\underline{2}?????\cdots_3 .$$

This means that $p \in RRLRLL$ while $q \in RRLRLR$. This means that any number x in the middle third of the interval $RRLRL$ is *not* in C and is between p and q. Therefore C is totally disconnected.

Cantor's set is perfect*

We say that a set S is *perfect* if every point in S is the limit of *other* points in S. Crudely, this means that the set can't be spread out too thinly. The unit interval $[0, 1]$ is an example of a perfect set.

To show that Cantor's set is perfect, we choose a point p in C, for example, $0.022020222\cdots_3$. We have to show that there is a sequence of points $p_1, p_2, \ldots$ which are *all* in C, are *all* different from p, and which converge to p. Here's how we do it. Let

$$p_1 = 0.0_3,$$

$$p_2 = 0.02_3,$$

$$p_3 = 0.022_3,$$

$$p_4 = 0.0220_3,$$

$$p_5 = 0.02202_3,$$

$$p_6 = 0.022022_3,$$

$$\vdots$$

In words, we let p_k be the ternary number obtained by truncating p after k digits. Now, it is clear that all the p_k's are in C and that $p_k \to p$ as $k \to \infty$. Finally, the p_k's are all different from p *unless* p has a terminating ternary expansion, such as $p = 0.202_3$. In this case, our trick of using more and more digits of p doesn't work, so we reach back into our bag and pull out another trick. If p's ternary representation is finite (ends with an infinite stream of 0's), then for, say $p = 0.202_3$, we let

$$p_1 = 0.2022_3,$$

$$p_2 = 0.20202_3,$$

$$p_3 = 0.202002_3,$$

$$p_4 = 0.2020002_3,$$

$$p_5 = 0.20200002_3,$$

$$\vdots$$

The trick here is to move the last, additional 2 farther and farther to the right. Notice that none of the p_k's is equal to p, all are in C, and $p_k \to p$ as $k \to \infty$.

Generalized Cantor sets*

Cantor sets are also called *fractal dust*.

Cantor's set C is *the* Cantor's set; it is also called *Cantor's middle thirds set*. However, there are other Cantor sets. Any set which is compact, totally disconnected, and perfect is called *a* Cantor set. For example, the set B from the previous chapter (points which are bounded under iteration of $f(x) = x^2 - 2.64$) is a Cantor set.

5.1.5   In what sense a fractal?

The Cantor set C has many interesting mathematical properties, but it is not exciting from a pictorial point of view. It looks (more or less) like the last row of Figure 5.1—pretty boring. Nevertheless, it is a good first example of a fractal.

The word *fractal* was coined by Mandelbrot. It is meant to convey the *fractured* look these sets have as well as the fact that they have *fractional* dimension.

What is a *fractal*? Rather than give a precise definition, we list the two important features of fractals: self-similarity and fractional dimension.

First, fractals look the same under a microscope as they do to the naked eye. By this we mean if we zoom in on a section of a fractal image, what we see looks very much (or exactly) like the set itself. For instance, we can look at the whole Cantor set C or just the tiny bit that lies between $\frac{2}{81}$ and $\frac{3}{81}$ (in the interval $LLLR$). The tiny portion is simply a $\frac{1}{81}$th scale model of the original! No matter how much we magnify a portion of C, what we observe looks exactly like C.

Second, fractals have *fractional* dimension. We make this precise in §5.6, but we can give a loose, intuitive idea here. A point is a zero-dimensional object: It has no length. A real interval, or a line, or a curve in space are examples of one-dimensional objects; they have length but no area. The interior of a square or the surface of a sphere are examples of two-dimensional objects, having area but no volume. Now, think about Cantor's set. It has no length, so it makes sense that its dimension is less than 1. On the other hand, there are *a lot* of points in Cantor's set and a lot of structure (it's perfect, just like [0, 1]). It doesn't seem fair to give it the lowly zero-dimensional rating. These comments are very vague and mushy, but they are meant only to convey an impression. In §5.6 we precisely work out the exact dimension of Cantor's set. That number, by the way, is $\log 2 / \log 3 \approx 0.6309$.

Problems for §5.1

◆1.   Which of the following numbers are in Cantor's set?

   (a) 0.

   (b) 1/9.

   (c) 1/10.

   (d) 3/4.

   (e) 8/3.

   (f) 19/26.

   (g) 7/8.

   (h) π.

◆2.   Consider a middle "three-fifths" version of Cantor's set. Start with the unit interval [0, 1]. Remove the open middle three-fifths, i.e., delete $\left(\frac{1}{5}, \frac{4}{5}\right)$. In each remaining interval, remove the middle three-fifths again. Continue this procedure *ad infinitum*.

   Describe the set of numbers in this middle-three-fifths version of Cantor's set. [Hint: work in base 5.]

◆3.   Consider the function

$$f(x) = -3\left|x - \frac{1}{2}\right| + \frac{3}{2}.$$

   When we iterate f we often have $f^k(x) \rightarrow -\infty$ (for example, try this starting at $x = \frac{1}{2}$). Let B be the set of starting values x_0 so that $f^k(x_0) \not\rightarrow -\infty$.

   Find *exactly* the set B and describe how f treats the values in B.

♦4. Which of the following sets are bounded? Which are closed? Which are compact?

  (a) The unit interval [0, 1].

  (b) A ray (with its endpoint) in the plane.

  (c) A half-plane including the boundary line.

  (d) A half-plane without the boundary line.

  (e) The real line.

  (f) The set of points in the plane which are distance at least 5 from the origin.

  (g) The set of points in the plane which are distance at most 5 from the origin.

  (h) The interior of a triangle.

  (i) Any set of 15 points in the plane.

  (j) The empty set.

♦5.* Give four examples of subsets of the line which are connected; make them as different as possible.

♦6.* Give three examples of subsets of the line which are totally disconnected. Can a subset of the real line be both connected and totally disconnected?

5.2 Biting out the middle in the plane

Cantor's set is constructed by repeated applications of the "bite out the middle" operation. The resulting set C is mathematically intriguing but, as it is a subset of the line, visually dull. Before we get on with the meat of this chapter, we thought we'd sneak a few bites (ha!) from the dessert tray.

We consider two extensions of the "bite out the middle" concept to sets in the plane, $\mathbf{R}^2$. The first gives us Sierpiński's triangle and the second Koch's snowflake.

5.2.1 Sierpiński's triangle

Biting out the middle of a triangle.

The raw material for Cantor's set was the unit interval. To build Sierpiński's triangle, we start with a filled-in triangle in the plane. (By *filled-in* we mean that we include the interior.)

The basic construction step for Cantor's set was the removal of the middle part of the interval. For Sierpiński's triangle, the basic step is to remove the middle of the triangle. What do we mean by the *middle of the triangle*? Given a triangle, we can form another triangle by joining the

Figure 5.2. Construction of Sierpiński's triangle by removing middle triangles.

midpoints of the three sides. This smaller triangle, covering $\frac{1}{4}$ the area of the original, is what we mean by the middle triangle. We now delete the *interior* of the middle triangle. What we are left with is three half-size copies of the original triangle which are just-touching each other at their corners; see Figure 5.2. We now repeat this "delete the middle triangle" operation on each of the three remaining triangles. (See the figure.) We continue this process *ad infinitum*, and the result is the Sierpiński triangle (Figure 5.3).

Sierpiński's triangle has the same self-similarity properties that Cantor's set has. Each of the three subtriangles (in the three regions left after the first bite) is a half-size copy of the original.

Each bite removes $\frac{1}{4}$ of the area left, so after k bites the area remaining is $(\frac{3}{4})^k$ times the original area. Thus, Sierpiński's triangle has zero area. On the other hand, it is clearly fatter than a one-dimensional curve, so it would seem appropriate to award Sierpiński's triangle with a dimension between 1 and 2. Indeed, we show (in §5.6) that its dimension is $\log 3/\log 2 \approx 1.585$.

5.2.2   Koch's snowflake

We now present a second construction akin to that of Cantor's set. We start with an equilateral triangle (just the triangle without its interior). We want to bite out the middle third of each of its sides, but instead of deleting that section, we bend it out to make two sides of a smaller equilateral triangle; see Figure 5.4. We do this basic step to push out each side of the equilateral triangle. The result (in the upper left of Figure 5.5) looks

Pushing out the middle.

Figure 5.3. Sierpiński's triangle.

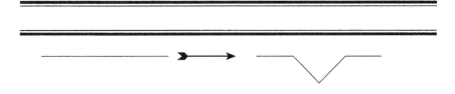

Figure 5.4. The basic step in the construction of the Koch snowflake.

like a Star of David. Next we perform the basic operation on each side of the Star of David: We replace the middle third by pushing out two sides of an equilateral triangle. The next three stages in the construction are shown in Figure 5.5. The Koch snowflake, which looks very much like the lower right image in Figure 5.5, is the result of repeating this operation *ad infinitum*.

At each stage, the length of the curve increases by a factor of $\frac{4}{3}$. Thus the Koch snowflake has infinite length but zero area. Each section of the snowflake has the same basic appearance as the whole, regardless of how closely you might zoom in on a section.

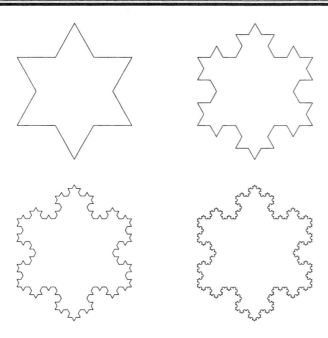

Figure 5.5. Steps 1, 2, 3, and 4 of the Koch snowflake construction.

Problems for §5.2

◆1. Recall Pascal's triangle, which displays the binomial coefficients. The nth row of Pascal's triangle are the coefficients of $(x + y)^n$ (binomial coefficients):

Draw a picture (a computer is helpful here) of Pascal's triangle in which every odd entry is replaced by a dot, and every even entry is left blank. What is the result?

◆2. Start with the unit square, i.e., the set $\{(x, y) : 0 \le x, y \le 1\}$. Delete the open square of side length $\frac{1}{3}$ from the middle of this square. This leaves eight closed squares of side length $\frac{1}{3}$ remaining. Now repeat this procedure on each of these pieces, and continue forever; see Figure 5.6.

Which ordered pairs of real numbers (x, y) are in this set?

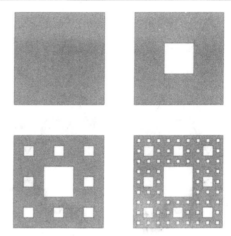

Figure 5.6. Repeatedly deleting the middle third of a square.

5.3   Contraction mapping theorems

In this section we lay the theoretical foundation for understanding how fractals are attractive fixed "points" of dynamical systems. This material is used to justify the fractal algorithms we discuss later.

5.3.1   Contraction maps

If $|f'(x)| < 1$, then f shrinks distances.

Let $\tilde{x}$ be a fixed point of f, i.e., $f(\tilde{x}) = \tilde{x}$. Recall that if $|f'(\tilde{x})| < 1$, then $\tilde{x}$ is an attractive fixed point. Geometrically, the condition $|f'(\tilde{x})| < 1$ means that the graph of f crosses the line $y = x$ at $x = \tilde{x}$ with a gentle slope. We now present another geometric view of this situation.

Suppose a and b are values near $\tilde{x}$. What do we mean by *near*? Let us suppose that f' is continuous,[6] and therefore $f'(x)$ doesn't vary wildly near $\tilde{x}$. Thus we know that $f'(x)$ stays between $\pm s$ for some positive number $s < 1$ whenever x is in some open interval J around $\tilde{x}$. We assume a and b are in this open interval J within which $|f'(x)| \leq s < 1$. What do we know about the distance between $f(a)$ and $f(b)$? This distance is $|f(b) - f(a)|$. By the mean value theorem (see §A.3.1),

$$f(b) - f(a) = f'(c)[b - a]$$

---

[6]In order to even *talk* about f' we need f to be differentiable. We also assume that this derivative is a continuous function.

for some c between a and b. We can rewrite this as

$$|f(b) - f(a)| = |f'(c)| \cdot |b - a| \leq s|b - a|.$$

In other words, the distance between a and b shrinks by a factor of s (or better) after we apply f. In particular, the distance between a and $\tilde{x}$ (and b and $\tilde{x}$) also shrinks by at least a factor of s, so $f(a)$ and $f(b)$ move closer to $\tilde{x}$. What happens after two iterations? We have

$$|f^2(b) - f^2(a)| \leq s|f(b) - f(a)| \leq s^2|b - a|$$

so in two iterations of f, the distance between a and b has shrunk by at least a factor of s^2. We see that after k iterations,

$$|f^k(b) - f^k(a)| \leq s^k|b - a|$$

and since $s^k \to 0$ as $k \to \infty$ (because $0 < s < 1$) we know that the distance between points is shrinking to zero as we iterate.

The important feature we have found is that (near the fixed point $\tilde{x}$) the function f is *contractive*: It shrinks distances. Let's make this more precise.

Let $f: \mathbf{R} \to \mathbf{R}$. We say that f is a *contraction mapping* provided there is a number s, with $0 < s < 1$, so that for any numbers a and b we have

Definition of contraction mapping.

$$|f(b) - f(a)| \leq s|b - a|.$$

If we wish to be more specific, we can say that f is a mapping with *contractivity s*.

For example, let $f(x) = \frac{1}{3}x - 1$. Let a and b be any numbers. Observe that

$$|f(b) - f(a)| = \left|(\frac{b}{3} - 1) - (\frac{a}{3} - 1)\right| = \frac{1}{3}|b - a|$$

so f is a contraction map with contractivity $\frac{1}{3}$.

Note that the condition $|f(b) - f(a)| \leq s|b - a|$ implies that if $b \to a$, then $f(b) \to f(a)$, i.e., f must be continuous. However, f need not be differentiable. (See problems 2 and 3 on page 258.)

5.3.2   Contraction mapping theorem on the real line

We now give a first version of the contraction mapping theorem.

Theorem (contraction mapping for R). Let $f : \mathbf{R} \to \mathbf{R}$ be a contraction mapping. Then f has a unique, stable fixed point, x^*. Furthermore, for any number a, we have $f^k(a) \to x^*$ as $k \to \infty$.

This value x^* turns out to be the unique fixed point of f. Does the infinite sum converge?

Let's see why this is true. We define

$$x^* = f(0) + \left[f^2(0) - f(0) \right] + \left[f^3(0) - f^2(0) \right] + \left[f^4(0) - f^3(0) \right] + \cdots.$$
$$(5.1)$$

This is an infinite sum, so our first worry is that this sum converges. The kth term of this sum is $f^{k+1}(0) - f^k(0)$, which we can also write as

$$f(f^k(0)) - f(f^{k-1}(0)).$$

We now use the fact that f is a contraction mapping to write

$$\left| f\left(f^k(0) \right) - f\left(f^{k-1}(0) \right) \right| \le s \left| f^k(0) - f^{k-1}(0) \right|$$

which we can rearrange to read

$$\left| \frac{f(f^k(0)) - f(f^{k-1}(0))}{f^k(0) - f^{k-1}(0)} \right| \le s.$$

What this says is that the absolute values of the ratios of the successive terms in equation (5.1) are bounded by $s < 1$. Hence by the ratio test, we know that the infinite sum defining x^* converges (indeed, converges absolutely). So the number x^* exists!

Here we show that $f(x^*) = x^*$.

Now, what is x^*? Let's look at the partial sums of equation (5.1):

One term $\to f(0)$.

Two terms $\to f(0) + [f^2(0) - f(0)] = f^2(0)$.

Three terms $\to f(0) + [f^2(0) - f(0)] + [f^3(0) - f^2(0)] = f^3(0)$.

$$\vdots$$

Because the summation in equation (5.1) telescopes,[7] we see that the partial sums are $f(0), f^2(0), f^3(0), \cdots$. In other words, we have shown that $f^k(0) \to x^*$ as $k \to \infty$. We are almost done with our first goal: finding a fixed point of f. Now we ask, What is $f(x^*)$? Since x^* is the limit of $f^k(0)$ (as $k \to \infty$) and f is continuous, we have

$$f(x^*) = \lim_{k \to \infty} f[f^k(x^*)] = \lim_{k \to \infty} f^{k+1}(x^*) = x^*.$$

---

[7] Yes, that's really what mathematicians say, because the successive terms fold into each other like a collapsing telescope.

Aha! We see that x^* is a fixed point. What's more, we can compute x^* simply by iterating f starting at 0.

Now we still have three tasks:

Three tasks remain.

(1) to show that there are no *other* fixed points of f,

(2) to show that x^* is an stable fixed point, and

(3) to show that $f^k(a) \to x^*$ as $k \to \infty$ for any a.

For (1) we note that if there were two (or more) distinct fixed points of f, say, a and b, then we would have the following contradiction to f's contractivity:

There are no other fixed points.

$$|f(a) - f(b)| = |a - b| \not\leq s|a - b|.$$

For (2) and (3) we choose any starting value a (not necessarily near x^*) and we repeatedly use f's contractivity to write

x^* is *very* stable.

$$|f^k(a) - x^*| = |f^k(a) - f^k(x^*)| \leq s^k|a - x^*|$$

which tends to 0 (since $s < 1$) as $k \to \infty$. Thus $f^k(a) \to x^*$ for *any* number a. This completes our justification of the contraction mapping theorem (for **R**).

5.3.3   Contraction mapping in higher dimensions

We plan to greatly extend the idea of contraction mapping and to present increasingly general versions of this theorem. Our first extension is modest: We move from **R** to $\mathbf{R}^n$.

Let $f: \mathbf{R}^n \to \mathbf{R}^n$, i.e., f is a function from n-vectors to n-vectors. Now we know exactly what we mean by the distance between vectors **x** and **y**. We have

We write $d(\mathbf{x}, \mathbf{y})$ to stand for the distance between **x** and **y**.

$$d(\mathbf{x}, \mathbf{y}) = |\mathbf{x} - \mathbf{y}| = \sqrt{(x_1 - y_1)^2 + (x_2 - y_2)^2 + \cdots + (x_n - y_n)^2}.$$

We can now express what it means for vector functions to be contraction mappings. A function $f: \mathbf{R}^n \to \mathbf{R}^n$ is a *contraction mapping* means that there is a number s, with $0 < s < 1$, so that for any vectors **x** and **y** we must have

$$d[f(\mathbf{x}), f(\mathbf{y})] \leq sd(\mathbf{x}, \mathbf{y}).$$

Theorem (contraction mapping for $\mathbf{R}^n$). Let $f: \mathbf{R}^n \to \mathbf{R}^n$ be a contraction mapping. Then f has a unique, stable fixed point, $\tilde{\mathbf{x}}$. Furthermore, for any $\mathbf{a} \in \mathbf{R}^n$ we have $f^k(\mathbf{a}) \to \tilde{\mathbf{x}}$ as $k \to \infty$.

The justification of this theorem is similar to that of the one-dimensional version.

5.3.4 Contractive affine maps: the spectral norm*

How to tell if an affine
function is contractive.

Later, we will use functions of the form $f(\mathbf{x}) = A\mathbf{x} + \mathbf{b}$ to build fractals. We want to know whether these functions are contraction maps. In the one-dimensional case, $f(x) = ax + b$, we just need $|a| < 1$. What do we need in the multidimensional case? If you said "all eigenvalues of A have absolute value less than 1" you are—I'm very sorry—wrong. But give yourself a big pat on the back; it was an *excellent* guess.

Let's see what's going on. We want $d[f(\mathbf{x}), f(\mathbf{y})] \le sd(\mathbf{x}, \mathbf{y})$, that is,

$$|(A\mathbf{x} - \mathbf{b}) - (A\mathbf{y} - \mathbf{b})| \le s|\mathbf{x} - \mathbf{y}|$$

$$\Rightarrow \qquad |A(\mathbf{x} - \mathbf{y})| \le s|\mathbf{x} - \mathbf{y}|$$

$$\Rightarrow \qquad \frac{|A(\mathbf{x} - \mathbf{y})|}{|\mathbf{x} - \mathbf{y}|} \le s.$$

For $\mathbf{z} = \mathbf{x} - \mathbf{y} \ne \mathbf{0}$, this becomes $|A\mathbf{z}|/|\mathbf{z}| \le s$ for any $\mathbf{z} \in \mathbf{R}^n$. Notice that since $A(r\mathbf{z}) = rA(\mathbf{z})$, we may as well assume that $\mathbf{z}$ is a unit vector (has length 1). In this case, we just want to be sure that $|A\mathbf{z}| \le s < 1$ for all unit vectors $\mathbf{z}$.

Let's consider an example. Let $A = \begin{bmatrix} 0.2 & 0.3 \\ 0.5 & -0.4 \end{bmatrix}$. We ask, Given that $\mathbf{z}$ is a unit vector, how large can $|A\mathbf{z}|$ be? Now, unit vectors are those with length 1 and correspond to points on the unit circle. Thus $\mathbf{z}$ can be written as $\mathbf{z} = \begin{bmatrix} \cos\theta \\ \sin\theta \end{bmatrix}$, where $0 \le \theta \le 2\pi$. So

$$A\mathbf{z} = \begin{bmatrix} 0.2 & 0.3 \\ 0.5 & -0.4 \end{bmatrix} \begin{bmatrix} \cos\theta \\ \sin\theta \end{bmatrix} = \begin{bmatrix} 0.2\cos\theta + 0.3\sin\theta \\ 0.5\cos\theta - 0.4\sin\theta \end{bmatrix},$$

and

$$|A\mathbf{z}| = \sqrt{(0.2\cos\theta + 0.3\sin\theta)^2 + (0.5\cos\theta - 0.4\sin\theta)^2}.$$

We can plot this last expression as a function of θ, as we have done in Figure 5.7. Note that $|A\mathbf{z}|$ varies roughly between 0.36 and 0.64. In any case, we see that $|A\mathbf{z}| < 0.65$ for any unit vector $\mathbf{z}$. This implies, by our preceding work, that a map of the form $f(\mathbf{x}) = A\mathbf{x} + \mathbf{b}$ (with $A = \begin{bmatrix} 0.2 & 0.3 \\ 0.5 & -0.4 \end{bmatrix}$) is a contraction map.

Now, the eigenvalues of A are 0.3899 and -0.5899; neither is the special value 0.64. What we really want to know is the maximum value of

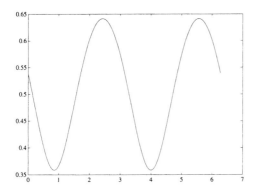

Figure 5.7. A plot of the possible values of $|A\mathbf{z}|$ where $\mathbf{z}$ is a unit vector.

$|A\mathbf{z}|$ for a unit vector $\mathbf{z}$. This number is called the *spectral norm* of A. Let us denote the spectral norm of A by $\|A\|$, i.e.,

$$\|A\| = \max_{|\mathbf{z}|=1} |A\mathbf{z}|.$$

The spectral norm of A is also called the *first singular value* of A. How do we calculate it? Advanced linear algebra texts give the full story,[8] but we shall be content if our computer can compute it. In MATLAB we use the norm function:

> The spectral norm, not the eigenvalues, of a matrix determines whether or not it is contractive.

> How to compute $\|A\|$.

```
a =
    0.2000    0.3000
    0.5000   -0.4000
>>norm(a)
ans =
    0.6414
```

Aha! There is the "magic" number we have been looking for. We that the maximum of $|A\mathbf{z}|$ (i.e., $\|A\|$) is about 0.6414. Since this number is less than 1, we see that that $A\mathbf{x} + \mathbf{b}$ is a contraction mapping.

(Other computer packages also enable you to compute spectral norm. In *Mathematica* you can use the SingularValues function to compute the spectral norm. You can create your own command by entering

---

[8]Here's a quick synopsis: Compute the square roots of the eigenvalues of $A^T A$; these numbers are the singular values of A. The largest singular value is the spectral norm.

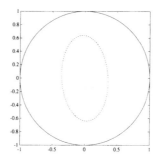

Figure 5.8. A plot of **z** (the unit circle) and $|A\mathbf{z}|$ (the enclosed ellipse) for all unit vectors **z**. The spectral norm of A is less than 1.

```
SpectralNorm[ matrix_ ] :=
  Max[ SingularValues[matrix][[2]] ]
```

To compute the singular value of a matrix such as $\begin{bmatrix} 0.2 & 0.3 \\ 0.5 & -0.4 \end{bmatrix}$ you would type

```
SpectralNorm[{{.2,.3},{.5,-.4}}]
```

and the computer would respond with 0.641421. You can also use *Maple* to compute the spectral norm. Use the Svd command or the `singularvals` command in the `linalg` package. The spectral norm is the largest of the singular values.)

A geometric view of spectral norm.

We can also look at this situation graphically. If we plot all the unit vectors **z** in $\mathbf{R}^2$, we get the unit circle. If we also plot $A\mathbf{z}$ for all **z**, we get (typically) an ellipse. If the ellipse lies entirely in the interior of the unit circle, then we know that $|A\mathbf{z}| \le s < 1$ for some number s. Such a plot is given in Figure 5.8. The circle of radius 1 represents the set of all unit vectors **z**. The ellipse represents the set of all vectors of the form $A\mathbf{z}$ (with **z** a unit vector). Since this ellipse is enclosed in the interior of the unit circle, we see that $\|A\| < 1$, so $A\mathbf{x} + \mathbf{b}$ is a contraction mapping.

A matrix whose eigenvalues have absolute value less than 1, but whose spectral norm is greater than 1.

Now let's do another example. Let $B = \begin{bmatrix} 0.8 & -0.9 \\ 0 & 0.9 \end{bmatrix}$. The eigenvalues of B are 0.8 and 0.9; both have absolute value less than 1. We compute the spectral norm of B using MATLAB:

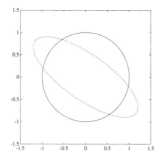

Figure 5.9. A plot of $\mathbf{z}$ (the unit circle) and $|B\mathbf{z}|$ (the ellipse) for all unit vectors $\mathbf{z}$. The spectral norm of B is greater than 1.

```
b =
      0.8000    -0.9000
           0     0.9000

>>norm(b)

ans =
      1.4145
```

Since $\|B\| > 1$, we know that $f(\mathbf{x}) = B\mathbf{x} + \mathbf{b}$ is *not* a contraction mapping. For example, let $\mathbf{x} = \begin{bmatrix} 1 \\ -1 \end{bmatrix}$ and $\mathbf{y} = \mathbf{0} = \begin{bmatrix} 0 \\ 0 \end{bmatrix}$. Now, $d(\mathbf{x}, \mathbf{y}) = |\mathbf{x}| = \sqrt{2} \approx 1.414$. However,

$$d(f(\mathbf{x}), f(\mathbf{y})) = \left\| \begin{bmatrix} 1.7 \\ -0.9 \end{bmatrix} \right\| \approx 1.9235,$$

so the distance between $\mathbf{x}$ and $\mathbf{0}$ *increases* after we apply f to each.

We can also see that $\|B\| > 1$ geometrically. We plot the unit circle (all $\mathbf{z}$ with $|\mathbf{z}| = 1$), and we also plot the points $B\mathbf{z}$; see Figure 5.9. Notice that the ellipse is not contained in the interior of the unit circle, but is a distance $\|B\| > 1$ from the origin at its farthermost point.

In summary, we can judge if an affine function $f(\mathbf{x}) = A\mathbf{x} + \mathbf{b}$ is contractive by computing the spectral norm of A. If $\|A\| < 1$, then f is contractive with contractivity $\|A\|$. Otherwise ($\|A\| \geq 1$), the function is not a contraction mapping.

5.3.5   Other metric spaces

What is distance? On the real line, the distance between numbers x and y is the absolute value of their difference. In the plane, the distance between two points (given their coordinates) can be computed using the Pythagorean theorem. We can generalize the idea of distance. We might want to talk about the distance between two functions or two compact sets in the plane. We want to extend the notion of distance.

Let $\mathcal{X}$ be a set of objects; in $\mathcal{X}$ we might have all points in the plane, or all continuous functions from the reals to the reals. A function d, defined for pairs of elements of $\mathcal{X}$, is called a *metric* or a *distance* provided the following three properties hold:

- For any $x, y \in \mathcal{X}, d(x, y) = d(y, x)$ is a nonnegative real number.

- For any $x, y \in \mathcal{X}$, we have $d(x, y) = 0$ if and only if $x = y$.

- For any $x, y, z \in \mathcal{X}$, we have $d(x, y) + d(y, z) \geq d(x, z)$.

Ordinary distance of points satisfies all three of these properties: The distance between two points is a nonnegative number which is zero just when the two points are the same. Ordinary distance also satisfies the third property (known as the *triangle inequality*), which states that the distance from x to z is never bigger than the distance from x to y plus the distance from y to z.

A set $\mathcal{X}$ which has a metric d is called a *metric space*. Now if f is a function from $\mathcal{X}$ to $\mathcal{X}$, we can ask, Is f a contraction mapping? This makes sense. We say that f is a contraction mapping provided there is a number s, with $0 < s < 1$, so that for any points $x, y \in \mathcal{X}$, we have $d(f(x), f(y)) \leq sd(x, y)$.

And now we can ask, Does the contraction mapping theorem hold for any metric space $(\mathcal{X}, d)$? Regrettably, the answer is no. We need one more technical condition: The metric space should be *complete*. I don't want to discuss what this means except in very vague terms. Very roughly, to be *complete*, sequences of points in $\mathcal{X}$ which you think ought to converge to a limit, in fact, do converge.

[In fact, we work with only one novel metric space: the set of nonempty compact sets. And we promise that this metric space is complete (we fully describe this metric space in the next section). The technical details of why that particular metric space is complete can be found in other books.]

Even though we haven't explained what *complete* means, we present our final version of the contraction mapping theorem.

Theorem (contraction mapping for complete metric spaces). Suppose $(\mathscr{X}, d)$ is a complete metric space. Let $f : \mathscr{X} \to \mathscr{X}$ be a contraction mapping. Then f has a unique, stable fixed point, $\tilde{x}$. Further, for any $a \in \mathscr{X}$ we have $f^k(a) \to \tilde{x}$ as $k \to \infty$.

This has been rather abstract. We now turn to considering a new metric space.

5.3.6   Compact sets and Hausdorff distance

Recall (page 238) that a compact set is a closed and bounded subset of $\mathbf{R}^n$. Let $\mathscr{H}^n$ denote the set of all nonempty compact sets in $\mathbf{R}^n$. Thus an element $X \in \mathscr{H}^n$ is an *entire set* of points in $\mathbf{R}^n$. For example, the unit circle, $\{(x, y) : x^2 + y^2 = 1\}$, is an element of $\mathscr{H}^2$.

We want to talk about the distance between two compact sets. In order for this distance to be a *metric* we need to satisfy the three properties listed above: (1) the distance is a nonnegative real number which (2) is zero exactly when the objects are the same, and (3) the distance function must obey the triangle inequality.

The definition of distance for $\mathscr{H}^n$ takes a bit of work to develop. We already know what the distance between points x and y is, and we write this distance as $d(x, y)$. Now we work to develop a notion of distance between nonempty compact sets $d(A, B)$ with $A, B \in \mathscr{H}^n$.

We proceed in steps.

Step 1: Distance between a point and a compact set.

The first step is to define the number $d(x, A)$, where x is a point and A is a nonempty compact set in $\mathbf{R}^n$. We define $d(x, A)$ to be the distance between x and a point in A which is closest to x, that is,

$$d(x, A) = \min\{d(x, y) : y \in A\}.$$

This is illustrated in Figure 5.10. The distance between x and A is the distance from x to a nearest point in A.

Notice that if $x \in A$, then $d(x, A) = d(x, x) = 0$, but if $x \notin A$, then $d(x, A) > 0$.

Step 2: Asymmetrical distance from one compact set to another.

The next step is to build an asymmetrical distance from a compact set A to a compact set B. We define $\vec{d}(A, B)$ to be the *largest* distance from a point in A to the set B, i.e.,

$$\vec{d}(A, B) = \max\{d(x, B) : x \in A\}.$$

(margin notes)

The distance between compact sets: the Hausdorff metric

$d(\text{point,set})$

$\vec{d}(\text{set,set})$

Figure 5.10. The distance between a point x and a compact set A is $d(x, A) = \min\{d(x, y) : y \in A\}$.

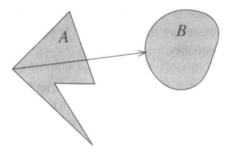

Figure 5.11. The asymmetrical distance $\vec{d}(A, B)$ from compact set A to compact set B.

This idea is illustrated in Figure 5.11. The function $\vec{d}$ is, unfortunately, not a metric because it is not symmetric: We can have $\vec{d}(A, B) \neq \vec{d}(B, A)$. For example, if A and B are distinct elements of $\mathcal{H}^n$ with $A \subset B$, then $\vec{d}(A, B) = 0$ (the point in A farthest from B is *still* in B), but $\vec{d}(B, A) > 0$ (there are points in B that aren't in A). This example ($A \subset B$) also shows that we can have $\vec{d}(A, B) = 0$ without $A = B$ (violating the second part of the definition of metric). So $\vec{d}$ is not a metric, but it's getting close. We now show that $\vec{d}$ satisfies the triangle inequality: $\vec{d}(A, B) + \vec{d}(B, C) \geq \vec{d}(A, B)$.

Why $\vec{d}$ satisfies the triangle inequality.

To see why, we select $A, B, C \in \mathcal{H}^n$. To compute $\vec{d}(A, C)$, we find a point $a \in A$ maximally distant from C and let x be a point in C to which a is closest. Thus $\vec{d}(A, C) = d(a, x)$; see Figure 5.12. Next, we find a point

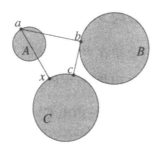

Figure 5.12. Understanding why $\vec{d}$ satisfies the triangle inequality.

$b \in B$ closest to a, so $d(a, B) = d(a, b)$. Finally, we find a point $c \in C$ which is closest to b, so $d(b, C) = d(b, c)$. We also know the following facts:

(1) $\vec{d}(A, B) \geq d(a, B) = d(a, b)$ because $\vec{d}(A, B)$ is the greatest distance of a point in A to B.

(2) Likewise, $\vec{d}(B, C) \geq d(b, C) = d(b, c)$ because $\vec{d}(B, C)$ is the greatest distance of a point in B to C.

(3) $d(a, c) \geq d(a, x)$ because x is a point of C closest to a.

Armed with these facts, we now reason

$$
\begin{aligned}
\vec{d}(A, B) + \vec{d}(B, C) \quad &\geq \quad d(a, b) + d(b, c) \qquad && \text{by (1) and (2),} \\
&\geq \quad d(a, c) && \text{by the triangle inequality,} \\
&\geq \quad d(a, x) && \text{by (3),} \\
&= \quad \vec{d}(A, C) && \text{by construction.}
\end{aligned}
$$

Thus $\vec{d}(A, B) + \vec{d}(B, C) \geq \vec{d}(A, C)$ as promised.

Step 3: The Hausdorff metric on $\mathscr{H}^n$. $d(\text{set,set})$

The last step in defining a metric on $\mathscr{H}^n$ is to fix the asymmetry in $\vec{d}$. We do this by defining

$$
d(A, B) = \max\{\vec{d}(A, B), \vec{d}(B, A)\},
$$

that is, $d(A, B)$ is the larger of the numbers $\vec{d}(A, B)$ and $\vec{d}(B, A)$. It is clear that $d(A, B) = d(B, A)$ is a nonnegative real number. We have

$d(A, A) = 0$, since $\vec{d}(A, A) = 0$. Also if $A \neq B$, then one of $\vec{d}(A, B)$ or $\vec{d}(B, A)$ is positive, so $d(A, B) > 0$. Lastly,

$$d(A, B) + d(B, C) \geq \vec{d}(A, B) + \vec{d}(B, C) \geq \vec{d}(A, C), \quad \text{and}$$

$$d(A, B) + d(B, C) \geq \vec{d}(B, A) + \vec{d}(C, B) \geq \vec{d}(C, A),$$

and therefore $d(A, B) + d(B, C) \geq d(A, C)$ verifying the triangle inequality. Thus d *is* a metric, called the *Hausdorff metric* for $\mathscr{H}^n$.

Now $(\mathscr{H}^n, d)$ is a complete metric space (we haven't explained or proved the "complete" part) and therefore the contraction mapping theorem applies: If $f: \mathscr{H}^n \to \mathscr{H}^n$ is a contraction mapping, then f must have a unique, stable fixed point, $X \in \mathscr{H}^n$, and for any $A \in \mathscr{H}^n$ we have $f^k(A) \to X$ as $k \to \infty$.

The stable fixed "point" of f is actually an entire compact set, also called the attractor of f.

It is important to understand that the fixed "point" of the function f is not a point in the usual sense. We are using the word "point" in a rather broad sense. In this setting, it simply means a member of $\mathscr{H}^n$, i.e., a compact set. Thus the fixed point of $f: \mathscr{H}^n \to \mathscr{H}^n$ need not be a single dot—it can be an intricate compact set.

Indeed, this is exactly how we build fractals. In the next section we create fractals by first constructing contraction maps and then finding their fixed "points."

Problems for §5.3

◆1.   Which of the following are contraction maps? For those that are, find their contractivities (a number $s < 1$ so that $|f(x) - f(y)| \leq s|x - y|$ for all real x and y).

    (a) $f(x) = 3x - 4$.

    (b) $f(x) = -\frac{x}{2} + 10$.

    (c) $f(x) = \frac{1}{3} - x$.

    (d) $f(x) = \cos x$.

    (e) $f(x) = \frac{1}{2} \sin x$.

    (f) $f(x) = \exp\{-x^2\}$.

◆2.* Let $f: \mathbf{R} \to \mathbf{R}$ be a contraction map. Prove that f must be continuous.

To do this, suppose the sequence $x_1, x_2, \ldots$ converges to x, i.e., the distance between the x_j's and x goes to 0. Show that $f(x_1), f(x_2), \ldots$ converges to $f(x)$.

◆3.   Let $f: \mathbf{R} \to \mathbf{R}$ be a contraction map. Show that f need not be differentiable. To do this, find an example of a function which (1) is a contraction map and (2) is not differentiable at all values x.

◆4.   Suppose $f: \mathbf{R} \to \mathbf{R}$ is differentiable, and let $0 < s < 1$. Show that if $|f'(x)| \leq s$ for all x, then f is a contraction mapping.
       [Hint: Use the mean value theorem.]

◆5.   Let's weaken the assumptions of the previous problem. Suppose $f: \mathbf{R} \to \mathbf{R}$ is differentiable, and $|f'(x)| < 1$ for all x. Show that f need not be a contraction map and that f need not have a fixed point.
       [Hint: Let $f(x) = x - \frac{1}{2} \log(e^x + 1)$.]

◆6.   Let J be the unit square and K be the unit disk, i.e.,

$$J = \{(x, y) : 0 \leq x, y \leq 1\}, \text{ and}$$

$$K = \{(x, y) : x^2 + y^2 \leq 1\}.$$

       Find $\vec{d}(J, K), \vec{d}(K, J)$, and $d(J, K)$.

◆7.   Let J be the unit *circle* and K be the unit *disk*, i.e.,

$$J = \{(x, y) : x^2 + y^2 = 1\}, \text{ and}$$

$$K = \{(x, y) : x^2 + y^2 \leq 1\}.$$

       Find $\vec{d}(J, K), \vec{d}(K, J)$, and $d(J, K)$.

◆8.*   The purpose of this problem is to give an example of a metric space in which the contraction mapping theorem doesn't hold. Thus this example is not a *complete* metric space.

       Recall that $\mathbf{Q}$ denotes the set of *rational numbers*, i.e., the set of all numbers of the form p/q, where p and q are integers and $q \neq 0$. Let $\mathscr{X} = \mathbf{Q} \cap [1, 2]$, i.e., $\mathscr{X}$ is the set of rational numbers between 1 and 2 inclusive.

       We can consider $\mathscr{X}$ to be a metric space by endowing it with the usual distance function, i.e., for $x, y \in \mathscr{X}$ we put $d(x, y) = |x - y|$.

       Now, we define $f: \mathscr{X} \to \mathscr{X}$ by

$$f(x) = x - \frac{x^2 - 2}{2x}.$$

       You should check that if $x \in \mathscr{X}$, then $f(x) \in \mathscr{X}$ as well. This means checking that if x is rational and in $[1, 2]$, then $f(x)$ is also rational and in $[1, 2]$.

       Please show (1) that f is a contraction mapping on $\mathscr{X}$, but (2) f does not have a fixed point.

       [Comments and hints: (1) Forget for a moment that f is a function defined on rational numbers and think about $f'(x)$ in the interval $[1, 2]$; how big can $|f'(x)|$ be? (2) If you think of f as a function from $[1, 2]$ to $[1, 2]$ (the full interval), then f *does* have a fixed point (what is it?); nevertheless, f—defined just on $\mathscr{X}$—does not have a fixed point.]

5.4   Iterated function systems

In this section we introduce the notion of *iterated function systems*, or IFSs, for short. The IFSs are, in fact, functions defined on $\mathscr{H}^n$. We develop methods to check if they are contraction maps and then create fractals as the fixed points of these maps. Our raw ingredients are affine functions.

5.4.1   From point maps to set maps

Recall that $\mathscr{H}^n$ stands for the set of all nonempty compact subsets of $\mathbf{R}^n$.

f(set)

There is a natural way to convert a function defined on $\mathbf{R}^n$ to a function defined on $\mathscr{H}^n$. Let $f : \mathbf{R}^n \to \mathbf{R}^n$ be a continuous function. If A is a nonempty compact set in $\mathbf{R}^n$ (i.e., $A \in \mathscr{H}^n$), then we define

$$f(A) = \{ f(a) : a \in A \}.$$

In other words, $f(A)$ is the set of all points of the form $f(a)$, where a ranges over the possible elements of A. For example, if $f(x) = x^2$ and $A = [-1, 3]$, then $f(A) = [0, 9]$. Why? As a ranges over the compact interval $[-1, 3]$, then $f(a) = a^2$ takes on all values from 0 (at the lowest) to 9 (at the greatest).

Here is another example. Let $f : \mathbf{R}^2 \to \mathbf{R}^2$ be defined by $f(\mathbf{x}) = A\mathbf{x}$, where A is a 2×2 matrix. Let C be the unit circle, a compact set and therefore an element of $\mathscr{H}^2$; then $f(C)$ is an ellipse (unless A is noninvertible).

Another example: A singleton set $\{a\}$ is a compact set. What is $f(\{a\})$? There is only one value in $\{a\}$, so, by definition, $f(\{a\})$ is the set $\{f(a)\}$.

It is a theorem of topology that if f is continuous and A is a compact set, then $f(A)$ is also compact.[9]

The double use of the letter f (as a function defined on points and as a function defined on sets) is undesirable but something we can tolerate. The crux is, if we have a continuous function defined from points to points ($f : \mathbf{R}^n \to \mathbf{R}^n$), it extends to a function defined on compact sets ($f : \mathscr{H}^n \to \mathscr{H}^n$).

Contraction implies contraction

Suppose $f : \mathbf{R}^n \to \mathbf{R}^n$ is a contraction mapping (with contractivity s). We know we can also view f as a function defined on $\mathscr{H}^n$, which is a metric

---

[9] This theorem is not, in fact, hard to prove. It is simply beyond the scope of this text.

space. Thus it makes sense to ask, Is f a contraction mapping on $\mathscr{H}^n$ as well? Happily, the answer is yes.

If $f: \mathbf{R}^n \to \mathbf{R}^n$ is a contraction map with contractivity s, then $f: \mathscr{H}^n \to \mathscr{H}^n$ is also a contraction map with contractivity s.

Let's prove this. We know that $d(f(a), f(b)) \le sd(a, b)$ for any a, b. We work to show that $d(f(A), f(B)) \le sd(A, B)$ for any $A, B \in \mathscr{H}^n$.

Step 1: $d(f(a), f(B)) \le sd(a, B)$ for any point $a \in \mathbf{R}^n$ and any set $B \in \mathscr{H}^n$.

We compute

$$
\begin{aligned}
d(f(a), f(B)) &= \min\{d(f(a), y) : y \in f(B)\} && \text{def'n } d(\text{point,set}) \\
&= \min\{d(f(a), f(b)) : b \in B\} && \text{def'n } f(B) \\
&\le \min\{sd(a, b) : b \in B\} && f \text{ is a contraction} \\
&= s\min\{d(a, b) : b \in B\} && \text{factor out } s \\
&= sd(a, B) && \text{def'n } d(\text{point,set}).
\end{aligned}
$$

Step 2: $\vec{d}(f(A), f(B)) \le s\vec{d}(A, B)$ for any two sets $A, B \in \mathscr{H}^n$.

We compute

$$
\begin{aligned}
\vec{d}(f(A), f(B)) &= \max\{d(x, f(B)) : x \in f(A)\} && \text{def'n } \vec{d} \\
&= \max\{d(f(a), f(B)) : a \in A\} && \text{def'n } f(A) \\
&\le \max\{sd(a, B) : a \in A\} && \text{by step 1} \\
&= s\max\{d(a, B) : a \in A\} && \text{factor out } s \\
&= s\vec{d}(A, B) && \text{def'n } \vec{d}.
\end{aligned}
$$

Step 3: $d(f(A), f(B)) \le sd(A, B)$ for any two sets $A, B \in \mathscr{H}^n$.

We compute

$$
\begin{aligned}
d(f(A), f(B)) &= \max\left\{\vec{d}(f(A), f(B)), \vec{d}(f(B), f(A))\right\} && \text{def'n } d(\text{set,set}) \\
&\le \max\left\{s\vec{d}(A, B), s\vec{d}(B, A)\right\} && \text{by step 3} \\
&= s\max\left\{\vec{d}(A, B), \vec{d}(B, A)\right\} && \text{factor out } s \\
&= sd(A, B) && \text{def'n } d(\text{set,set}).
\end{aligned}
$$

Thus we have shown that if $f: \mathbf{R}^n \to \mathbf{R}^n$ is a contraction map (as a pointwise function), then $f: \mathscr{H}^n \to \mathscr{H}^n$ is also a contraction map (as a setwise function).

Summary

We summarize the main points of this section:

- If f is a (pointwise) function from $\mathbf{R}^n$ to $\mathbf{R}^n$, then f can also be considered a (setwise) function from $\mathscr{H}^n$ to $\mathscr{H}^n$.

- If f is contractive as a pointwise function, then f is also contractive as a setwise mapping.

- If f is contractive, then f has a unique stable fixed point.

  - In the pointwise setting ($f: \mathbf{R}^n \to \mathbf{R}^n$) this is a vector $\tilde{\mathbf{x}}$ for which $f(\tilde{\mathbf{x}}) = \tilde{\mathbf{x}}$.

  - In the setwise setting ($f: \mathscr{H}^n \to \mathscr{H}^n$) this is a compact set $X \in \mathscr{H}^n$ for which $f(X) = X$.

We now know that if f is a *pointwise* contractive map, then f is also a *setwise* contractive map. Thus, by the contraction mapping theorem, f has a *setwise* fixed "point." What is the set for which $f(A) = A$? The answer is anticlimactic. We know that as a pointwise map, a contractive map f has a unique fixed point $\tilde{x}$, i.e., $f(\tilde{x}) = \tilde{x}$. Thus $f(\{\tilde{x}\}) = \{f(\tilde{x})\} = \{\tilde{x}\}$. Thus the singleton set $\{\tilde{x}\}$ is the setwise fixed "point" of f. Since the fixed point of a contraction map is unique, this can be the only one!

We said that we create our fractals as fixed points of contractive set maps. A one-point set is not an exciting fractal. What we need to do next is find an additional source of contractive maps defined on $\mathscr{H}^n$.

5.4.2   The union of set maps

Suppose f and g are both set maps, i.e., $f: \mathscr{H}^n \to \mathscr{H}^n$ and $g: \mathscr{H}^n \to \mathscr{H}^n$. We define a new function defined on $\mathscr{H}^n$ by taking the *union* of f and g, which we denote by $f \cup g$. The definition is simple. Let

$$(f \cup g)(A) = f(A) \cup g(A).$$

In words, the function $F = f \cup g$ is computed on a set A by taking the union of $f(A)$ and $g(A)$.

For example, let $f(x) = x^2$ and $g(x) = 2x + 5$. We can also think of f and g as functions on sets. Thus $f([-1, 1]) = [0, 1]$ and $g([-1, 1]) = [3, 7]$. Thus $(f \cup g)([-1, 1]) = [0, 1] \cup [3, 7]$.

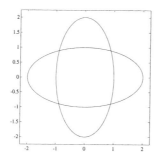

Figure 5.13. The result of $(f \cup g)(C)$ where f and g are simple linear transformations and C is the unit circle.

Another example: Suppose

$$f\begin{bmatrix} x \\ y \end{bmatrix} = \begin{bmatrix} 2 & 0 \\ 0 & 1 \end{bmatrix}\begin{bmatrix} x \\ y \end{bmatrix} \quad \text{and} \quad g\begin{bmatrix} x \\ y \end{bmatrix} = \begin{bmatrix} 1 & 0 \\ 0 & 2 \end{bmatrix}\begin{bmatrix} x \\ y \end{bmatrix}.$$

Let us compute $(f \cup g)(C)$, where C is the unit circle. Now f stretches the x-axis by a factor of 2, stretching C into a horizontal ellipse, while g stretches the y-axis by a factor of 2, transforming C into a vertical ellipse. The result of $(f \cup g)(C)$ is shown in Figure 5.13.

We want to use the idea of union of functions to build contraction maps. The union of contraction The question is, If f and g are contraction maps, is $f \cup g$ a contraction maps is a contraction map. map as well? The answer is yes. Here's the result.

If $f, g: \mathscr{H}^n \to \mathscr{H}^n$ are contraction maps with contractivities s and t respectively, then $F: \mathscr{H}^n \to \mathscr{H}^n$ defined by $F = f \cup g$ is also a contraction map with contractivity $\max\{s, t\}$.

To prove this, we begin by proving an important intermediate step. We claim that for any sets $A, B, C, D \in \mathscr{H}^n$ we have

$$d(A \cup B, C \cup D) \leq \max\{d(A, C), d(B, D)\}. \tag{5.2}$$

The verification of inequality (5.2) goes in three steps.

- **Step 1**: $\vec{d}(X \cup Y, Z) = \max\left\{\vec{d}(X, Z), \vec{d}(Y, Z)\right\}$.

  To see why this is true, we simply choose $x \in X$ so that $d(x, Z)$ is greatest, and we choose $y \in Y$ so that $d(y, Z)$ is greatest. Now we ask, What point in $X \cup Y$ is farthest from Z? It must be either x or y, thus

$$\vec{d}(X \cup Y, Z) = \max\{d(x, Z), d(y, Z)\} = \max\left\{\vec{d}(X, Z), \vec{d}(Y, Z)\right\}.$$

- **Step 2**: $\vec{d}(X, Y \cup Z) \leq \min\left\{\vec{d}(X, Y), \vec{d}(X, Z)\right\}$.

  Note that for *any* $x \in X$, we have $d(x, Y \cup Z) \leq d(x, Y)$ and $d(x, Y \cup Z) \leq d(x, Z)$. Thus if x is a point of X farthest from $Y \cup Z$, we have

  $$\begin{aligned}
  \vec{d}(X, Y \cup Z) &= d(x, Y \cup Z) \\
  &= \min\{d(x, Y), d(x, Z)\} \\
  &\leq \min\left\{\vec{d}(X, Y), \vec{d}(Y, Z)\right\}
  \end{aligned}$$

  verifying step 2.

- **Step 3**. We use steps 1 and 2 to complete the proof of inequality (5.2) as follows:

  $$\begin{aligned}
  d(A \cup B, C \cup D) &= \max\left\{\vec{d}(A \cup B, C \cup D), \vec{d}(C \cup D, A \cup B)\right\} \\
  &= \max\left\{\vec{d}(A, C \cup D), \vec{d}(B, C \cup D), \right. \\
  &\qquad\qquad \left. \vec{d}(C, A \cup B), \vec{d}(D, A \cup B)\right\} \\
  &\leq \max\left\{\vec{d}(A, C), \vec{d}(B, D), \vec{d}(C, A), \vec{d}(D, B)\right\} \\
  &= \max\{d(A, C), d(B, D)\},
  \end{aligned}$$

  verifying inequality (5.2).

We now complete the proof that $F = f \cup g$ is a contraction mapping. We let A and B be nonempty compact sets, then we use inequality (5.2) to compute

$$\begin{aligned}
d(F(A), F(B)) &= d(f(A) \cup g(A), f(B) \cup g(B)) \\
&\leq \max\{d(f(A), f(B)), d(g(A), g(B))\} \\
&\leq \max\{sd(A, B), td(A, B)\} \\
&= \max\{s, t\}d(A, B).
\end{aligned}$$

Thus F is a contraction map with contractivity $\max\{s, t\} < 1$.

Summary

In this section we introduced the idea of the union of functions. If f and g are functions from $\mathscr{H}^n$ to $\mathscr{H}^n$, then $(f \cup g)$ is also a function from $\mathscr{H}^n$ to $\mathscr{H}^n$ defined by $(f \cup g)(A) = f(A) \cup g(A)$.

Furthermore, if f and g are contraction maps, then so is $f \cup g$.

5.4.3 Examples revisited

Cantor's set, again

Let's see how the idea of the union of functions gives rise to a world of fractals. Our first goal is to re-create the Cantor set.

Let

$$f(x) = \frac{1}{3}x \qquad \text{and} \qquad g(x) = \frac{1}{3}x + \frac{2}{3}.$$

Both f and g are affine functions with constant derivative $\frac{1}{3}$ (which has absolute value less than 1). Hence both are contractive mappings: both as pointwise mappings and as functions on $\mathscr{H}^1$. The fixed point of f is 0 and the fixed point of g is 1. As setwise mappings, the fixed point of f is $\{0\}$ and the fixed point of g is $\{1\}$.

We now ask, What is the fixed "point" of the setwise map $F = f \cup g$? Since f and g are contraction maps with contractivity $\frac{1}{3}$, so is F. Thus F has a unique fixed point which satisfies $F(X) = f(X) \cup g(X) = X$. What is this fixed point? A reasonable guess might be the interval $[0, 1]$. This is incorrect, and let's see why. We have $f([0, 1]) = [0, \frac{1}{3}]$, and $g(0, 1) = [\frac{2}{3}, 1]$, so

Remember: The fixed "point" of F is not a *point* at all; rather, it is a compact set S for which $F(S) = S$.

$$F([0, 1]) = \left[0, \frac{1}{3}\right] \cup \left[\frac{2}{3}, 1\right].$$

Since $F([0, 1]) \neq [0, 1]$ we know that $[0, 1]$ is not the unique fixed point of F. However, we see something interesting: the first step in the construction of Cantor's set. What happens if we apply F again? Let's work it out:

$$f\left(\left[0, \frac{1}{3}\right]\right) = \left[0, \frac{1}{9}\right],$$

$$g\left(\left[0, \frac{1}{3}\right]\right) = \left[\frac{2}{3}, \frac{7}{9}\right],$$

$$f\left(\left[\frac{2}{3}, 1\right]\right) = \left[\frac{2}{9}, \frac{1}{3}\right],$$

$$g\left(\left[\frac{2}{3}, 1\right]\right) = \left[\frac{8}{9}, 1\right],$$

so

$$F^2([0, 1]) = \left[0, \frac{1}{9}\right] \cup \left[\frac{2}{9}, \frac{1}{3}\right] \cup \left[\frac{2}{3}, \frac{7}{9}\right] \cup \left[\frac{8}{9}, 1\right],$$

which is the *second* step in the construction of Cantor's set. Go ahead. Check out that $F^3([0, 1])$ gives just what you want: the third step in the construction. It would seem that $F^k([0, 1])$ is converging to the Cantor set. That's exactly right; let's see why.

Let C be the Cantor set. We ask, What is $F(C)$? Now, $f(C)$ shrinks C by a factor of 3, and therefore $f(C)$ is the portion of C between 0 and $\frac{1}{3}$. Similarly, $g(C)$ shrinks C by factor of 3 and then slides the result to the right a distance of $\frac{2}{3}$. Thus $g(C)$ is the portion of C between $\frac{2}{3}$ and 1. Together, $f(C) \cup g(C)$ is simply C. In other words, $F(C) = C$. Aha! Cantor's set is a fixed point of F. Indeed, by the contraction mapping theorem, it is the unique, stable fixed point of F, and if we start with any nonempty compact set $A \in \mathcal{H}^1$, we have $F^k(A) \to C$ as $k \to \infty$.

How did we find f and g?

At first blush the choice of f and g might seem mysterious. You might wonder, How did you ever think to pick those particular functions? Actually, the choice was rather simple. When we look at Cantor's set, we see two third-sized copies of Cantor's set sitting inside. So we choose f to map the whole of Cantor's set onto the left portion and we choose g to map the whole of Cantor's set onto the right. Because we were shrinking, we knew that f and g would be contraction maps. We also rigged it so that $C = f(C) \cup g(C)$. Let's see if we can recreate Sierpiński's triangle by the same method.

Sierpiński's triangle, again

We can build a Sierpiński triangle S by iteratively biting out the middle of a triangle. For this example, we use the isosceles triangle with corners at $(0, 0)$, $(1, 0)$, and $(\frac{1}{2}, 1)$. See Figure 5.14, in which we show the first step in the construction. When the construction is complete, we know we have three half-sized copies of the original Sierpiński triangle in regions A, B, and C. What we need to do is to find three contraction maps for which $f_1(S) = A$, $f_2(S) = B$, and $f_3(S) = C$. Notice that the basic steps we need are rescaling and translation. Copy A is the simplest, so we begin with it. If we simply shrink the x- and y-coordinates of all the points in S by a factor of 2, we will get exactly A. So let us put

$$f_1 \begin{bmatrix} x \\ y \end{bmatrix} = \begin{bmatrix} \frac{1}{2} & 0 \\ 0 & \frac{1}{2} \end{bmatrix} \begin{bmatrix} x \\ y \end{bmatrix}.$$

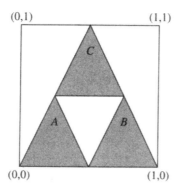

Figure 5.14. Sierpiński's triangle contains three half-sized copies of itself: A, B, and C.

Next we want to make copy B. Notice that B is the same as A but just shifted to the right a distance $\frac{1}{2}$. To do this, we let

$$f_2\begin{bmatrix} x \\ y \end{bmatrix} = \begin{bmatrix} \frac{1}{2} & 0 \\ 0 & \frac{1}{2} \end{bmatrix}\begin{bmatrix} x \\ y \end{bmatrix} + \begin{bmatrix} \frac{1}{2} \\ 0 \end{bmatrix}.$$

Copy C is the trickiest, but not really any harder. We want to take A and move it to the right a distance of $\frac{1}{4}$ and up a distance of $\frac{1}{2}$. Thus we let

$$f_3\begin{bmatrix} x \\ y \end{bmatrix} = \begin{bmatrix} \frac{1}{2} & 0 \\ 0 & \frac{1}{2} \end{bmatrix}\begin{bmatrix} x \\ y \end{bmatrix} + \begin{bmatrix} \frac{1}{4} \\ \frac{1}{2} \end{bmatrix}.$$

Notice that f_1, f_2, and f_3 are all contraction maps with contractivity $\frac{1}{2}$. Thus $F = f_1 \cup f_2 \cup f_3$ is also a contraction map with contractivity $\frac{1}{2}$. Thus F has a unique fixed point in $\mathscr{H}^2$, which we hope is our Sierpiński triangle S. Let's see if that's correct. By construction, we know

$$f_1(S) = A, \qquad f_2(S) = B, \qquad \text{and} \qquad f_3(S) = C;$$

therefore

$$F(S) = f_1(S) \cup f_2(S) \cup f_3(S) = A \cup B \cup C = S.$$

Thus S is a fixed point of F and, by the contraction mapping theorem, it's the *only* fixed point of F. Further, it's a stable fixed point. If we begin

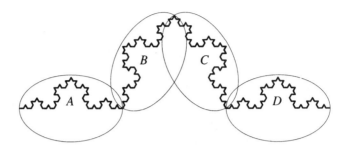

Figure 5.15. The top of the Koch snowflake. Notice that there are four sections which are miniatures of the original.

with *any* nonempty compact set X and iterate F, i.e., compute $F^k(X)$, we converge to the Sierpiński triangle S.

(Now you may be wondering about how we compute—on a computer— the set $F^k(X)$. We show how to do these computations in the next section.)

Koch's snowflake, again

There is no subset of the Koch snowflake K (Figure 5.5 on page 245) which is a miniature copy of K. Rather, we look at just the top portion of the snowflake. The top portion is what we get when we iterate the basic Koch step on a horizontal line segment (with the tent portion sticking upward); see Figure 5.15. Let us call the top of Koch's snowflake T. Observe that within T there are four third-sized versions of T, which are labeled A through D in the figure. Let us construct four contraction maps, f_1 through f_4, for which $f_1(T) = A$, $f_2(T) = B$, $f_3(T) = C$, and $f_4(T) = D$. Let's assume the origin $(0, 0)$ is situated at the lower left corner of the diagram and the far right of the diagram is located at $(1, 0)$.

The first contraction, f_1, is the simplest. We want to contract the whole set T by a factor of 3. So we simply let

$$f_1 \begin{bmatrix} x \\ y \end{bmatrix} = \begin{bmatrix} \frac{1}{3} & 0 \\ 0 & \frac{1}{3} \end{bmatrix} \begin{bmatrix} x \\ y \end{bmatrix}.$$

Thus $f_1(T) = A$.

Let's skip to D since that's the next easiest. We want to contract by a factor of 3 and then slide the result to the right a distance of $\frac{2}{3}$. The following works:

$$f_4 \begin{bmatrix} x \\ y \end{bmatrix} = \begin{bmatrix} \frac{1}{3} & 0 \\ 0 & \frac{1}{3} \end{bmatrix} \begin{bmatrix} x \\ y \end{bmatrix} + \begin{bmatrix} \frac{2}{3} \\ 0 \end{bmatrix}.$$

This gives $f_4(T) = D$.

Now let's tackle B. We want to first shrink by $\frac{1}{3}$, then rotate $60° = \frac{\pi}{3}$, and then slide a distance $\frac{1}{3}$ to the right. To do this, we let

$$f_2 \begin{bmatrix} x \\ y \end{bmatrix} = \begin{bmatrix} \cos\frac{\pi}{3} & -\sin\frac{\pi}{3} \\ \sin\frac{\pi}{3} & \cos\frac{\pi}{3} \end{bmatrix} \begin{bmatrix} \frac{1}{3} & 0 \\ 0 & \frac{1}{3} \end{bmatrix} \begin{bmatrix} x \\ y \end{bmatrix} + \begin{bmatrix} \frac{1}{3} \\ 0 \end{bmatrix}.$$

Let's take this apart piece by piece. First we multiply $\begin{bmatrix} x \\ y \end{bmatrix}$ by $\begin{bmatrix} \frac{1}{3} & 0 \\ 0 & \frac{1}{3} \end{bmatrix}$; this shrinks the vector $\begin{bmatrix} x \\ y \end{bmatrix}$ by a factor of 3. Next, we multiply by the matrix $\begin{bmatrix} \cos\frac{\pi}{3} & -\sin\frac{\pi}{3} \\ \sin\frac{\pi}{3} & \cos\frac{\pi}{3} \end{bmatrix}$; this is a rotation matrix corresponding to a rotation through an angle of $\frac{\pi}{3}$. Thus the shrunken vectors are rotated $60°$ counterclockwise. Finally, we add $\begin{bmatrix} \frac{1}{3} \\ 0 \end{bmatrix}$ to shift to the right a distance $\frac{1}{3}$. Thus $f_2(T) = B$.

Finally, we work out f_3, the most complicated of the lot. To make the C section from the original, we want to (1) shrink to $\frac{1}{3}$ size, (2) rotate through $-60° = -\frac{\pi}{3}$, (3) slide to the right a distance $\frac{1}{2}$, and (4) slide upward a distance $\sqrt{3}/6$ (that's the height of an equilateral triangle of side length $\frac{1}{3}$.) To do this, we let

$$f_3 \begin{bmatrix} x \\ y \end{bmatrix} = \begin{bmatrix} \cos(-\frac{\pi}{3}) & -\sin(-\frac{\pi}{3}) \\ \sin(-\frac{\pi}{3}) & \cos(-\frac{\pi}{3}) \end{bmatrix} \begin{bmatrix} \frac{1}{3} & 0 \\ 0 & \frac{1}{3} \end{bmatrix} \begin{bmatrix} x \\ y \end{bmatrix} + \begin{bmatrix} \frac{1}{2} \\ \frac{\sqrt{3}}{6} \end{bmatrix}.$$

Although it is a bit messy, it does give $f_3(T) = C$.

Summarizing, we have created the four affine transformations f_1 through f_4 which are contractive maps (with contractivity $\frac{1}{3}$). Thus $F = f_1 \cup f_2 \cup f_3 \cup f_4$ is also a contractive map on $\mathscr{H}^2$. We know that

$$F(T) = f_1(T) \cup f_2(T) \cup f_3(T) \cup f_4(T) = A \cup B \cup C \cup D = T;$$

therefore T is the unique, stable fixed point of F.

Summary

In each of the preceding examples, we exploited the self-similar nature of fractals to find functions $f_1, f_2, \ldots$ so that the fractal is the unique stable fixed point of the setwise map $F = f_1 \cup f_2 \cup \cdots$. In particular, since the fractal contains miniature copies of itself, we devise $f_1, f_2, \ldots$ to map the entire fractal onto all the smaller copies it contains of itself. Because the embedded copies are smaller than the whole, we know that the f_i's are contraction maps. Because we cover all the copies, we know that the fractal is the fixed point of F. We call the fractal the *attractor* of the IFS.

5.4.4   IFSs defined

We have seen that a single pointwise contraction map f, when considered setwise, does not lead to interesting fractals. However, by combining two or more by the union operation, $f \cup g$, we create new setwise contraction maps with interesting fixed points: fractals!

We give a formal name to this idea. Suppose $f_1, \ldots, f_k$ are contraction maps defined on $\mathbf{R}^n$ (and therefore on $\mathscr{H}^n$ as well). We call the union of these functions, $F = f_1 \cup f_2 \cup \cdots \cup f_k$, an *iterated function system* or IFS for short.

For the most part we are interested in working in the plane, $\mathbf{R}^2$. The reasons are both practical and aesthetic. The practical rationale is that fractals in $\mathbf{R}^3$ are hard to draw, and the aesthetic rationale is that fractals in $\mathbf{R}^1$ are dull. However, in $\mathbf{R}^2$ we are *able* to produce figures that are *visually appealing*.

In the previous section we gave examples of iterated function systems which give rise to Cantor's set, Sierpiński's triangle, and the top of the Koch snowflake. In each case the individual functions making up the IFSs are affine functions: $f(x) = ax + b$ for Cantor's set, and $f(\mathbf{x}) = A\mathbf{x} + \mathbf{b}$ for the others.

For the fractals we draw in $\mathbf{R}^2$, we use affine functions, i.e., functions of the form

$$g\begin{bmatrix} x \\ y \end{bmatrix} = \begin{bmatrix} a & b \\ c & d \end{bmatrix}\begin{bmatrix} x \\ y \end{bmatrix} + \begin{bmatrix} e \\ f \end{bmatrix}.$$

To specify this function g, we need to know just the six numbers a through f. For example, the third function for Sierpiński's triangle is

$$f_3\begin{bmatrix} x \\ y \end{bmatrix} = \begin{bmatrix} \frac{1}{2} & 0 \\ 0 & \frac{1}{2} \end{bmatrix}\begin{bmatrix} x \\ y \end{bmatrix} + \begin{bmatrix} \frac{1}{4} \\ \frac{1}{2} \end{bmatrix}$$

Function	a	b	c	d	e	f
#1	0.5	0	0	0.5	0	0
#2	0.5	0	0	0.5	0.5	0
#3	0.5	0	0	0.5	0.25	0.5

Table 5.1. IFS for Sierpiński's triangle

Function	a	b	c	d	e	f
#1	0.3333	0	0	0.3333	0	0
#2	0.1667	−0.2887	0.2887	0.1667	0.3333	0
#3	0.1667	0.2887	−0.2887	0.1667	0.5	0.2887
#4	0.3333	0	0	0.3333	0.6667	0

Table 5.2. IFS for top of Koch's snowflake.

To specify this function, we can list just the six numbers $(a, b, c, d, e, f) =$ $(0.5, 0, 0, 0.5, 0.25, 0.5)$. The full recipe for Sierpiński's triangle consists of listing the three lists of six numbers. We can package these recipes (called IFS codes) into tables such as Tables 5.1 and 5.2.

5.4.5   Working backward

Given a fractal K, can we find an IFS for which K is the attractor, i.e., can we find pointwise contraction maps $f_1, \ldots, f_k$ so that K is the stable fixed point of the union $F = f_1 \cup \cdots \cup f_k$?

Given a fractal, find its IFS.

Often the answer is yes, but finding the maps f_j can be tricky. With some practice, you can develop this art. Let's start with a simple fractal and try to find an IFS for it. Look at the fractal in Figure 5.16 (don't look at the table of numbers below it yet). You should observe five third-size copies of the original (four corners and center). This tells us we want five affine functions $g_1, \ldots, g_5$. Each is of the form

$$g_j \begin{bmatrix} x \\ y \end{bmatrix} = \begin{bmatrix} \frac{1}{3} & 0 \\ 0 & \frac{1}{3} \end{bmatrix} \begin{bmatrix} x \\ y \end{bmatrix} + \begin{bmatrix} e_j \\ f_j \end{bmatrix}.$$

Now we have to figure out the offsets (the e's and f's). It is helpful to imagine the fractal as sitting inside the unit square: $\{(x, y) : 0 \le x, y \le 1\}$.

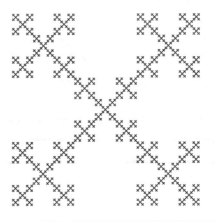

Function	a	b	c	d	e	f
#1	0.3333	0	0	0.3333	0	0
#2	0.3333	0	0	0.3333	0.6667	0
#3	0.3333	0	0	0.3333	0	0.6667
#4	0.3333	0	0	0.3333	0.6667	0.6667
#5	0.3333	0	0	0.3333	0.3333	0.3333

Figure 5.16. A fractal and its associated IFS.

The first third-sized copy is not displaced at all ($e_1 = f_1 = 0$). For the three other corners, we want to displace $\begin{bmatrix} 2/3 \\ 0 \end{bmatrix}$, $\begin{bmatrix} 0 \\ 2/3 \end{bmatrix}$, and $\begin{bmatrix} 2/3 \\ 2/3 \end{bmatrix}$. And for the center copy we want to shift $\begin{bmatrix} 1/3 \\ 1/3 \end{bmatrix}$. The IFS code below the figure gives these values (with the numbers in approximate decimal form).

Let's look at a somewhat more complicated image. Consider the fractal in Figure 5.17. Let's call this fractal W. Notice there are four quarter-sized copies of W in the top half of the image. In the bottom half of the image, we see a copy of the whole, but it is distorted (stretched a bit wide).

This can be seen more clearly in Figure 5.18. Portions A, B, C, and D are the quarter-sized copies of the whole fractal W. In portion E we see a copy of W which has been reduced by a factor of $\frac{3}{4}$ in the horizontal direction and by a factor of $\frac{1}{2}$ in the vertical. Examine the IFS code (Figure 5.17) to understand which function gives rise to each portion. The five functions are

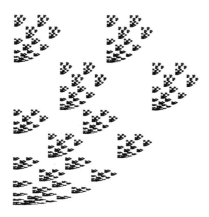

Function	a	b	c	d	e	f
#1	0.25	0	0	0.25	0	0.75
#2	0.25	0	0	0.25	0.25	0.5
#3	0.25	0	0	0.25	0.5	0.75
#4	0.25	0	0	0.25	0.75	0.5
#5	0.75	0	0	0.5	0	0

Figure 5.17. A fractal W and its associated IFS.

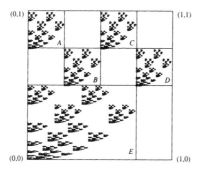

Figure 5.18. Understanding the fractal W from Figure 5.17.

Function	a	b	c	d	e	f
#1	0.3333	−0.3333	0.3333	0.3333	0.3333	0
#2	0.6667	0	0	0.6667	0.3333	0.3333
#3	0.3333	0	0	0.3333	0.6667	0

Figure 5.19. A fractal J and its associated IFS.

clearly contraction mappings, and we have $F(W) = g_1(W) \cup \cdots \cup g_5(W) = A \cup \cdots \cup E = W$.

Finding IFSs from a fractal image can get rather complicated. As a modest example, consider the fractal J in Figure 5.19. If you look carefully, you can find a third-sized copy in the lower right, a two-third-sized copy in the upper right, and a shrunken and *rotated* copy in the lower left. This is made clearer in Figure 5.20. The affine transformations to make the B and C portions are easy. They are, respectively,

$$f_2 \begin{bmatrix} x \\ y \end{bmatrix} = \begin{bmatrix} \frac{2}{3} & 0 \\ 0 & \frac{2}{3} \end{bmatrix} \begin{bmatrix} x \\ y \end{bmatrix}, \quad \text{and}$$

$$f_3 \begin{bmatrix} x \\ y \end{bmatrix} = \begin{bmatrix} \frac{1}{3} & 0 \\ 0 & \frac{1}{3} \end{bmatrix} \begin{bmatrix} x \\ y \end{bmatrix} + \begin{bmatrix} \frac{2}{3} \\ 0 \end{bmatrix}.$$

It is easy to see that $f_2(J) = B$, and $f_3(J) = C$.

The affine transformation for which $f_1(J) = A$ is harder. We know

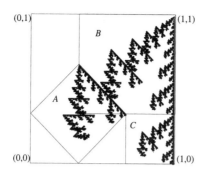

Figure 5.20. Understanding the fractal J from Figure 5.19.

that f_1 has the form

$$f_1 \begin{bmatrix} x \\ y \end{bmatrix} = \begin{bmatrix} a & b \\ c & d \end{bmatrix} \begin{bmatrix} x \\ y \end{bmatrix} + \begin{bmatrix} e \\ f \end{bmatrix}.$$

Further, we see that

$$f_1 \begin{bmatrix} 0 \\ 0 \end{bmatrix} = \begin{bmatrix} \frac{1}{3} \\ 0 \end{bmatrix}, \quad f_1 \begin{bmatrix} 1 \\ 0 \end{bmatrix} = \begin{bmatrix} \frac{2}{3} \\ \frac{1}{3} \end{bmatrix}, \quad \text{and} \quad f_1 \begin{bmatrix} 0 \\ 1 \end{bmatrix} = \begin{bmatrix} \frac{1}{3} \\ \frac{2}{3} \end{bmatrix}.$$

Now, since $f_1 \begin{bmatrix} 0 \\ 0 \end{bmatrix} = \begin{bmatrix} \frac{1}{3} \\ 0 \end{bmatrix}$, we know that $e = \frac{1}{3}$ and $f = 0$. Working from the other two equations, we get

$$a = \frac{1}{3}, \quad b = -\frac{1}{3}, \quad c = \frac{1}{3}, \quad \text{and} \quad d = \frac{1}{3}.$$

This confirms the first entry in the table in Figure 5.19.

In general, the affine transformations can be much more complicated. They might reflect and rotate the image and then skew and shrink it. The transformations might overlap one another. Finding IFSs to generate desired fractal images is an art. It's a good idea to develop an affine transformation tool (see problem 2) to assist you.

On the other hand, once you have an IFS, you naturally want to see what it looks like. To do this, there is no substitute for a computer. In the next section we discuss algorithms for drawing fractal images.

Problems for §5.4

◆1. Let f and g be functions and let X be a compact set. Compute $f(X)$, $g(X)$, and $(f \cup g)(X)$ when f, g, and X are:

(a) $f(x) = x^2$, $g(x) = x + 2$, and $X = [-1, 1]$.

(b) $f(x) = x/2$, $g(x) = 2x$, and $X = [1, 2]$.

(c) $f(x) = \sqrt{x}$, $g(x) = -x$, and $X = [0, 1]$.

(d) $f(x) = x/3$, $g(x) = (x + 2)/3$, and X is Cantor's set.

(e) $f(\mathbf{x}) = \begin{bmatrix} 1 & 0 \\ 0 & 2 \end{bmatrix} \mathbf{x}$, $g(\mathbf{x}) = \begin{bmatrix} 2 & 0 \\ 0 & 1 \end{bmatrix} \mathbf{x}$, and X is the unit square, i.e., $X = \{(x, y) : 0 \le x, y \le 1\}$.

(f) $f(\mathbf{x}) = \begin{bmatrix} 0 & -1 \\ 1 & 0 \end{bmatrix} \mathbf{x}$, $g(\mathbf{x}) = \begin{bmatrix} 0 & 1 \\ -1 & 0 \end{bmatrix} \mathbf{x}$, and X is the unit square.

◆2. Develop a computer tool which will help you find affine transformations of $\mathbf{R}^2$, $g(\mathbf{x}) = A\mathbf{x} + \mathbf{b}$, where A is a 2×2 matrix and $\mathbf{b}$ is a fixed vector in $\mathbf{R}^2$. Specifically, if it is told the action of g on three points,

$$g(\mathbf{x}) = \mathbf{x}^*, \quad g(\mathbf{y}) = \mathbf{y}^*, \quad \text{and} \quad g(\mathbf{z}) = \mathbf{z}^*,$$

your program should return the six numbers a, b, c, d, e, and f so that

$$g \begin{bmatrix} x \\ y \end{bmatrix} = \begin{bmatrix} a & b \\ c & d \end{bmatrix} \begin{bmatrix} x \\ y \end{bmatrix} + \begin{bmatrix} e \\ f \end{bmatrix}.$$

◆3. Let $0 < t < 1$ be a real number. Let C_t be constructed in the same manner as Cantor's set, but instead of removing the middle $\frac{1}{3}$, remove the middle t section. In other words, begin with the unit interval $[0, 1]$ and delete the open interval $\left(\frac{1-t}{2}, \frac{1+t}{2}\right)$. Now repeat this on each of the two remaining subintervals, and then on the remaining four, etc.

[The fractal in problem 2 on page 241 is $C_{3/5}$.]

Find an IFS for which C_t is the attractor.

◆4. Find an IFS for which the fractal of problem 2 on page 245 (see Figure 5.6) is the attractor.

◆5. Let $f(x) = x/2$ and let $g(x) = (x + 1)/2$. What is the unique, stable fixed "point" of $f \cup g$.

5.5 Algorithms for drawing fractals

In this section we develop two algorithms for drawing a fractal given its IFS code. The first is a deterministic algorithm based on the contraction mapping theorem. It is slow and memory intensive.

The second is a randomized algorithm (an algorithm which makes random choices), and we discuss its theoretical underpinnings. This latter algorithm is quicker and requires much less memory.

It is well worth your while to actually implement these algorithms on a computer. Besides being good programming exercises, they give you an opportunity to explore your own IFSs and generate delightful fractal pictures.[10]

5.5.1 A deterministic algorithm

Let $F = f_1 \cup f_2 \cup \cdots \cup f_k$ be an iterated function system where the f_i's are contractive affine functions of the plane.

By the contraction mapping theorem, we know that if A is any compact set, then $F^k(A)$ tends to a unique, stable fixed point, i.e., the attractor of F. In this case the "point" is the fractal we want to draw. The idea of our first fractal drawing algorithm is quite simple. We start with an arbitrary compact set A and iterate F.

Now, computers can easily store numbers; how do we put a *compact set* into a computer? Anything we draw on a computer screen can be thought of as a compact set. The screen is actually an array of tiny dots, and we turn on those dots which are part of the image and turn off those dots which are not. Thus we can think of the computer screen as a large matrix of 0's and 1's. The 0's represent points that are not in the compact set, while the 1's represent points that are in it.

Representing a compact set in a computer as a matrix.

We assume that the fractal we want to draw sits inside the unit square, i.e., all of its x- and y-coordinates are between 0 and 1. We store the image in an $n \times n$ array. Now the tricky part is that the i, j entry of our array corresponds to a point in the plane. We make the following correspondences between m_{ij} (an entry in our matrix) and $\begin{bmatrix} x \\ y \end{bmatrix}$ (a point in the plane):

The notation $\lfloor \cdot \rfloor$ means "round down to the nearest integer"; it's called the floor function.

$$m_{i,j} \to \begin{bmatrix} \frac{i-1}{n-1} \\ \frac{j-1}{n-1} \end{bmatrix}, \quad \text{and}$$

$$\begin{bmatrix} x \\ y \end{bmatrix} \to m_{i,j} \text{ with } i = \lfloor x(n-1) \rfloor + 1, \text{ and } j = \lfloor y(n-1) \rfloor + 1.$$

In this way, for each entry in M we have a point in the unit square, and each point in the unit square corresponds (approximately) to an entry in M.

---

[10]See the `fractory` and `fraclite` programs in the accompanying software.

Deterministic Fractal Drawing Algorithm

Inputs:

- A $6 \times k$ matrix IFSDATA containing a through f for each affine function in the iterated function system,

- a positive integer n (the resolution), and

- a positive integer NITS (number of iterations).

Procedure:

1. Initialize. Let M be an $n \times n$ matrix of all 1's. (We assume our initial compact set is the entire unit square.)

2. Do the following set of instructions NITS times:

   (a) A be an $n \times n$ matrix of all zeros.

   (b) For $i = 1, 2, \ldots, n$, and $j = 1, 2, \ldots, n$ do:

      i. Let $\begin{bmatrix} x \\ y \end{bmatrix}$ be the point in the unit square corresponding to $m_{i,j}$.

      ii. If $m_{i,j} = 1$, then for $p = 1, 2, \ldots, k$ do:

         A. Let $\begin{bmatrix} x' \\ y' \end{bmatrix}$ be $f_p \begin{bmatrix} x \\ y \end{bmatrix}$, where f_p is the pth affine transformation whose parameters (a through f) are stored in the pth row of IFSDATA.

         B. Let i', j' be the matrix position corresponding to $\begin{bmatrix} x' \\ y' \end{bmatrix}$.

         C. Let $a_{i',j'} = 1$.

   (c) Let $M = A$.

3. Display the matrix M on the computer screen.

Notice that $\begin{bmatrix} 0 \\ 0 \end{bmatrix}$ corresponds to $m_{1,1}$, and $\begin{bmatrix} 1 \\ 1 \end{bmatrix}$ corresponds to $m_{n,n}$.

The last step of our algorithm is to display the matrix M on the computer screen. How you do this depends very much on the kind of computer and the programming environment you use. We present this algorithm on page 278.

What's going on in this algorithm? The matrix M corresponds to the compact set S which is inside the unit square. All the work of this algorithm occurs in step 2(b), where we compute $F(S)$, saving the result in a temporary matrix A. Each point of S (corresponding to each '1' entry in M) is mapped to several points when we compute $F(S) = f_1(S) \cup \cdots \cup f_p(S)$; this happens in step 2(b)ii. In step 2(c) we copy the result of the computation of $F(S)$ (stored in A) back over S (stored in M) and then repeat the whole procedure.

This algorithm is quite slow. For the most basic image we would like n to be around 200. This will give acceptable resolution in a modest-sized picture. The bad news is, when $n = 200$, the matrices M and A contain 40,000 entries each. Each iteration of the algorithm looks at all these entries and can take quite a lot of time even on a fast machine.

If you decide to implement this algorithm, here are some suggestions. First, you need *not* start M as a matrix filled entirely with 1's. You might choose to have only the boundary entries set equal to 1, or fill in about 10% of the entries at random with 1's. This will speed things up quite a bit in most cases. Second, you might like to display the current state of the matrix M each time through the main loop. The program can then ask the user whether or not to continue on to the next iteration. Third, you might want to test if $A = M$ each time through the main loop; if it does, your fractal has converged, and no further iterations will change the image. Finally, if your computer has color, you might want to think about how you can use color to better illustrate how the algorithm operates.

5.5.2 Dancing on fractals

The following experiment is fun. *You absolutely must try it!* Get a large sheet of paper, a marker, a ruler, and a die (singular of dice). Mark three points A, B, and C at the corners of a large triangle on your paper (more or less an equilateral triangle, but it doesn't matter). Position your marker at A. This is your *current point*.

Now roll the die and do the following depending on your roll:

- If you roll a 1 or a 2, find the point halfway from where you are now to A. This is your new current point; draw a dot there.

- If you roll a 3 or a 4, find the point halfway from where you are now to B. This is your new current point; draw a dot there.

- If you roll a 5 or a 6, find the point halfway from where you are now to C. This is your new current point; draw a dot there.

Repeat the steps above a few hundred times. What do you expect to see? What *do* you see?

Do it!! If you're a sloppy person like me (and you'd end up getting marker all over the floor), do your drawing with a computer program. There's a ready-to-run MATLAB program called dance.m in the accompanying software. A listing for this program is given in §B.2 on page 365.

In any case, I absolutely insist you try this. Don't read another word (I'll wait).

$$\boxed{\mathscr{P}lease\ do\ it\ now.}$$

Did you do it? Yes, good. (No, well do it!)

You should see Sierpiński's triangle. Neat! Now, why?

Let's say you put the points A, B, and C at the following locations:

$$A = \begin{bmatrix} 0 \\ 0 \end{bmatrix}, \quad B = \begin{bmatrix} 1 \\ 0 \end{bmatrix}, \quad \text{and} \quad C = \begin{bmatrix} \frac{1}{2} \\ 1 \end{bmatrix}.$$

First, let's work out what "jump halfway to A" (or B or C) means analytically. Suppose we are at a point $\begin{bmatrix} x \\ y \end{bmatrix}$. "Jump halfway to A" means find the point $\left(A + \begin{bmatrix} x \\ y \end{bmatrix} \right) /2$, but since $A = \begin{bmatrix} 0 \\ 0 \end{bmatrix}$, this simply means

$$\begin{bmatrix} x \\ y \end{bmatrix} \mapsto \begin{bmatrix} \frac{1}{2} & 0 \\ 0 & \frac{1}{2} \end{bmatrix} \begin{bmatrix} x \\ y \end{bmatrix}.$$

Next, consider "jump halfway to B." This means $\begin{bmatrix} x \\ y \end{bmatrix}$ becomes $\left(\begin{bmatrix} x \\ y \end{bmatrix} + B \right) /2$ and since $B = \begin{bmatrix} 1 \\ 0 \end{bmatrix}$, we compute

$$\begin{bmatrix} x \\ y \end{bmatrix} \mapsto \begin{bmatrix} \frac{1}{2} & 0 \\ 0 & \frac{1}{2} \end{bmatrix} \begin{bmatrix} x \\ y \end{bmatrix} + \begin{bmatrix} \frac{1}{2} \\ 0 \end{bmatrix}.$$

Finally, "jump halfway to C" with $C = \begin{bmatrix} \frac{1}{2} \\ 1 \end{bmatrix}$ works out to be

$$\begin{bmatrix} x \\ y \end{bmatrix} \mapsto \begin{bmatrix} \frac{1}{2} & 0 \\ 0 & \frac{1}{2} \end{bmatrix} \begin{bmatrix} x \\ y \end{bmatrix} + \begin{bmatrix} \frac{1}{4} \\ \frac{1}{2} \end{bmatrix}.$$

Aha! These are the three affine transformations f_1, f_2, f_3 in the IFS for Sierpiński's triangle (see Table 5.1 on page 271). The "choose a corner at random and jump halfway there" instruction can be rephrased, "choose among f_1, f_2, or f_3 at random and apply that transformation."

Now, let's see why this procedure draws the fractal. In Figure 5.14 (on page 267) we broke Sierpiński's triangle up into three sections: A, B, and C (it was no accident we used these names for the three corners above). Applying f_1 to any point of Sierpiński's triangle, S, puts us into region A, applying f_2 puts us into B, and applying f_3 puts us into C.

We can refine our labeling method, just as we refined the L and R labels for Cantor's set. We can divide each of the three triangles A, B, and C into three subtriangles each, resulting in nine regions: AA, AB, AC, BA, BB, BC, CA, CB, and CC. Each of these we can break down again into a total of 27 little triangles named AAA through CCC; see Figure 5.21. Now, for each infinite sequence of the letters A, B, and C, such as $ABACCAB\cdots$ we have a nested sequence of compact sets:

A symbolic addressing system for Sierpiński's triangle.

$$A \supset AB \supset ABA \supset ABAC \supset ABACC \supset \cdots.$$

And, as with intervals, the intersection of a nested sequence of compact sets is nonempty. The intersection of this chain of sets is a point of Sierpiński's triangle. Indeed, every point in Sierpiński's triangle can be given a code name (address) using this ABC labeling method.

Note that $AAAAA\cdots$ names the lower leftmost point of the triangle (the point we had been calling simply A). The other corners are clearly $BBBBB\cdots$ and $CCCCC\cdots$. Some points in the triangle have two equivalent addresses. Notice in the figure the point $CBBBB\cdots$ is the same as the point $BCCCC\cdots$.

We want to know how to do calculations with the addresses? In particular, given a point X specified by its address, what are $f_1(X)$, $f_2(X)$, and $f_3(X)$?

Understanding how the f_i's work using the ABC addresses.

To begin, what is $f_1(BB)$? Since f_1 shrinks the entire triangle into the A section, we see that the BB section becomes the ABB section. Likewise, the $BBCA$ section gets shrunk to become the $ABBCA$ section. So if we have a point $X = BBCACA\cdots$, then $f_1(X) = ABBCACA\cdots$. In other words, $f_1(X)$ prepends the symbol A to the address of X. We check that f_2 and f_3 work in similar ways. Since f_2 compresses everything into the lower right section, $f_2(X)$ is computed by prepending a B to X's address. Likewise, as f_3 packs everything into the upper triangle, $f_3(X)$ prepends a

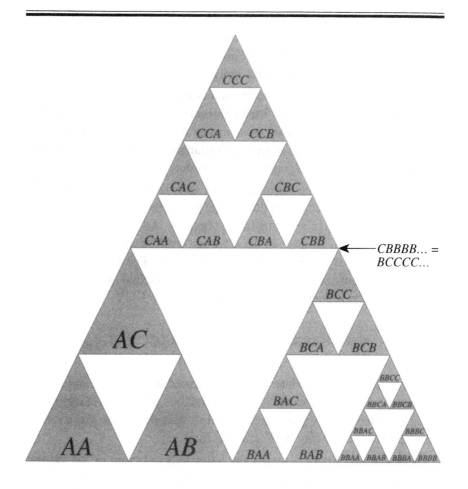

Figure 5.21. The ABC labeling system for Sierpiński's triangle.

C. Thus,

$$f_i(X_1 X_2 X_3 X_4 \cdots) = \begin{cases} A X_1 X_2 X_3 X_4 \cdots & \text{if } i = 1, \\ B X_1 X_2 X_3 X_4 \cdots & \text{if } i = 2, \text{ and} \\ C X_1 X_2 X_3 X_4 \cdots & \text{if } i = 3. \end{cases}$$

Understanding the triangle dance using the ABC encoding.

We can use this symbolic representation of the points in Sierpiński's triangle and the action of the f_i's to understand how the triangle dance works. We begin at the origin, i.e., at $AAAA \cdots$ and then randomly apply f_1, f_2, or f_3. Equivalently, we randomly put the symbol A, B, or C at the beginning of the current point. We now see why the algorithm draws Sierpiński's triangle. First, each step of the algorithm produces another point in the triangle. Second, given a point in the triangle, say $X = ABAAC \cdots$, we note that there is a $1/243 = 1/3^5$ chance that the next four symbols we pick are (in order) C, A, A, B, A. In that case (which is likely to happen, since we are plotting thousands of points) we arrive at a point Y whose initial symbols are $ABAAC$. Now, Y might not be *exactly* the same as X, but it is very close. Thus the successive points we plot dance about on Sierpiński's triangle and get near every element of it. In this way, the fractal is drawn on our screen.

We hope it is clear now why our triangle dance draws out Sierpiński's triangle. Before we proceed to the more general case, we'd like to consider what would happen if we began with a point (such as $X = \begin{bmatrix} 1 \\ 1 \end{bmatrix}$) which is *not* in S. Now, X and any $f_i(X)$ are not in S, so as we start drawing, we'll be making a mess, i.e., we'll be plotting points not in S. However, after several iterations, the points we plot are close to, and for all practical purposes in, S. The reason is the contraction mapping theorem. Let's think about the one point set $\{X\}$. This is a compact set, so we know that after several iterations of $F = f_1 \cup f_2 \cup f_3$, the result, $F^k(\{X\})$ will look very much like the Sierpiński triangle S. The technical meaning is that the point of $F^k(\{X\})$ which is furthest from S is, in fact, very close to S. In other words, all points of $F^k(\{X\})$ are very near S. What are the points of $F^k(\{X\})$? They are exactly the points of the form

$$f_{i_1} \circ f_{i_2} \circ f_{i_3} \circ \cdots \circ f_{i_k}(X)$$

where $i_1, i_2, \ldots, i_k \in \{1, 2, 3\}$. Now in the random dance, we choose the f_i's at random. What our analysis shows is that *no matter how we choose the f_i's* the resulting point is (after a few dozen iterations) incredibly close to being in S. Indeed, the computer's arithmetic (and its screen's resolution) is not precise enough to tell the difference!

Thus if we do our random triangle dance starting at any point but not plotting the first dozen points or so, the result will be the same: we draw Sierpiński's triangle. And...

5.5.3   A randomized algorithm

...this works in general! (This section is inexorably linked to the previous.)

Suppose we have an IFS: $F = f_1 \cup f_2 \cup \cdots \cup f_k$, where each f_i is a contraction map (and therefore F is). We want to draw the fractal K which

Extending the triangle dance method to draw other fractals.

Randomized Fractal Drawing Algorithm

Inputs:

- A $6 \times k$ matrix IFSDATA containing a through f for each affine function in the iterated function system, and

- NPTS, a positive integer (or ∞) indicating how many points to plot.

Procedure:

1. Let $X = \begin{bmatrix} 0 \\ 0 \end{bmatrix}$ (or any other staring value you prefer).

2. For $i = 1, 2, \ldots,$ NPTS $+ 50$, do:

   (a) Choose an integer j at random between 1 and k (inclusive). (See the discussion below.)

   (b) Let $X = f_j(X)$, where f_j is the affine transformation from row j of IFSDATA.

   (c) If $i > 50$, plot the point X on the screen.

---

is the unique, attractive fixed point of F. One way to do this (and this was the deterministic algorithm) is to compute $F^t(A)$ for some starting compact set A, such as the unit square. Alternatively, we can begin our algorithm with a singleton set $A = \{X\}$ for some point X (such as the origin). After many iterations, we have $F^t(\{X\})$, which is the set of all points of the form

$$f_{i_1} \circ f_{i_2} \circ f_{i_3} \circ \cdots \circ f_{i_t}(X),$$

where the i_j's are in $\{1, 2, \ldots, k\}$.

Now instead of plotting all k^t such points, we can plot representative points, i.e., choose points in the fractal at random. To do this, we pick the f's we apply at random. This is exactly the rationale behind the Randomized Fractal Drawing Algorithm shown on page 284.

Several comments are in order. First, this algorithm is *much simpler* (and therefore easier to implement) than the deterministic fractal algorithm. Second, we ignore the issue of how to "plot the point X on the screen." This depends on your computing environment. Third, we discard the first 50 iterations. Fifty is probably more than necessary, but computing successive points is fast and cheap, so there's no harm done. Fourth (and most important), is a more careful discussion of step 2(a): "Choose an integer j at random between 1 and k (inclusive)."

Picking an integer at random

How do we pick an integer from the set $\{1, 2, \ldots, k\}$ at random? One interpretation of this statement (and not the one you should use) is to choose each of the numbers with *equal* probability (i.e., uniformly). We do this by using the computer's random number generator[11] to create a (real) value between 0 and 1, say x. We then multiply x by k and round up. This gives a random number in the set $\{1, \ldots, k\}$, i.e., we compute $\lceil kx \rceil$. There is nothing wrong with this method per se, but if you try it (and you should) you should notice something interesting.

Picking f_i uniformly at random may result in non-uniform darkening of the fractal image.

If you run your program on the IFS for Sierpiński's triangle (Table 5.1 on page 271) all should be fine (indeed, you will be doing the triangle dance). However, if you try the fractal J from Figure 5.19 (page 274), you should notice that the picture's density is uneven. One-third of the points you plot go into section C in the lower right (see Figure 5.20 on page 275). This makes sense, since we use transformation #3 one-third of the time on average. However, because this part of the picture is smaller than the other parts, it darkens more quickly than, say, section B, which covers four times as much area.

Uneven darkening.

To correct this uneven drawing, we need to use transformation #2 more frequently than #3. Indeed, we want to choose #2 four times as often as #3. The general principle is, we choose a transformation proportional to the area it covers. Now, in theory, the fractals have zero area, which seems to make this entire discussion vacuous. However, in reality, the computer image *does* cover area on the screen. Thus if the fractal K we are drawing counts for one unit of area, how much area does, say, the $f_1(K)$ portion cover? We are working with affine functions, so f_1 is of the form $f_1(\mathbf{x}) = A\mathbf{x} + \mathbf{b}$. From linear algebra we know that transforming a region by multiplying by a matrix A changes the region's area by a factor of $|\det A|$. Now, if $A = \begin{bmatrix} a & b \\ c & d \end{bmatrix}$, we simply want to choose this affine transformation with probability proportional to $|ad - bc|$. Thus we should choose functions f_i not with uniform probability but, rather, with probability proportional to $|a_i d_i - b_i c_i|$.

Choose the transformations with probability proportioanl to their determinants . . .

---

[11] As we discussed in Chapter 1, the random numbers generated by the computer are actually not random at all! Nonetheless, it is safe in this situation to trust the randomness of the numbers the computer generates.

For example, consider the fractal in Figure 5.19 (page 274). The determinants of the three transformations are

$$\det \begin{bmatrix} \frac{1}{3} & -\frac{1}{3} \\ \frac{1}{3} & \frac{1}{3} \end{bmatrix} = -\frac{2}{9}, \ \det \begin{bmatrix} \frac{2}{3} & 0 \\ 0 & \frac{2}{3} \end{bmatrix} = \frac{4}{9}, \ \text{and } \det \begin{bmatrix} \frac{1}{3} & 0 \\ 0 & \frac{1}{3} \end{bmatrix} = \frac{1}{9}.$$

Now $\frac{2}{9} + \frac{4}{9} + \frac{1}{9} = \frac{7}{9}$, so we should choose transformation #1 with probability $\frac{2}{7}$, transformation #2 with probability $\frac{4}{7}$, and transformation #3 with probability $\frac{1}{7}$. In this way, all sections will darken at the same rate, giving a nice picture.

... but give transformations with zero determinant a break.

This is good, but suppose one of the affine transformations has a matrix with *zero* determinant? Our rule says that we should *never* apply this transformation, and this would change the IFS. So let's break our rule in this case; after all, on a computer screen a line does cover some area. We can fix a minimum probability (say, 2% or whatever you choose) with which a transformation must be chosen.

Problems for §5.5

◆1.   On the real line, plot a point at 0. Now flip a coin. If you get heads, jump $\frac{2}{3}$ of the way to 0 and if you get tails, jump $\frac{2}{3}$ of the way to 1 and plot a new point. Repeat this procedure several times. What do you get? Explain.

◆2.   Suppose $f_1, f_2, f_3 : \mathbf{R}^2 \to \mathbf{R}^2$ are contraction maps, and let $F = f_1 \cup f_2 \cup f_3$. We know that F has an attractor (stable fixed point) $K \in \mathscr{H}^2$.

Since f_1, f_2, and f_3 are pointwise contraction maps, we know that they have fixed points $\mathbf{x}_1, \mathbf{x}_2, \mathbf{x}_3 \in \mathbf{R}^2$ (respectively).

Explain why $\mathbf{x}_1, \mathbf{x}_2, \mathbf{x}_3 \in K$.

◆3.   In the randomized fractal drawing algorithm we suggest not plotting the first 50 points for fear that these points are not in (or close to) the fractal we are drawing.

Suppose the contractivities of the affine transformations are $s_1, s_2, \ldots, s_k$ (all less than 1, of course). Suppose further that the distance from the initial point to the fractal is 1.

Provide a formula which bounds the distance between the 50th point generated and the fractal.

Is the number 50 reasonable most of the time?

Find contractivities $s_1, \ldots, s_k$ for which 50 points may not be sufficient to guarantee that the points you are plotting are essentially in the fractal. Instead of ignoring the first 50 points, suppose you skip the first 500 points. Is this sufficient for the contractivities you found?

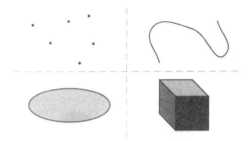

Figure 5.22. Common geometric objects have whole-number dimensions. The finite set of points has dimension zero, the curve is one-dimensional, the filled ellipse is two-dimensional, and the solid cube is three-dimensional.

5.6   Fractal dimension

Sierpiński's triangle is fatter than a curve but skinnier than a disk (a circle with its interior). In this section we make this rather vague, intuitive statement precise. One of the reasons fractals are called *frac*tals is that they have *fractional* dimension.

There are certain objects whose dimension you can understand intuitively. For example, a curve, even if it twists and winds through space, is a one-dimensional object. A filled-in ellipse and the surface of a sphere are two-dimensional objects. A solid cube and a ball (a sphere and its interior) are three-dimensional objects. A finite collection of points is a zero-dimensional set. See Figure 5.22.

We now turn to understanding what it means for an object to have dimension $\frac{1}{2}$ or any other number!

5.6.1   Covering with balls

We use the common word *ball* as a technical term. A *ball* is defined to be a set of points within a given distance of a fixed point. Specifically, let

Balls.

$$B(\mathbf{x}, r) = \{\mathbf{y} : d(\mathbf{x}, \mathbf{y}) \leq r\},$$

i.e., the set of all points $\mathbf{y}$ at a distance r or less from $\mathbf{x}$. In three-dimensional space ($\mathbf{R}^3$) a ball is a sphere together with its interior. In the plane ($\mathbf{R}^2$) a ball is the same thing as a disk, i.e., a circle (centered at $\mathbf{x}$ with radius r)

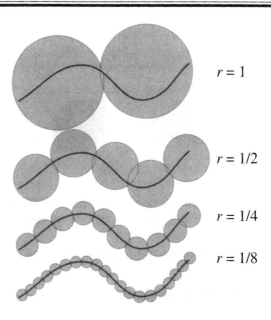

Figure 5.23. Covering a curve with balls of smaller and smaller radius.

together with its interior. Finally, on the line ($\mathbf{R}^1$) a ball is simply a closed
interval $[x - r, x + r]$, where x is the midpoint.

$N(K, r)$.

The central idea in calculating the dimension of an object K is *covering*
K with balls of a given radius. This means we seek a collection of balls so
that K is contained in their union. And we want to do it with as few balls
as possible! Define $N(K, r)$ to be the minimum number of balls of radius
r needed to cover K.

Covering a curve:
$N(K, r) \propto 1/r$.

For example, let K be the curve (drawn several times) of Figure 5.23.
The top part of the figure shows that we can cover the curve with two balls
of radius 1, so $N(K, 1) = 2$. Next, we see that when we cover K with balls
of radius $\frac{1}{2}$ we need 5, so $N(K, \frac{1}{2}) = 5$. In the next portion, we see that
$N(K, \frac{1}{4}) = 10$, and finally, $N(K, \frac{1}{8}) = 20$. As we halve the radius of the
balls, the number we need (more or less) doubles. This makes sense, since
a ball of radius r covers about $2r$ of the length of the curve. The smaller
the balls, the more accurate this estimation technique becomes. Thus we
can write that $N(K, r) \propto r^{-1}$.

Covering a square:
$N(K, r) \propto 1/r^2$.

Now, suppose we want to cover a square (with interior) with balls. How
many balls of radius r does it take to cover a (filled-in) square of side length

Figure 5.24. A square of side length r can be covered by a single ball of radius r. Likewise, a cube of side length 1 can be covered by a ball of radius 1. However, see problem 9 on page 307.

1? I haven't the vaguest idea! However, we can estimate. A ball of radius r covers an area of πr^2, so we need *at least* $1/(\pi r^2)$ balls. We certainly need more, since we can't cover the area of the square with 100% efficiency. On the other hand, we can break up the unit square into $\frac{1}{r} \times \frac{1}{r}$ little squares (assuming $\frac{1}{r}$ is a whole number, but if not, we won't be too far off). Each of these little squares of side length r can be covered by a single ball[12] of radius r; see Figure 5.24. Thus we can certainly cover the square with at most $1/r^2$ balls. Summarizing, we know

$$\frac{1}{\pi} r^{-2} \le N(K, r) \le r^{-2}$$

or more succinctly, $N(K, r) \propto r^{-2}$.

Let K be a finite collection of points. What is $N(K, r)$? Naturally, this number depends on K and on r, but once r is quite small (less than half the distance between any two of the points), then $N(K, r)$ is constant (it's exactly the number of points in K).

Covering a finite collection of points: $N(K, r) \propto 1$.

We write this as $N(K, r) \propto 1$ (i.e., $N(K, r)$ is proportional to a constant) and, we can write "1" in a rather bizarre fashion: r^{-0}. Thus for a finite collection of points, we have $N(K, r) \propto r^{-0}$.

As a last example, let's jump to $\mathbf{R}^3$ and consider how many balls of radius r we need to cover the unit cube (a solid cube with side length 1). Since the volume of a sphere is $\frac{4}{3}\pi r^3$, we know that $N(K, r) \ge 3/(4\pi r^3)$. By chopping the unit cube into $1/r^3$ subcubes of side length r, we find that $N(K, r) \le 1/r^3$. Thus $N(K, r) \propto r^{-3}$.

Covering a cube: $N(K, r) \propto 1/r^3$.

We have considered four examples of compact sets: a finite collection

Summary of examples.

---

[12]This doesn't work in higher dimensions; see problem 9 on page 307.

Set K	Dimension K	$N(K, r) \propto$
Finite set of points	0	r^{-0}
Curve	1	r^{-1}
Filled square	2	r^{-2}
Solid cube	3	r^{-3}

Table 5.3. Common geometric objects, their dimensions, and the number of balls needed to cover them.

of points, a curve, a square (filled), and a cube (solid). Table 5.3 summarizes these results. I hope the pattern is clear: For an object K of dimension d, we have $N(K, r) \propto r^{-d}$. We can use this pattern to make a definition.

5.6.2 Definition of dimension

Recall that $N(K, r)$ is the minimum number of balls of radius r needed to cover a compact set K. We observed that $N(K, r) \propto r^{-d}$ for some objects whose dimension we know. Our next aim is to convert this intuition into a good theoretical definition of dimension and then to convert our theoretical definition into a form that can be used in practice.

First, let's make the "is proportional to" sign ($\propto$) a bit more specific. Given functions f and g of a real number r, when we write $f(r) \propto g(r)$ we roughly mean

$$f(r) = g(r) \times \text{(less significant terms involving } r\text{)} .$$

To make this rigorous, we take logs[13] of both sides of the equation to get

$$\log f(r) = \log g(r) + \log \text{(lower order terms)} .$$

Now, what do we mean by *lower order terms*? We mean terms that are very much smaller than $f(r)$ or $g(r)$. So let us divide both sides by $\log g(r)$ to get

$$\frac{\log f(r)}{\log g(r)} = 1 + \frac{\log \text{(lower order terms)}}{\log g(r)} .$$

---

[13]The base of the logarithm is unimportant. We get the same eventual answer regardless of which logarithm we use. But if you prefer, then the natural choice is the natural log: base e. See problem 10 on page 307.

Since the lower order terms are very much smaller than $g(r)$, the term after "1+" must be small.

Thus we formalize the notation $f(r) \propto g(r)$ to mean

Formalizing $\propto$.

$$\lim_{r \downarrow 0} \frac{\log f(r)}{\log g(r)} = 1.$$

For example, earlier we noted that if K is a filled square, then

$$\frac{1}{\pi} r^{-2} \leq N(K, r) \leq r^{-2}.$$

Taking logs and dividing by $\log \left(r^{-2}\right) = -2 \log r$, which is positive if r is small, we have

$$\frac{\log(1/\pi)}{-2 \log r} + 1 \leq \frac{\log N(K, r)}{-2 \log r} \leq 1.$$

Therefore, as $r \downarrow 0$ we have

$$\frac{\log N(K, r)}{\log \left(r^{-2}\right)} \to 1,$$

verifying that $N(K, r) \propto r^{-2}$.

We can use the logarithm trick to extract the 2 from r^{-2}. Observe that when $N(K, r) \propto r^{-2}$, then

$$\frac{\log N(K, r)}{(-\log r)} \to 2 \quad \text{as } r \downarrow 0.$$

We now are ready to present the definition of dimension. (The $-\log r$ in the denominator can also be written as $\log(1/r)$. Since r is a small number—nearly zero—we know that $\log(1/r)$ is positive.)

Let K be a compact set. The *dimension*[14] of K is defined to be

The formal definition of dimension.

$$\dim K = \lim_{r \downarrow 0} \frac{\log N(K, r)}{\log(1/r)}.$$

It is a bit slippery to define something in terms of a limit because the limit might not exist. In such cases, we say the dimension of K is undefined. However, for fractals arising as the attractors of iterated function systems of contractive affine maps, the limit always exists.

---

[14]There is more than one way to define the dimension of a compact set. The definition we use is known as the Kolmogorov definition. Another definition, which we do not consider, is the Hausdorff-Besicovitch dimension.

5.6.3   Simplifying the definition

Now that we have a definition of dimension, let's apply it to various fractals such as Cantor's set or Sierpiński's triangle and compute their dimensions. The difficulty is, the number $N(K, r)$ is very hard to compute. Fortunately, we can make some simplifications which do not alter the meaning of the definition but make it easier to apply and compute.

Square boxes instead of balls: $N_\square$.

Let K be a square of side length 1. To compute the dimension of K we want to know $N(K, r)$, but as we discussed above, we don't know how to get this number exactly. However, we don't *need* to know this number exactly; our approximations are good enough to confirm dim $K = 2$. We used a trick: Instead of covering by balls, we covered by squares. Let's see that this trick is actually a technique[15] we can use in other situations.

We define $N_\square(K, r)$ to be the minimum number of axis-parallel square boxes of side length r we need to cover K. By *square box* we mean a square together with its interior; thus square box is to square as disk is to circle. By *axis-parallel* we mean that the bottom sides of the squares are all parallel to the x-axis of the plane.

There are two nice things about $N_\square(K, r)$: First, it is often easier to compute than $N(K, r)$. Second, we can replace N with $N_\square$ in the definition of dimension. The key to seeing why is the following:

$$N(K, r) \leq N_\square(K, r) \leq 4N(K, r). \tag{5.3}$$

In words, these inequalities say:

- If we can cover K with $n = N_\square(K, r)$ boxes, then we can cover K with n or fewer disks.

- If we can cover K with $n = N(K, r)$ disks, then we can cover K with $4n$ or fewer boxes.

To see why these statements are true, look at Figure 5.25. For the first claim, look at the left portion of the figure. If we can cover K with n boxes of side length r, simply replace each box with a disk of radius r concentric with the box. This gives a cover of K by n disks, but perhaps we can do better. So $N(K, r) \leq n = N_\square(K, r)$.

For the second claim, suppose we cover K with $n = N(K, r)$ disks of radius r; then we can replace each disk with four square boxes, as shown in the right-hand portion of Figure 5.25. This gives a cover of K by $4n$ boxes,

---

[15]In mathematics, a technique is a trick that works more than once.

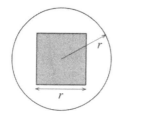 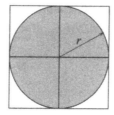

Figure 5.25. Why $N(K, r) \leq N_\square(K, r) \leq 4N(K, r)$.

but perhaps we can do better. Thus $N_\square(K, r) \leq 4n = 4N(K, r)$, and we have verified the two claims justifying inequality (5.3).

Now, we take logs of all terms in inequality (5.3) and divide by $\log(1/r)$ to get

$$\frac{\log N(K, r)}{\log(1/r)} \leq \frac{\log N_\square(K, r)}{\log(1/r)} \leq \frac{\log 4 + \log N(K, r)}{\log(1/r)}.$$

Finally, we let $r \downarrow 0$. Notice that the outer terms both tend to $\dim K$, and therefore the term sandwiched between them also converges to $\dim K$.

In conclusion, we have an equivalent definition of dimension whether we use $N(K, r)$ (cover by balls) or $N_\square(K, r)$ (cover by axis-parallel square boxes):

$$\dim K = \lim_{r \downarrow 0} \frac{\log N(K, r)}{\log(1/r)} = \lim_{r \downarrow 0} \frac{\log N_\square(K, r)}{\log(1/r)}.$$

Whether we use square boxes or disks, computation of $N(K, r)$ (or $N_\square(K, r)$) can get nasty if the number r doesn't fit the geometry of the situation. For example, if $r = \frac{1}{2}$, it is easy to see that we need exactly three boxes to cover Sierpiński's triangle, S. Likewise, $N_\square(S, \frac{1}{4}) = 9$. However, $N_\square(S, \frac{1}{5})$ is not so easy to compute. In short, it's typically very hard to get a formula for $N_\square(K, r)$ which works for *all* values of r. [margin: Choose a convenient subsequence.]

This brings us to our next simplification. If the dimension of K is defined, then we know $\log N_\square(K, r) / \log(1/r) \to \dim K$ as $r \downarrow 0$. What's more, for *any* sequence of numbers $r_1 > r_2 > r_3 > \cdots$ which tends to zero the sequence

$$\frac{\log N_\square(K, r_1)}{\log(1/r_1)}, \ \frac{\log N_\square(K, r_2)}{\log(1/r_2)}, \ \frac{\log N_\square(K, r_3)}{\log(1/r_3)}, \ldots$$

also converges to dim K.

Let's see how this works with Sierpiński's triangle, S. We noted that $N_\square(S, \frac{1}{2}) = 3$ and $N_\square(S, \frac{1}{4}) = 9$. Continuing in this fashion, we have

$$N_\square(S, \frac{1}{2}) = 3,$$

$$N_\square(S, \frac{1}{4}) = 9,$$

$$N_\square(S, \frac{1}{8}) = 27,$$

$$N_\square(S, \frac{1}{16}) = 81,$$

$$\vdots$$

$$N_\square(S, \frac{1}{2^k}) = 3^k.$$

So instead of worrying about $N_\square(S, r)$ for *all* values of r, we instead use only a convenient sequence of values of r, namely $\frac{1}{2}, \frac{1}{4}, \frac{1}{8}, \cdots$. This is a decreasing sequence which tends to 0. Thus

$$\dim S = \lim_{r \downarrow 0} \frac{\log N_\square(S, r)}{\log(1/r)}$$

$$= \lim_{k \to \infty} \frac{\log N_\square\left(S, 2^{-k}\right)}{\log\left(1/2^{-k}\right)} \quad (k = 1, 2, 3, \ldots)$$

$$= \lim_{k \to \infty} \frac{\log 3^k}{\log 2^k}$$

$$= \lim_{k \to \infty} \frac{k \log 3}{k \log 2}$$

$$= \frac{\log 3}{\log 2} = \log_2 3 \approx 1.585.$$

Let's use these same ideas to compute the fractal dimension of Cantor's set, C. It is clear that $N_\square(C, \frac{1}{3}) = 2$: the two intervals (one-dimensional boxes) of side length $\frac{1}{3}$ we need are L and R. Next, $N_\square(C, \frac{1}{9}) = 4$ (the four boxes are $LL, LR, RL,$ and RR). Continuing in this fashion, we have $N_\square(C, \frac{1}{27}) = 8$ and $N_\square(C, \frac{1}{81}) = 16$. In general,

$$N_\square\left(C, \frac{1}{3^k}\right) = 2^k.$$

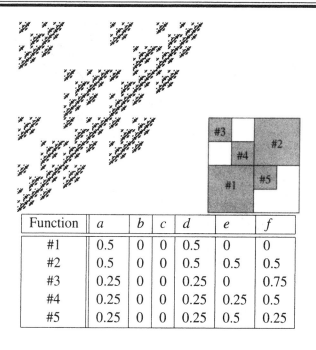

Function	a	b	c	d	e	f
#1	0.5	0	0	0.5	0	0
#2	0.5	0	0	0.5	0.5	0.5
#3	0.25	0	0	0.25	0	0.75
#4	0.25	0	0	0.25	0.25	0.5
#5	0.25	0	0	0.25	0.5	0.25

Figure 5.26. A fractal T and its associated IFS.

With the preceding formula we compute the dimension of C using the convenient sequence $\frac{1}{3}, \frac{1}{9}, \frac{1}{27}, \ldots$ as follows

$$\dim C = \lim_{r \downarrow 0} \frac{\log N_\square(C, r)}{\log(1/r)}$$

$$= \lim_{k \to \infty} \frac{\log N_\square(C, 3^{-k})}{\log(1/3^{-k})}$$

$$= \lim_{k \to \infty} \frac{\log 2^k}{\log 3^k}$$

$$= \lim_{k \to \infty} \frac{k \log 2}{k \log 3}$$

$$= \frac{\log 2}{\log 3} = \log_3 2 \approx 0.6309.$$

Now let's try a more difficult example. Consider the fractal T depicted in Figure 5.26. This fractal sits just inside the unit square. We now want

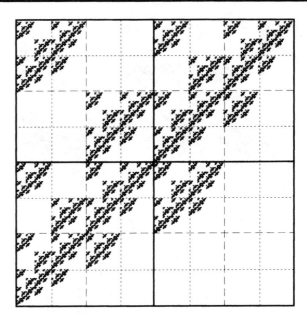

Figure 5.27. The fractal T (from Figure 5.26) overlaid with an 8×8 grid.

to count the number of square boxes of side length r we need to cover T. We choose the numbers r to be convenient.

To assist us in this process, we show a larger picture of T in Figure 5.27 in which we have overlaid T with a grid. From the figure we directly count:

$$N_\square(T, 1) = 1,$$

$$N_\square(T, \frac{1}{2}) = 4,$$

$$N_\square(T, \frac{1}{4}) = 11,$$

$$N_\square(T, \frac{1}{8}) = 34, \quad \text{and}$$

$$N_\square(T, \frac{1}{16}) = 101.$$

(To get the last one, mentally subdivide each $\frac{1}{8} \times \frac{1}{8}$ square into fourths and count *very* carefully.)

Do you see the pattern? It's not obvious until you observe that

$$11 = 2 \times 4 + 3 \times 1,$$

$$34 = 2 \times 11 + 3 \times 4, \quad \text{and}$$

$$101 = 2 \times 34 + 3 \times 11.$$

If we believe this pattern, we predict that

$$N_\square(T, \frac{1}{32}) = 2 \times 101 + 3 \times 34 = 274.$$

If you have a large computer screen you might like to make a large drawing of this fractal, overlay a 32×32 grid, and count.

Let's explain why the pattern we observe is correct. To assist our reasoning, we introduce a simpler notation. For this problem only, let

$$x(k) = N_\square(T, 2^{-k}).$$

So far, we know $x(0) = 1$, $x(1) = 4$, $x(2) = 11$, $x(3) = 34$, and $x(4) = 101$. Our guess is that

$$x(k + 2) = 2x(k + 1) + 3x(k). \tag{5.4}$$

Now look again at Figure 5.26. Notice that within T there are two half-size and three quarter-size copies of T, and in equation (5.4) we also have a 2 and a 3. This is not a coincidence! Let's understand why we need 101 boxes of side length $\frac{1}{16}$ to cover T. To cover sections #1 and #2 with $\frac{1}{16}$ boxes is exactly the same as covering the entire image with $\frac{1}{8}$ boxes, because sections #1 and #2 are half-size copies. To cover sections #3, #4, and #5 with $\frac{1}{16}$ boxes is exactly the same as covering the entire image with $\frac{1}{4}$ boxes because each of these sections is a quarter-sized copy. Thus, in total, we need $2 \times N_\square(K, \frac{1}{8}) + 3 \times N_\square(K, \frac{1}{4})$ boxes of side length $\frac{1}{16}$ to cover K.

This idea is not special to boxes of side length $\frac{1}{16}$. In general, to cover sections #1 and #2 with boxes of side length $1/2^{k+2}$ is the same as covering the entire fractal with boxes of side length $1/2^{k+1}$. And to cover sections #3, #4, and #5 is the same as covering the entire image with boxes of side length $1/2^k$. Thus

$$N_\square(K, 1/2^{k+2}) = 2N_\square(K, 1/2^{k+1}) + 3N_\square(K, 1/2^k)$$

$$\Rightarrow \quad x(k + 2) = 2x(k + 1) + 3x(k),$$

verifying equation (5.4).

The next step is to get some idea of how big $x(k)$ is as k grows. To

We are using the
reduction-of-order technique
from problem 13 on page 34.

assist us, put $y(k) = x(k+1)$. Then equation (5.4) can be rewritten as

$$x(k+1) = y(k),$$

$$y(k+1) = x(k+2) = 2x(k+1) + 3x(k) = 3x(k) + 2y(k),$$

or in matrix notation,

$$\begin{bmatrix} x(k+1) \\ y(k+1) \end{bmatrix} = \begin{bmatrix} 0 & 1 \\ 3 & 2 \end{bmatrix} \begin{bmatrix} x(k) \\ y(k) \end{bmatrix}.$$

Aha! This is a two-dimensional linear discrete time dynamical system. Now we can work out an exact formula for $x(k)$ and/or $y(k)$ using the methods of Chapter 2. Let A be the matrix $\begin{bmatrix} 0 & 1 \\ 3 & 2 \end{bmatrix}$. The eigenvalues of A are -1 and 3, so we know that $x(k) = a3^k + b(-1)^k$ for some constants a and b. Since $x(0) = a + b = 1$ and $x(1) = 3a - b = 4$, we work out that $a = \frac{5}{4}$ and $b = -\frac{1}{4}$. Hence

$$N_\square\left(T, \frac{1}{2^k}\right) = \frac{5}{4} \cdot 3^k - \frac{1}{4} \cdot (-1)^k.$$

Therefore, for k large, $N_\square(T, 1/2^k) \approx \frac{5}{4}3^k$. We are now ready to compute the dimension of this fractal. Subsituting into the definition, we compute

$$\dim T = \lim_{r \downarrow 0} \frac{\log N(K, r)}{\log(1/r)}$$

$$= \lim_{k \to \infty} \frac{\log N_\square(K, 1/2^k)}{\log(2^k)}$$

$$= \lim_{k \to \infty} \frac{\log\left(\frac{5}{4} \cdot 3^k - \frac{1}{4} \cdot (-1)^k\right)}{\log(2^k)}$$

$$= \lim_{k \to \infty} \frac{\log \frac{5}{4} + k \log 3}{k \log 2}$$

$$= \frac{\log 3}{\log 2} \approx 1.5850.$$

That was a lot of work. In the next section we will see how to compute the dimension of this fractal less painfully.

The boxes we used in computing the dimension of T (the fractal in Figure 5.26) were nicely lined up in a grid. I think this is the best way to cover J with boxes of side length $\frac{1}{4}$, $\frac{1}{8}$, $\frac{1}{16}$, etc., but I am not 100% sure. Nonetheless, we can have 100% confidence in our dimension computation. Here's why.

We know that in the definition of dimension we may use either N or $N_\square$ and obtain the same result. The axis-parallel square boxes we use in computing $N_\square(K, r)$ just have to have side length r and be parallel to the x-axis. Otherwise, they are free to float about the plane. Once you find a good covering of K by these boxes, you may still worry that you have not found the very best (minimum size) cover. It would be a lot easier to simply draw some graph paper over the image and use the boxes that touch the fractal. Not only is it easier, it's correct!

To make "draw graph paper over the fractal" more precise, we introduce the idea of a grid box. Let n be a positive integer. A *grid box* of side length $1/n$ is defined to be a square box of side length $1/n$ whose corners' coordinates are rational numbers with n in the denominator. In other words, they are a set of the form

$$\left\{ (x, y) : \frac{i}{n} \le x \le \frac{i+1}{n}, \ \frac{j}{n} \le y \le \frac{j+1}{n} \right\},$$

where i and j are integers.

Let $N'_\square(K, 1/n)$ be the number of *grid* boxes of side length $1/n$ which intersect the fractal K. Reread the previous sentence! We did *not* say "... the *minimum* number of..."! The question is, Can we use $N'_\square$ in place of $N_\square$ (or N) in the definition of dimension? The answer is yes. To see why, we claim the following inequality:

$$N_\square(K, 1/n) \le N'_\square(K, 1/n) \le 4N_\square(K, 1/n). \tag{5.5}$$

The first inequality is easy: If the fractal K is covered by $N'_\square(K, 1/n)$ grid boxes, then the best covering with *any* boxes of side length $1/n$ is no larger.

To see the second part, consider a best possible cover of K by boxes of side length $1/n$. Replace each of these $N_\square(K, 1/n)$ boxes by four grid boxes as shown in Figure 5.28. This gives a cover of K by $4N_\square(K, 1/n)$ grid boxes. It might be the case that the fractal doesn't touch all these grid boxes—no matter, inequality (5.5) is correct.

Now we take logs across inequality (5.5) and divide by $\log(1/\frac{1}{n}) = \log n$ to get

$$\frac{\log N_\square(K, 1/n)}{\log n} \le \frac{\log N'_\square(K, 1/n)}{\log n} \le \frac{\log 4 + \log N_\square(K, 1/n)}{\log n}$$

Counting grid boxes: $N'_\square$.

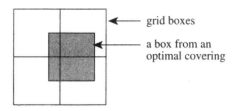

Figure 5.28. Replacing free-floating boxes with grid boxes.

and then we let $n \to \infty$ to find

$$\dim K \le \lim_{n\to\infty} \frac{\log N'_\square(K, 1/n)}{\log n} \le \dim K.$$

Therefore, we may use $N'_\square$ in place of either $N_\square$ or N in the definition of fractal dimension.

5.6.4 Just-touching similitudes and dimension

The dimension of the fractal T in Figure 5.26 (page 295) was difficult to compute. The structure of that fractal, however, is rather special and allows us to find its dimension in a simpler manner. The IFS of which T is the attractor consists of *just-touching similitudes*. Let's explore what they are.

Similitudes

Similitudes: translations, rotations, reflections, and dilations.

An affine transformation of the plane is a function of the form

$$h \begin{bmatrix} x \\ y \end{bmatrix} = \begin{bmatrix} a & b \\ c & d \end{bmatrix} \begin{bmatrix} x \\ y \end{bmatrix} + \begin{bmatrix} e \\ f \end{bmatrix}.$$

Now, if A is a compact set, then $h(A)$ is another compact set which looks like A. The transformation from A to $h(A)$ might rotate, reflect, squash, and skew the set A, as in Figure 5.29. The salient feature of affine transformations is that straight lines remain straight. Angles between lines, however, can change. See §A.1.6 on page 344.

A *similitude* is a rather special type of affine transformation: It preserves angles. A *similitude* is any combination of the following actions:

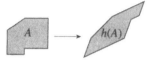

Figure 5.29. A typical affine transformation distorts the shape of sets.

• *a translation*: move points by a fixed vector, i.e.,

$$h\begin{bmatrix} x \\ y \end{bmatrix} = \begin{bmatrix} x \\ y \end{bmatrix} + \begin{bmatrix} e \\ f \end{bmatrix},$$

or

• *a rotation*: rotate points around the origin through an angle θ, i.e.,

$$h\begin{bmatrix} x \\ y \end{bmatrix} = \begin{bmatrix} \cos\theta & -\sin\theta \\ \sin\theta & \cos\theta \end{bmatrix}\begin{bmatrix} x \\ y \end{bmatrix},$$

or

• *a reflection*: reflect points through a line, as in

$$h\begin{bmatrix} x \\ y \end{bmatrix} = \begin{bmatrix} -1 & 0 \\ 0 & 1 \end{bmatrix}\begin{bmatrix} x \\ y \end{bmatrix} = \begin{bmatrix} -x \\ y \end{bmatrix},$$

or

• *a dilation*: multiply the coordinates of all points by a constant, i.e.,

$$h\begin{bmatrix} x \\ y \end{bmatrix} = \begin{bmatrix} c & 0 \\ 0 & c \end{bmatrix}\begin{bmatrix} x \\ y \end{bmatrix} = \begin{bmatrix} cx \\ cy \end{bmatrix}.$$

These basic actions are illustrated in Figure 5.30. When h is a similitude, then A and $h(A)$ have the same shape but not necessarily the same size. Observe that angles are not distorted by similitudes.

Now, a similitude can be a contraction mapping provided it includes a dilation which shrinks objects, i.e., it includes an action of the form $\begin{bmatrix} x \\ y \end{bmatrix} \mapsto \begin{bmatrix} cx \\ cy \end{bmatrix}$ with $|c| < 1$. Indeed, in a similitude all distances are treated alike, i.e., $d[h(\mathbf{x}), h(\mathbf{y})] = cd(\mathbf{x}, \mathbf{y})$, where c is a constant that

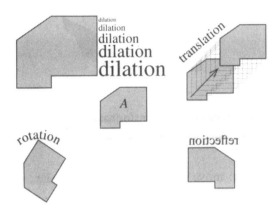

Figure 5.30. The four basic similitudes: dilation, translation, rotation, and reflection. Notice that the shape of A is the same as that of $h(A)$.

doesn't depend on **x** or **y**. We call c the *dilation factor* for the similitude, and a similitude is a contraction map if and only if its dilation factor is less than 1.

Observe that the five affine transformation in T's IFS (refer to Figure 5.26 on page 295) are all similitudes. Their dilation factors are $\frac{1}{2}$, $\frac{1}{2}$, $\frac{1}{4}$, $\frac{1}{4}$, and $\frac{1}{4}$.

Just-touching

The other feature of the five similitudes, f_1 through f_5, defining T (Figure 5.26 on page 295) is that they are *just-touching*. The precise meaning of *just-touching* is that the intersection $f_i(T) \cap f_j(T)$ is of smaller dimension than T. This is not very helpful, since we want to *use* the idea of just-touching in computing dimension.

Fortunately, there's an easier way to tell if the images $f_1(T)$ through $f_5(T)$ are just-touching. Draw a box around each of $f_1(T)$ through $f_5(T)$. If these boxes do not overlap—if they meet only along their boundaries or not at all—then the regions are just-touching. This is the case for the fractal T. We see in Figure 5.26 the five regions #1 through #5 which verify that the IFS has the just-touching property.

There are other examples of just-touching IFSs in this chapter: Cantor's set, Sierpiński's triangle, the top of Koch's snowflake in Figure 5.15 (page 268), the X-shaped fractal of Figure 5.16 (page 272), the fractal W of

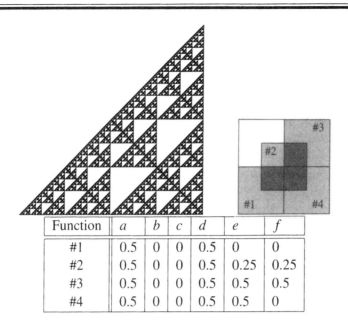

Function	a	b	c	d	e	f
#1	0.5	0	0	0.5	0	0
#2	0.5	0	0	0.5	0.25	0.25
#3	0.5	0	0	0.5	0.5	0.5
#4	0.5	0	0	0.5	0.5	0

Figure 5.31. Notice that the four half-size copies of this fractal are not just-touching.

Figure 5.17 (page 273), and the fractal J of Figure 5.19 (page 274). However, consider the fractal in Figure 5.31. The image contains four half-size copies of the original, but these four copies are *not* just-touching.

A formula for just-touching similitudes

Of the just-touching IFSs we have seen in this chapter (listed in the preceding paragraph), all except W consist completely of similitudes. We now give a simple formula for the dimension of any fractal K which is the attractor of an iterated function system of just-touching similitudes.

Suppose $F = f_1 \cup f_2 \cup \cdots \cup f_k$ is an IFS of just-touching similitudes with dilation factors $c_1, c_2, \ldots, c_k$, respectively. Let K be its attractor and let $d = \dim K$. Then

$$c_1^d + c_2^d + \cdots + c_k^d = 1. \tag{5.6}$$

Solve this equation for d to find the dimension of a fractal generated by just-touching similitudes.

Equation (5.6) is a beautiful relation between the dilation factors and the dimension, but is often hard to solve for d. Before we explain why equation (5.6) works, let's look at some examples.

Consider the fractal X in Figure 5.16. Its IFS consists of five similitudes all with dilation factor $\frac{1}{3}$. We therefore know that its dimension, d, satisfies

$$\left(\frac{1}{3}\right)^d + \left(\frac{1}{3}\right)^d + \left(\frac{1}{3}\right)^d + \left(\frac{1}{3}\right)^d + \left(\frac{1}{3}\right)^d = 1,$$

or $5\left(\frac{1}{3}\right)^d = 1$. We can solve this equation for d as follows:

$$5\left(\frac{1}{3}\right)^d = 1$$

$$\left(\frac{1}{3}\right)^d = \frac{1}{5}$$

$$\log\left(\frac{1}{3}\right)^d = \log\frac{1}{5}$$

$$-d\log 3 = -\log 5$$

$$\therefore \dim X = d = \frac{\log 5}{\log 3} \approx 1.4650.$$

Let's do a more complicated example: the fractal T of Figure 5.26 on page 295). The IFS for T consists of five just-touching similitudes. Their dilation factors are $\frac{1}{2}, \frac{1}{2}, \frac{1}{4}, \frac{1}{4}$, and $\frac{1}{4}$. Thus by equation (5.6), $d = \dim T$ satisfies

$$\left(\frac{1}{2}\right)^d + \left(\frac{1}{2}\right)^d + \left(\frac{1}{4}\right)^d + \left(\frac{1}{4}\right)^d + \left(\frac{1}{4}\right)^d = 1. \tag{5.7}$$

This equation looks difficult to solve for d, but a trick makes it tractable. Let $x = 2^d$. So $\left(\frac{1}{2}\right)^d = 1/x$, and $\left(\frac{1}{4}\right)^d = 1/x^2$. Now equation (5.7) simplifies to

$$\frac{2}{x} + \frac{3}{x^2} = 1 \quad \Rightarrow \quad 2x + 3 = x^2 \quad \Rightarrow \quad x^2 - 2x - 3 = 0.$$

The roots of $x^2 - 2x - 3 = 0$ are $x = 3$ and $x = -1$. Since $x = 2^d$, it is impossible for x to be negative, so we know $x = 3$. Thus $2^d = 3$ and therefore $d = \log_2 3$.

Why the just-touching formula works.

A formal proof of equation (5.6) has many dirty details. However, the main idea is quite accessible. Let's see what's at the heart of equation (5.6).

Let K be the attractor of $F = f_1 \cup \cdots \cup f_k$, where the f_i's are just-touching similitudes, and the dilation factor of f_i is c_i. To compute $\dim K$

we would like to know $N(K, r)$, the minimum number of balls of radius r which cover K.

Now, K is the attractive fixed point of F, so

$$K = F(K) = f_1(K) \cup f_2(K) \cup \cdots \cup f_k(K).$$

Thus

$$N(K, r) \approx N(f_1(K), r) + N(f_2(K), r) + \cdots + N(f_k(K), r). \quad (5.8)$$

We do *not* have $=$ because a ball which covers part of $f_1(K)$ may also cover part of $f_2(K)$. We *do* have $\approx$ because this ball sharing can net us only a very modest savings. The reason is, the regions $f_i(K)$ are *just-touching* and have no significant overlap. (In a formal proof, we would need to make this precise.)

Next, we recall that $f_i(K)$ is a shrunken copy of K. Indeed, $f_i(K)$ is simply a c_i-scale miniature. So the number of balls of radius r we need to cover $f_i(K)$ is exactly the same as the number of balls of radius r/c_i we need to cover all of K. In equation form,

$$N(f_i(K), r) = N(K, r/c_i) \quad (5.9)$$

(and this time we do mean exactly equals).

Now, if K has dimension d, this means that $N(K, r) \approx ar^{-d}$ for some constant a. Using this relation in equation (5.9), we have

$$N(f_i(K), r) \approx a \left(\frac{r}{c_i} \right)^{-d},$$

which we substitute into equation (5.8) to get

$$ar^{-d} \approx a \left(\frac{r}{c_1} \right)^{-d} + a \left(\frac{r}{c_2} \right)^{-d} + \cdots + a \left(\frac{r}{c_k} \right)^{-d}.$$

As $r \downarrow 0$ the approximations become better and better, and if we divide the preceding equation through by ar^{-d}, we finally obtain our goal [equation (5.6)]:

$$1 = c_1^d + c_2^d + \cdots + c_k^d.$$

A final example: Let L be the fractal depicted in Figure 5.32. Observe that L is the attractor of an IFS of four just-touching similitudes with dilation factors $\frac{1}{2}, \frac{1}{2}, \frac{1}{3}$, and $\frac{1}{4}$. Thus dim $L = d$ satisfies

When we can't solve equation (5.6).

$$\left(\frac{1}{2} \right)^d + \left(\frac{1}{2} \right)^d + \left(\frac{1}{3} \right)^d + \left(\frac{1}{4} \right)^d = 1.$$

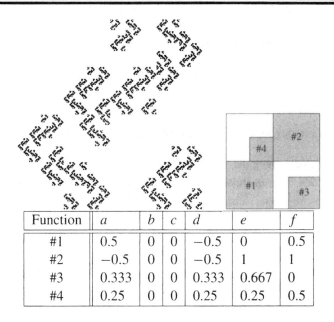

Function	a	b	c	d	e	f
#1	0.5	0	0	−0.5	0	0.5
#2	−0.5	0	0	−0.5	1	1
#3	0.333	0	0	0.333	0.667	0
#4	0.25	0	0	0.25	0.25	0.5

Figure 5.32. A fractal L and its associated IFS.

I don't know how to solve this equation for d, so we'll need to be content with a numerical solution. Happily, this is available by computer. Using *Mathematica* we type and the computer responds:

```
FindRoot[(1/2)^d+(1/2)^d+(1/3)^d+(1/4)^d==1,{d,1}]
```

```
{d -> 1.52844}
```

Hence dim $L \approx 1.52844$.

Problems for §5.6

◆1. What is the dimension of the fractal of problem 2 on page 245 (see Figure 5.6).

◆2. Let C_t be the fractal defined in problem 3 on page 276. What is its dimension?

◆3. Find a fractal with dimension exactly $\frac{1}{2}$.

◆4. Let K be a compact subset of the plane. Show that in the definition of fractal dimension of K we can replace $N(K, r)$ with $N_\Delta(K, r)$: the minimum number of equilateral triangles of side length r and with bottom side parallel to the x-axis needed to cover K. (In this problem, *triangle* means the triangle together with its interior.)

◆5. Let F be a compact set in the plane with fractal dimension d. Suppose we rescale F by doubling its size (in both dimensions). Find, in terms of d, the fractal dimension of the rescaled version of F.

◆6. What is the fractal dimension of the Koch snowflake?

◆7. Suppose K_1 and K_2 are compact sets with fractal dimensions d_1 and d_2 respectively. What is the fractal dimension of $K_1 \cup K_2$? What can you say about the fractal dimension of $K_1 \cap K_2$?

◆8. Suppose $K_1, K_2, K_3, \ldots$ and K are compact subsets of the plane and $d(K_i, K) \to 0$ as $i \to \infty$.

Suppose further that dim $K_i = d_i$ and dim $K = d$.

Is it true that
$$\lim_{i \to \infty} d_i = d?$$

◆9. A square of side length 1 can be covered by a single disk of radius 1. The same is true in $\mathbf{R}^3$: a cube of side length 1 fits inside a ball of radius 1.

The same is also true in $\mathbf{R}^4$, but false in $\mathbf{R}^5$. Prove both of these assertions. Namely,

   (a) Prove that a ball of radius 1 centered at the origin in $\mathbf{R}^4$ includes all points of the form $\left(\pm\frac{1}{2}, \pm\frac{1}{2}, \pm\frac{1}{2}, \pm\frac{1}{2}\right)$: the corners of a hypercube of side length 1.

   (b) Prove that a ball of radius 1 centered at the origin in $\mathbf{R}^5$ does not contain any of the points $\left(\pm\frac{1}{2}, \pm\frac{1}{2}, \pm\frac{1}{2}, \pm\frac{1}{2}, \pm\frac{1}{2}\right)$.

◆10. The definition of the dimension of a set K is
$$\lim_{r \downarrow 0} \frac{\log N(K, r)}{\log(1/r)}.$$

Explain why the result we get does not depend on the base of the logarithm.

Hint: If a, b, x, y are positive numbers, show that
$$\frac{\log_a x}{\log_a y} = \frac{\log_b x}{\log_b y}.$$

5.7 Examplification: Fractals in nature

From the moment you awaken in the morning until you nod off to sleep at night, it is hard *not* to see fractals. We know how to compute the fractal dimension of *mathematical* fractals. In this section we discuss how to compute the fractal dimension of real-world fractals and discuss how knowing the fractal dimension can be useful.

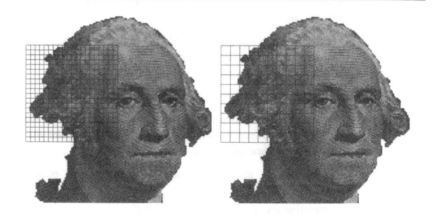

Figure 5.33. Overlaying a fractal image with different grids.

5.7.1 Dimension of physical fractals

The key to computing the dimension of real-world fractals is the box-counting method (see page 299). Here is what we need to do.

Draw a fractal F on very fine graph paper (or greatly magnify the image and plot it on ordinary graph paper). For example, see the left portion of Figure 5.33. Count how many 1×1 boxes touch the boundary of the image. Next, group the boxes in pairs (ignore every other vertical and horizontal line) so the graph paper is twice as coarse as before (right side of the figure). Count the number of 2×2 boxes which touch the boundary of the image. Continue in this fashion, counting how many 3×3 boxes touch the boundary, how many 4×4 boxes, and so on. Compile the data into a table, such as Table 5.4.

We expect the number of $r \times r$ boxes touched by the fractal to be proportional to r^{-d}, where d is the dimension. In symbols,

$$N'_\square(F, r) \approx C r^{-d},$$

where C is a constant. Taking logarithms, we have

$$\log N'_\square(F, r) \approx -d \log r + \log C.$$

This says that $\log N'_\square(F, r)$ should behave like an affine function of $\log r$. Thus if we plot $\log N'_\square(F, r)$ versus $\log r$, we expect to see a straight line with slope of $-d$. A convenient way to do this is to plot $N'_\square(F, r)$ versus

Box size $r \times r$	Number of boxes $N'_\square(F, r)$
1×1	938
2×2	408
3×3	250
4×4	177
5×5	135
6×6	109
7×7	90
8×8	77
9×9	67
10×10	59
12×12	47
14×14	39
16×16	33
18×18	29
20×20	25

Table 5.4. Box-counting data from a hypothetical fractal image, F.

r on log-log graph paper. We have done this in Figure 5.34 using the data from Table 5.4. The slope of the line in the graph is computed as follows:

$$\frac{\Delta \log N'_\square(F, r)}{\Delta r} = \frac{\log 25 - \log 938}{\log 20 - \log 1} \approx \frac{1.3979 - 2.9722}{1.3010 - 0} \approx -1.21;$$

therefore, the dimension of the hypothetical fractal image is roughly 1.2.

The data in Table 5.4 are contrived, and the resulting graph (Figure 5.34) is too perfect, so let's look at some real data.

A photograph of a fractal image is scanned into a computer and stored as a matrix. The matrix can be thought of as graph paper upon which the image is drawn, and each entry in the matrix is a pixel, i.e., a box of the graph paper. We then count the number of boxes (of various sizes) which touch the boundary of the image and assemble the data into a table; see Table 5.5. (These data come from the analysis of a fractal which is similar to the image in Color Plate 1; the fractal curve is the boundary between different colors, say, light and dark blue.) Then we plot the log of the number of boxes versus the log of the box size; see Figure 5.35. The dots in the figure represent the data in Table 5.5. The line drawn is a reasonable

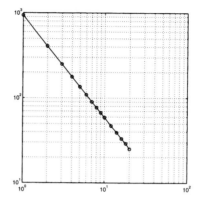

Figure 5.34. Log-log plot of r versus $N'_\square(F, r)$ using the data from Table 5.4.

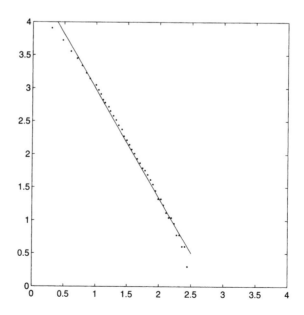

Figure 5.35. Log-log plot of box size versus box counts from the data in Table 5.5.

| Box | | Log of box | |
Size	Number	Size	Number
2	8110	0.3010	3.9090
3	5272	0.4771	3.7220
4	3589	0.6021	3.5550
5	2812	0.6990	3.4490
6	2188	0.7782	3.3400
7	1698	0.8451	3.2299
8	1396	0.9031	3.1449
10	1117	1.0000	3.0481
11	940	1.0414	2.9731
12	817	1.0792	2.9122
13	670	1.1139	2.8261
14	605	1.1461	2.7818
16	521	1.2041	2.7168
17	450	1.2304	2.6532
19	381	1.2788	2.5809
21	329	1.3222	2.5172
23	275	1.3617	2.4393
26	238	1.4150	2.3766
28	188	1.4472	2.2742
31	164	1.4914	2.2148
34	140	1.5315	2.1461
37	119	1.5682	2.0755
41	103	1.6128	2.0128
45	86	1.6532	1.9345
50	74	1.6990	1.8692
55	62	1.7404	1.7924
60	57	1.7782	1.7559
66	49	1.8195	1.6902
73	41	1.8633	1.6128
80	35	1.9031	1.5441
88	28	1.9445	1.4472
97	21	1.9868	1.3222
107	21	2.0294	1.3222
117	17	2.0682	1.2304
129	13	2.1106	1.1139
142	11	2.1523	1.0414
156	11	2.1931	1.0414
172	9	2.2355	0.9542
189	6	2.2765	0.7782
208	6	2.3181	0.7782
229	4	2.3598	0.6021
252	4	2.4014	0.6021
277	2	2.4425	0.3010

Table 5.5. Box-counting data from a real-world fractal, such as the one in Color Plate 1. (Data courtesy Professor K. R. Sreenivasan, Yale University.)

approximation of the data by a straight line; [16] the slope of that line is about -1.67, so the fractal dimension of the image is approximately $\frac{5}{3}$.

5.7.2 Estimating surface area

Many objects (such as automobiles) have smooth surfaces. Others, such as clouds or sandpaper, have textured surfaces which are better described by fractals. A pretty example of a physical system with fractal surfaces is shown in Color Plate 1. This photograph shows one fluid being injected at moderate speed into a surrounding fluid. What is the surface area of the boundary between regions of different concentrations (different colors in Color Plate 1) of the injected liquid? One way to estimate this area is to approximate the shape of the jet stream with a cone. This is a rather poor approximation because it ignores the intricate folding of this surface. However, using the box-counting method, we can make a much better approximation of the surface area of the turbulent jet stream.

The surface area is of interest because the rate at which material in the jet and in the surrounding medium mix is proportional to the surface area.

When we compute the fractal dimension of a physical fractal F we cannot take the definition too literally. If we take r extremely small, then $N'_\square(F, r)$ counts the number of *atoms* in the fractal. Since this is a finite number, $N'_\square(F, r)$ becomes more or less constant as r enters the atomic scale, suggesting that the dimension of F is 0.

Don't take the definition of fractal dimension to its illogical extreme.

Typically, one does not have to drop to the atomic scale for the fractal nature of physical objects to vanish. For example, the liquid jet's (as in Color Plate 1) fractal nature vanishes on a scale around 1 mm; the surface of the turbulence is fairly smooth if we look at only a square millimeter (or smaller) portion of the surface.

The cutoff threshold.

Thus there is a cutoff in fractal dimension (which we denote by $r_{\min}$) below which the fractal nature of the physical object disappears. For the jet in Color Plate 1, $r_{\min} \approx 1$ mm. This means that the formula $N'_\square(F, r) \approx Cr^{-d}$ works down to only $r_{\min}$, and below that point the approximation breaks down.

For a graphic illustration of the cutoff phenomenon see Figure 5.36. For values of r above $r_{\min}$, the log-log plot is steeper than below. At scales below $r_{\min}$ the image is smooth and therefore of lower dimension than at the fractal scale. (Although the data for Figure 5.36 are contrived, the same phenomenon can be seen in the real data presented in Figure 5.35.)

A surface of constant concentration inside a turbulent jet stream has fractal dimension around $\frac{8}{3}$.

Let us consider the inner regions of a turbulent jet stream. Experiments

---

[16]The line was computed using linear regression, a technique for finding a line which best fits approximately linear data.

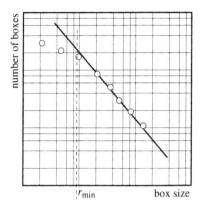

Figure 5.36. For box size r below r_{min}, the relation $N'_\square(F, r) \approx Cr^{-d}$ deteriorates.

show that surfaces of constant concentration have fractal dimension[17] around $\frac{8}{3}$, and the fractal behavior breaks down at $r_{min} \approx 1\text{mm}$. We can use this information to estimate the surface area.

To begin, recall that $N'_\square(F, r) \approx Cr^{-d}$ for $r \geq r_{min}$. It is reasonable to estimate $N'_\square(F, r)$ for a large value of r (say, around $r_{max} = 10\,\text{cm} = 100\,\text{mm}$) by direct observation of the jet. Since $N'_\square(F, r_{max}) \approx Cr_{max}^{-8/3}$, we can solve for $C \approx r_{max}^{8/3} N'_\square(F, r_{max})$.

Once we have estimated C, we can estimate

$$N'_\square(F, r_{min}) \approx Cr_{min}^{-8/3} \approx N'_\square(F, r_{max}) \left(\frac{r_{max}}{r_{min}} \right)^{8/3}.$$

Finally, since the surface is reasonably smooth on the scale of r_{min}, a box of side length r_{min} accounts for approximately r_{min}^2 of the surface area. Thus the surface area is roughly $N'_\square(F, r_{min})r_{min}^2$, or

$$\text{surface area} \approx N'_\square(F, r_{max})r_{max}^{8/3}r_{min}^{-2/3}.$$

In conclusion, because the surfaces of many objects appear to be fractals on all but the smallest scales (where they are smooth), we can estimate their

---

[17]The dimension is computed as follows. A cross-sectional photograph of the jet is taken by illuminating the liquid with a plane of light. The fractal boundary surface becomes a fractal boundary curve whose dimension is $\frac{5}{3}$ (see the data in Table 5.5, which are plotted in Figure 5.35). The fractal dimension of the surface is $1 + \frac{5}{3} = \frac{8}{3}$. This is just like slicing a smooth two-dimensional surface and getting a $2 - 1 = 1$-dimensional smooth curve.

surface areas by knowing the fractal dimension and then performing a quick box count on a coarse scale.

5.7.3   Image analysis

Consider the aerial photograph in the upper portion of Figure 5.37. In this picture we can see runways, roads, grass, trees, a bridge, buildings, and so on. You can recognize these structures because you have a brain which can interpret these images. Let's see how we can use fractal dimension to help a machine interpret a scene.[18] Methods akin to the ones described here are being developed to read mammograms and to detect cancerous tissue [21].

Converting a picture into a surface.

The first step is to convert a picture into a surface. To do this, we first notice that a picture is actually a large rectangular array of numbers. Each point in the picture has a corresponding gray-scale value between 0 (black) and 255 (white). To render the photograph, we put a dot of the appropriate shade of gray at each point in the photo. Alternatively, we could raise each point above the plane of the photo by an amount proportional to its gray-scale value; see Figure 5.38 which shows the "gray-scale surface" formed from a small portion of the photograph in the lower left where a cluster of trees has a fairly straight boundary with a patch of grass.

Notice that toward the front of the diagram the surface is crinkly, while to the rear the surface is smoother. Indeed, the fractal dimension of the surface near the front (tree tops) is about 2.6, whereas the dimension of the portion in back (grass) is around 2.3.

In the lower portion of Figure 5.37 we present an image derived from the original photograph by estimating the fractal dimension at each point in the photograph. Points of lower fractal dimension (near 2) are dark, and points of higher fractal dimension (near 3) are light. Notice that the river comes up quite dark in the image because its surface is fairly flat and smooth. The roads, runways, and bridge in the photograph (which are also quite flat) emerge as dark double bands in the dimension image (the double-banding is an artifact of how the fractal dimension was computed).

---

[18]There are much more sophisticated methods for analyzing image data. The fractal dimension method we discuss here is useful because although it cannot resolve fine details, it may be used as a preprocess for more sophisticated (but slower) methods. If we know the fractal dimension of the kind of object we seek, then a first pass with this method can greatly narrow the scope of a search with the more time consuming methods.

Figure 5.37. An aerial photograph (above) and an image representing the fractal dimension at each point in the scene (lower). Points of lower fractal dimension are darker. (Images courtesy Naval Surface Warfare Center, Dahlgren, Virginia.)

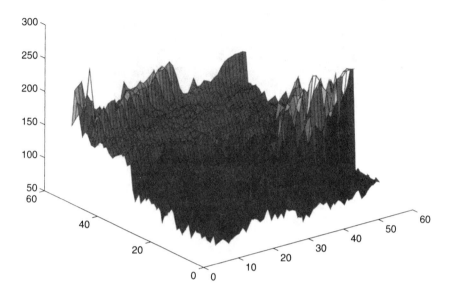

Figure 5.38. A portion of the photograph of Figure 5.37 rendered as a surface.

Problems for §5.7

◆1. When computing the dimension of a physical fractal, you should apply box counting to only the boundary of the object. For example, if you were to compute the dimension of George Washington's hair (see Figure 5.33) you should count only the boxes that are part empty and part covered by the head. Why? What would happen if you computed the dimension using *all* the boxes touched by the head?

◆2. The formula we derived for surface area (at the end of §5.7.3) depended on the dimension of the fractal being $\frac{8}{3}$. Rederive this formula for an arbitrary fractal of dimension d with d between 2 and 3.

Suppose the physical fractal is flat (lives in plane) and you wish to estimate its perimeter. Derive a formula for the perimeter based on the dimension d and the cut off threshold r_{min}.

Chapter 6

Complex Dynamical Systems

In most of our work we have been using real numbers. The occasional complex numbers which arose when we were finding eigenvalues (in Chapter 2) were quickly whisked away to become sine and cosine terms.

We now invite the return of complex numbers to our study of dynamical systems. We explore discrete time dynamical system in one *complex* variable. In other words, we ask, What happens when we iterate a function $f(x)$ where x may be a complex number?

Since we are working with complex numbers, you might like to review §A.2.

6.1   Julia sets

6.1.1   Definition and examples

In §4.2.3 we introduced the family of functions $f_a(x) = x^2 + a$. In §4.2.5 we studied the set of values B for which iterations of f_a (with $a = -2.64$) remains bounded. Let's consider what happens for other values of a, not just $a = -2.64$.

Let us define B_a to be the set of all values x for which the iterates $f_a^k(x)$ stay bounded, and let U_a be the set of x for which $f_a^k(x)$ explodes. In symbols,

$$B_a = \left\{ x : |f_a^k(x)| \nrightarrow \infty \text{ as } k \to \infty \right\}, \quad \text{and}$$

$$U_a = \left\{ x : |f_a^k(x)| \to \infty \text{ as } k \to \infty \right\}.$$

Note that B_a and U_a are complementary sets. The boundary between these sets is denoted J_a. The set J_a is called the *Julia set* of the function f_a, and

Julia set.

317

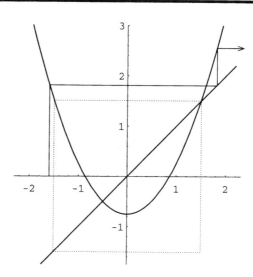

Figure 6.1. Plot of the function $f(x) = x^2 - \frac{3}{4}$. If $|x| \leq 1.5$, then $f^k(x)$ stays bounded. If $|x| > 1.5$, then $f^k(x)$ goes to infinity.

the set B_a is called the *filled-in Julia set* of f_a.

For example, when $a = -2.64$ we see (§4.2.5) that most numbers x are in U_a, and what remains in B_a forms a set with the same structure as Cantor's set.

Let's do another example: Let $a = -\frac{3}{4}$. There isn't much special about this value, except that it's between -2 and $\frac{1}{4}$. The fixed points of $f_{-3/4}(x)$ are $-\frac{1}{2}$ and $\frac{3}{2}$. Notice that if $-1.5 \leq x \leq 1.5$, then $f_a(x)$ is also between ± 1.5; this is most clearly seen in the plot of f_a shown in Figure 6.1. (We actually see that if x is between ± 1.5, then $f_a(x) \in [-0.75, 1.5] \subset [-1.5, 1.5]$.) However, if $|x| > 1.5$, then observe that $f_a^k(x) \to \infty$ as $k \to \infty$. Again, we can see this most clearly from the figure. Thus we see that $B_a = [-1.5, 1.5]$, and $U_a = (-\infty, -1.5) \cup (1.5, \infty)$. The boundary between these is the two-point set $J_a = \{-1.5, 1.5\}$.

The significance of the range $-2 \leq a \leq \frac{1}{4}$ is explained later—see, for example, Figure 6.13 on page 328.

That's all correct but not what we intended. Our intention is not to work just with real numbers (**R**) but to work with complex values (**C**) as well. The function $f_{-3/4}(x) = x^2 - \frac{3}{4}$ is also a function of a complex number. For example, if $x = 1 + 2i$, then $f_{-3/4}(x) = (1 + 2i)^2 - \frac{3}{4} = (-3 + 4i) - \frac{3}{4} = -3.75 + 4i$. All we have determined thus far is the *real part* of the sets B_a, U_a, and J_a. We want to know the full story. Since the

Figure 6.2. Filled-in Julia set B_a for $a = -\frac{3}{4}$.

real part of B_a is very simple (an interval centered at the origin), we might suppose that the full B_a is equally simple, say, a circle of radius 1.5 about the origin. This is a good guess, but the set B_a is not a circle. We're in for a delightful surprise.

We can make a picture of the set B_a as follows. Every complex number z corresponds to a point in the plane. We can plot a point for every element of B_a and thereby produce a two-dimensional depiction of B_a. This is exactly what we have done in Figure 6.2. The set $B_{-3/4}$ is symmetrical with respect to both the real (x) and the imaginary (y) axes. It runs from -1.5 to 1.5 on the real axis and roughly[1] between ± 0.9 on the imaginary.

A geometric view of Julia sets.

Notice that $B_{-3/4}$ is a rather bumpy set. These bumps don't go away as we look closer. In Figure 6.3 we greatly magnify the bump attached to the upper right part of the main section of $B_{-3/4}$. Notice that we see the same structure as we do in the whole. The set is a fractal!

What do other B_a sets look like?

For example, we can take a to be a pure imaginary number such as $-i/2$, as in Figure 6.4. The filled-in Julia set $B_{-i/2}$ looks like an island.

We can take a to be a complex number such as $-0.85 + 0.18i$. The result is the image in Figure 6.5. The set B_a looks like dancing flames. The image sits between (roughly) ± 1.7 on the real axis and ± 0.9 on the

---

[1]Actually, we can work out exactly how far up and down the y-axis $B_{-3/4}$ runs. Let $z = iy$ (where y is real). Then $f_{-3/4}(z) = -y^2 - \frac{3}{4}$, which is real (and negative). So $iy \in B_{-3/4}$ if and only if $-y^2 - \frac{3}{4} \in B_{-3/4}$. For this to happen, we need $|-y^2 - \frac{3}{4}| \le \frac{3}{2}$, i.e., $y^2 \le \frac{3}{4}$, or $|y| \le \sqrt{3/4} \approx 0.866$.

Figure 6.3. A close-up of one of the bumps on the main section of $B_{-3/4}$.

Figure 6.4. The filled-in Julia set $B_{-i/2}$.

Figure 6.5. The filled-in Julia set $B_{-0.85+0.18i}$.

Figure 6.6. Close-up of $B_{-0.85+0.18i}$.

imaginary.

Let's take a closer look at the region where the main flame is holding hands with the somewhat smaller flame on the right. We zoom in on this portion in Figure 6.6. This close-up view runs from about 0.46 to 0.67 on the real axis and from -0.17 to -0.02 on the imaginary. Again, we see that the two tips of the flames are just touching in the center of the picture, and so we zoom in even closer. Figure 6.7 is an extreme close-up of the

Julia sets look the same at all levels of magnification.

Figure 6.7. An extreme close-up of $B_{-0.85+0.18i}$.

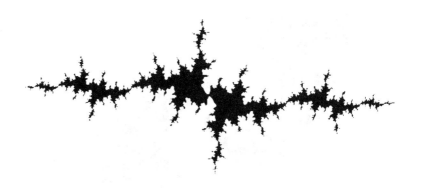

Figure 6.8. $B_{-1.24+0.15i}$.

middle of Figure 6.6. Notice that we don't see anything new! Now, this
extreme close-up runs from 0.55 to 0.56 on the real axis and from −0.09 to
−0.08 on the imaginary, and is about seven times bigger than the previous
close-up. Indeed, if you were to examine a $100\times$ magnification of this
same portion, you would see the same structure over and over again. The
Julia sets are fractals!

The Julia sets can have many wonderful shapes. Further examples are
given in Figures 6.8 through 6.11.

Figure 6.9. $B_{-0.16+0.74i}$.

Figure 6.10. $B_{0.375+0.333i}$.

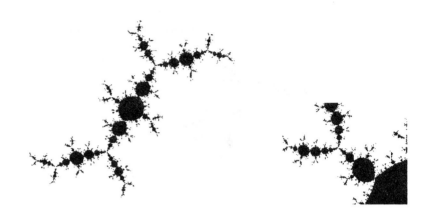

Figure 6.11. $B_{-0.117-0.856i}$. The inset on the right shows a close-up of a portion which appears disconnected in the main image.

6.1.2 Escape-time algorithm

How do we compute pictures of Julia sets (such as Figures 6.2 through 6.11)? The method is fairly straightforward.

We fix a complex number a and ask, Is a given complex number z in B_a? To answer this, we pick a big number k and check if $|f^k(z)|$ is large. The issue is, what do *big* and *large* mean in the previous sentence?

Large means bigger than 2.   First, let's tackle *large*. We say $|f^k(z)|$ is *large* if $|f^k(z)|$ is larger than both 2 and $|a|$. Typically (but not always) we have $|a| \leq 2$, so we simply require $|f^k(z)| > 2$. Now your reaction probably is, "2 isn't very *large*!", so let us justify this choice. We claim that if $|z| > 2$ and $|z| \geq a$, then $f^k(z)$ explodes as $k \to \infty$ (we prove this in a moment). If as we iterate f starting at z we ever reach a point whose absolute value is bigger than 2 (and at least $|a|$), then we are 100% sure that $|f^k(z)| \to \infty$. Now let's see why this works. Suppose $|z| > 2$ (and $|z| \geq |a|$). Since $|z| > 2$, we know that $|z| \geq 2 + \varepsilon$ for some fixed positive number ε. We now compute

$$
\begin{aligned}
|f_a(z)| &= |z^2 + a| & &\text{by definition of } f_a \\
&\geq |z^2| - |a| & &\text{since } |z^2 + a| + |-a| \geq |z^2| \text{ (triangle ineq.)} \\
&\geq |z^2| - |z| & &\text{since } |z| \geq |a| \\
&= |z|(|z| - 1) & &\text{factoring} \\
&\geq |z|(1 + \varepsilon) & &\text{since } |z| \geq 2 + \varepsilon.
\end{aligned}
$$

It now follows by repeated use of this reasoning that

$$
|f_a^k(z)| \geq |z|(1 + \varepsilon)^k \to \infty,
$$

and therefore $z \in U_a$.

Thus to test if $z \in B_a$, we compute $f_a^k(z)$ for $k = 1, 2, 3, \ldots$, and if we ever have $|f_a^k(z)| > \max\{2, |a|\}$, then we know that $z \in U_a$. There is a problem with this test. Suppose $z \in B_a$: How many iterations k of f_a should we compute before we are bored and decide that $z \notin U_a$? This is a harder question to answer. The answer depends on which you, the person making the computation, value more: speed or accuracy.

If speed is very important, then a modest value of k (such as 20) gives   *Speed versus accuracy.*
acceptable results. There will be some points which you will misidentify (you will think they are in B_a, but are they are really in U_a), but you will get a large percentage correct.

If accuracy is very important, then a large value of k (such as 1000) gives good results. You will make far fewer mistakes, but your computer will crank away much longer, and the pictures will take more time to produce.

Related to this issue is the magnification/resolution you are trying to   *Magnification versus*
achieve. If you simply want a coarse overview picture, then a small number   *resolution.*
of iterations suffices. On the other hand, if you are trying to produce a zoomed-in, high-resolution image, then a large value of k is important.

We assemble these ideas into an algorithm for computing Julia set images (page 326). In step 1 we say "For *all* values of $z = x + yi \ldots$" This should not be taken too literally. If the width of the screen is 200 pixels, then we want to take 200 values of x evenly spaced between $x_{\min}$ and $x_{\max}$. To prevent distortion, the shape of the plotting rectangle should be in proportion to the dimensions of the region ($x_{\max} - x_{\min}$ by $y_{\max} - y_{\min}$).

In step 1(d) we say "plot the point"; exactly how this is done depends on the computer system you are using.

Adding color

There is a natural and aesthetically pleasing way to add color to your plots of Julia sets. When you compute the iterations $f^k(z)$, record the first number k for which $f^k(z)$ is out of bounds, i.e., has large absolute value. That   *Remember: large means*
number k is called the *escape-time* of the point z. Now, if $z \in B_a$ we never   *bigger than 2.*
have $|f^k(z)|$ large, so its escape-time is infinite (or, effectively, MAXITS).

We now want to plot a point on the screen for *all* points z in the rectangular region ($x_{\min} \le x \le x_{\max}$ and $y_{\min} \le y \le y_{\max}$). For points in B_a (those whose escape-time we compute to be MAXITS) we plot a black dot. For all other points z (those z in U_a) we plot a color point depending on z's escape-time. For example, we could follow the usual spectrum from red to violet. When $k = 0$, we plot a red point, and as k increases to MAXITS -1, we plot various colors of the rainbow through to violet. On a gray-scale

Escape-Time Algorithm for Julia Sets

Inputs:

- The complex number a for $f_a(z) = z^2 + a$.

- The four numbers $x_{\min}$, $x_{\max}$, $y_{\min}$, $y_{\max}$ representing the region of the complex plane in which you are doing your computations. The lower-left corner is at $x_{\min} + y_{\min}i$, and the upper-right corner is at $x_{\max} + y_{\max}i$.

- A positive integer MAXITS. This is the maximum number of iterations of f_a you compute before you assume the point in question is in B_a.

Procedure:

1. For all values of $z = x + yi$ with $x_{\min} \leq x \leq x_{\max}$ and $y_{\min} \leq y \leq y_{\max}$ do the following steps:

   (a) Let $z^* = z$ (a temporary copy).

   (b) Let $k = 0$ (iteration counter).

   (c) While $|z^*| \leq 2$ and $|z^*| \leq |a|$ and $k < $ MAXITS do the following steps:

       i. Let $z^* = f_a(z^*) = (z^*)^2 + a$ (compute the next iterate).

       ii. Let $k = k + 1$ (increment the step counter).

   (d) IF $|z^*| \leq 2$ and $|z^*| \leq |a|$ (i.e., if you have fully iterated without breaking out of bounds)

   THEN plot the point z (i.e., (x, y)) on the screen.

monitor, we can assign various shades of gray to the possible escape-times. Try experimenting with applying different colors to different escape-times to create beautiful pictures. See Color Plate 2 for an example.

6.1.3   Other Julia sets

Using functions besides $f_a(x) = x^2 + a$.

There is no reason the functions $f_a(x) = x^2 + a$ must be used in the definition of Julia sets or in the escape-time algorithm. For any function $f: \mathbf{C} \to \mathbf{C}$ we can define B_f to be the set of complex numbers z which remain bounded when iterating f, U_f to be the set of those z for which $f^k(z)$ tends to infinity, and J_f to be the boundary between these regions. For example, we might let $f(z) = -0.3z^3 + 0.6z^2 - 0.4$. The resulting filled-in Julia set, B_f, is shown in Figure 6.12.

Figure 6.12. The filled-in Julia set B_f for $f(z) = -0.3z^3 + 0.6z^2 - 0.4$.

Besides polynomial functions f, you can also try the sine, cosine, and exp functions (see §A.2 to see how these are defined for complex numbers).

Problems for §6.1

◆1.   What is the filled-in Julia set B_0? Hint: You do not need a computer.

◆2.   Consider the set B_{-6}. Show that $2 \in B_{-6}$. Find several other values in B_{-6}.

◆3.   Find some points in B_{-1+3i}.

6.2   The Mandelbrot set

6.2.1   Definition and various views

If you create a variety of Julia sets (either using your own program, or using one of the many commercial and/or public-domain packages available), you may note that for some values of a the set B_a is fractal dust (as we saw in the case $a = -2.64$), and for some values of a the set B_a is a connected region. There is a simple way to decide which situation you are in: Iterate f_a starting at 0. If $f_a^k(0)$ remains bounded, then the set B_a will be connected, but if $|f_a^k(0)| \to \infty$, then B_a will be fractal dust. The justification of this fact is beyond the scope of this text.

Recall that $f_a(z) = z^2 + a$.

This leads to a natural question, For which values a does $f_a^k(0)$ remain bounded and for which values of a does it explode? The *Mandelbrot set*, denoted by $\mathscr{M}$, is the set of values a for which $f_a^k(0)$ remains bounded, i.e.,

$$\mathscr{M} = \left\{ a \in \mathbf{C} : |f_a^k(0)| \not\to \infty \right\}.$$

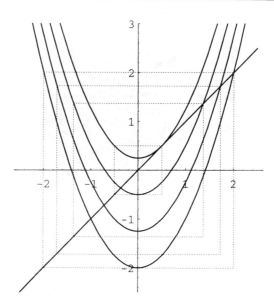

Figure 6.13. When a is between -2 and $\frac{1}{4}$, and $|x| \le (1 + \sqrt{1 - 4a})/2$, then $|f_a(x)| \le (1 + \sqrt{1 - 4a})/2$ as well.

Thus our question can be restated: Which numbers are in $.//?$ Let's start on the real line. When $-2 \le a \le \frac{1}{4}$, then $f_a(x) = x^2 + a$ has a fixed point at $\frac{1}{2}(1 + \sqrt{1 - 4a})$. Now, if $|x| \le \frac{1}{2}(1 + \sqrt{1 - 4a})$, graphical analysis shows that $|f(x)| \le \frac{1}{2}(1 + \sqrt{1 - 4a})$ as well; see Figure 6.13. Thus we certainly have $f_a^k(0)$ bounded whenever $-2 \le a \le \frac{1}{4}$.

Now, if $a > \frac{1}{4}$, we know that $f_a(x)$ has no (real) fixed point and that for any x (including $x = 0$) we have $f_a^k(x) \to \infty$. If $a < -2$, then $|f(0)| = |a| > \frac{1}{2}(1 + \sqrt{1 - 4a})$, and subsequent iterations of f_a explode; see Figure 6.14.

These computations show that for real a we have $f_a^k(0)$ bounded exactly for those $a \in \left[-2, \frac{1}{4}\right] = .//\cap \mathbf{R}$.

Now what about *complex values* of a? Again, it might be reasonable to expect that $.//$ has a simple appearance. Instead, we are startled to see that $.//$ looks like the image in Figure 6.15. This is an amazing set. Let's explore various parts of it.

The antenna sticking out to the left runs along the real (x) axis to -2. There appears to be a little blip toward the left end of the antenna. In

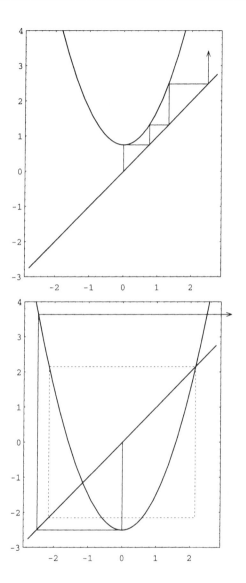

Figure 6.14. When $a > \frac{1}{4}$ (upper) or $a < -2$ (lower), we see that $f_a^k(0)$ explodes.

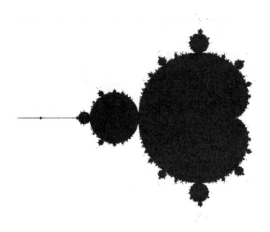

Figure 6.15. The Mandelbrot set . //.

Figure 6.16. The 'blip' in . //'s 'antenna' (near $a = -1.75$) looks like a small copy of . //.

Figure 6.16 we magnify this portion only to find another little copy of . // itself. (The antenna seems to have disappeared; this is an artifact of the software and the resolution of the image.)

Toward the top of the image in Figure 6.15 are some loose dots. In Figure 6.17 we zoom in on the topmost and see a small version . //. The indentation on the heart-shaped portion on the right is at $\frac{1}{4}$. In Figure 6.18 we extremely magnify a portion of . // in this region.

Next let's point our microscope at the region between the heart-shaped portion on the right and the largest circular portion in the middle. Fig-

Figure 6.17. Close-up near the top of $\mathcal{M}$ (around $a = -0.158 + 1.03i$).

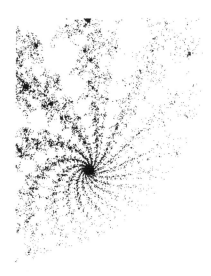

Figure 6.18. A portion of $\mathcal{M}$ near $a = 0.29 + 0.016i$.

Figure 6.19. A portion of $\mathscr{M}$ near $a = -0.748 + 0.0725i$.

ure 6.19 shows $\mathscr{M}$'s beautiful curlicues in this region.

6.2.2 Escape-time algorithm

There is clearly a lot to explore. It is worth your while to obtain commercial or public-domain software for drawing the Mandelbrot set, or else to write your own computer program. The idea is the same as the escape-time algorithm for computing Julia sets. First, we note that if $|a| > 2$, then $a \notin \mathscr{M}$. To see why, note that $f_a(0) = a$, and now we are iterating f_a at a point (a) whose absolute value is at least $|a|$ and greater than 2. Thus by the reasoning of §6.1.2 we conclude that $|f_a^k(0)| \to \infty$. If for some k we have $|f_a^k(0)| > 2$, then we know that $a \notin \mathscr{M}$. On the other hand, if $|f_a^k(0)| \leq 2$ for a large value of k, it is reasonable to suppose that $a \in \mathscr{M}$. We again have the speed/accuracy trade off. For coarse, quick pictures, you can iterate f a modest number of times, but for high-resolution/magnification images, a large number of iterations should be performed. The escape-time algorithm for computing images of the Mandelbrot set is presented on page 333. Note that this algorithm is virtually identical with the algorithm for computing Julia sets. If we assign colors to points in the plane based on their escape time, we can create beautiful pictures; see Color Plate 3.

Problems for §6.2

♦1.   Notice that $\mathscr{M}$ is symmetric about the x-axis. Explain why.

Escape-Time Algorithm for the Mandelbrot Set

Inputs:

- The four numbers x_{min}, x_{max}, y_{min}, y_{max} representing the region of the complex plane in which you are doing your computations. The lower-left corner is at $x_{min} + y_{min}i$, and the upper-right corner is at $x_{max} + y_{max}i$.

- A positive integer MAXITS. This is the maximum number of iterations of f_a (starting at 0) you compute before you assume the point in question is in $\mathcal{M}$.

Procedure:

1. For all values of $a = x + yi$ with $x_{min} \leq x \leq x_{max}$ and $y_{min} \leq y \leq y_{max}$ do the following steps:

   (a) Let $z^* = 0$ (a temporary copy).

   (b) Let $k = 0$ (iteration counter).

   (c) While $|z^*| \leq 2$ and $k <$ MAXITS, do the following steps:

      i. Let $z^* = f_a(z^*) = (z^*)^2 + a$ (compute the next iterate).

      ii. Let $k = k + 1$ (increment the step counter).

   (d) IF $|z^*| \leq 2$ (i.e., if you have fully iterated without breaking out of bounds)

      THEN plot the point z (i.e., (x, y)) on the screen.

6.3 Examplification: Newton's method revisited

In §1.2.9 on page 26 we discussed Newton's method for solving equations of the form $g(x) = 0$. In §3.4 on page 146 we analyzed this method and saw that roots of the equation $g(x) = 0$ are stable fixed points of the Newton step $f(x) = x - g(x)/g'(x)$. Let's take one more look at Newton's method.

Solving for a complex root using Newton's method.

Suppose we wish to solve the equation $x^3 - 1 = 0$ (i.e., $g(x) = 0$, where $g(x) = x^3 - 1$). Now, of course, we know that $x = 1$ is a root of this equation, but it is not the only root. The polynomial $x^3 - 1$ factors:

$$x^3 - 1 = (x - 1)(x^2 + x + 1),$$

so the other two roots are the solutions to the quadratic equation $x^2 + x + 1 =$

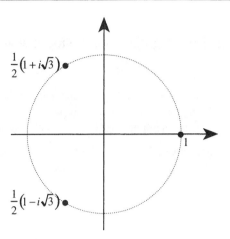

Figure 6.20. The three roots of the equation $z^3 - 1 = 0$. Notice they are evenly spaced around the unit circle.

0, namely, $\left(-1 \pm i\sqrt{3}\right)/2 \approx -0.5 \pm 0.866i$. The locations of these roots are shown in Figure 6.20. Notice that they are evenly spaced around the unit circle.

To find complex roots, start with a complex guess.

The question is, Can Newton's method find these complex roots? The answer is: Yes! We just have to give an appropriate complex starting guess.

When $g(x) = x^3 - 1$, we want to iterate $f(x) = x - (x^3 - 1)/(3x^2)$. Now, suppose we start with $x = i/2$. We compute $f(i/2)$, $f^2(i/2)$, $f^3(i/2)$, etc. Here's what we get (computer output):

```
     0 + 0.5000i
-1.3333 + 0.3333i
-0.7332 + 0.3053i
-0.1165 + 0.5786i
-0.9601 + 0.7560i
-0.5877 + 0.7210i
-0.4695 + 0.8580i
-0.5009 + 0.8655i
-0.5000 + 0.8660i
```

We see that $f^k(i/2)$ is converging to the root $-0.5 + 0.866i$. This makes sense: Of the three roots of $z^3 - 1 = 0$, the one closest to $0.5i$ is $-0.5 + 0.866i$.

We can ask, Given a point z in the complex plane, to which root of

$z^3 - 1 = 0,$

$$1, \quad \frac{-1 + i\sqrt{3}}{2}, \quad \text{or} \quad \frac{-1 - i\sqrt{3}}{2},$$

does Newton's method converge? A good guess is, To whichever root is closest. This guess is reasonable but, unfortunately, wrong. For example, suppose we start at $0.4i$ instead of $0.5i$. This new starting value is closest to $-0.5 + 0.866i$, but when we use Newton's method we get

Newton's method does not necessarily find the nearest root.

```
      0 + 0.4000i
-2.0833 + 0.2667i
-1.3158 + 0.1968i
-0.6971 + 0.1863i
 0.0901 + 0.4436i
-1.4376 - 0.3390i
-0.8217 - 0.2943i
-0.2097 - 0.4740i
-0.9744 - 1.2342i
-0.6809 - 0.9539i
-0.5328 - 0.8655i
-0.5005 - 0.8651i
-0.5000 - 0.8660i
```

We see that Newton's method is converging to $-0.5 - 0.866i$, the root farthest from $0.4i$. To make matters worse, if we start at $x = i/3$, we get

```
      0 + 0.3333i
-3.0000 + 0.2222i
-1.9636 + 0.1536i
-1.2242 + 0.1157i
-0.5996 + 0.1185i
 0.4257 + 0.4185i
 0.2998 - 0.6564i
-0.2194 + 0.0462i
 5.9240 + 2.7052i
 3.9545 + 1.7975i
 2.6479 + 1.1850i
 1.7917 + 0.7605i
 1.2556 + 0.4437i
 0.9833 + 0.1777i
 0.9683 + 0.0016i
 1.0010 - 0.0001i
 1.0000 - 0.0000i
```

Thus starting at $i/3$ leads to the root 1. What is going on here?

To gain perspective we need to draw a picture. For each complex number $z = x + iy$ (with x and y between, say, ± 2) we iterate the Newton step $f(z) = z - (z^3 - 1)/(3z^2)$. We color each point in the plane depending on to which of the three roots $(1, -0.5 + 0.866i,$ or $-0.5 - 0.866i)$ the method converges. The result is in Color Plate 4.

One of the interesting features of the diagram is the boundaries between the three regions. One can show that every boundary point is actually next to all *three* regions!

Problems for §6.3

◆1. We use Newton's method to solve the equation $z^3 - 1 = 0$. If our starting guess is $z = 0$, the method does not work. Why?

Find other starting guesses for which Newton's method fails to solve the equation $z^3 - 1 = 0$.

◆2. Consider the equation $x^2 - 3 = 0$. We want to solve this equation using Newton's method. Suppose we begin at $x = i$. What happens? What happens if we begin at $x = 1.01i$? Explain using the language of Chapter 4. What happens if we begin at $x = i\sqrt{3}$?

6.4 Examplification: Complex bases

6.4.1 Place value revisited

Recall that when we studied Cantor's set it was natural to work in ternary (base 3). Every number in Cantor's set can be expressed in the form $0.ddd \cdots_3$, where the d's are 0 or 2. Our goal is to extend this idea to *complex* number bases.

To begin, notice that if we divide all the number in Cantor's set by 2, the resulting numbers are exactly those which can be expressed in the form $0.ddd \cdots_3$, where the d's are either 0 or 1. Geometrically, we are simply looking at a half-size copy of Cantor's set.

Now we are ready to switch the number base. Instead of working in base 3, we can work in base b, where b is any number—real or complex.[2]

Definition of $C(b)$.   Let $C(b)$ denote the set of all numbers we can express in the form $0.ddd \cdots_b$, where the d's are 0's or 1's. This makes sense when b is 3 or 10, but what does it mean when $b = 2 + i$? Let's be a bit more careful. Base 10 is most familiar, so we start there. Let $x = 0.d_1d_2d_3d_4 \cdots_{10}$. What does this mean? It means

$$x = \frac{d_1}{10} + \frac{d_2}{10^2} + \frac{d_3}{10^3} + \frac{d_4}{10^4} + \cdots,$$

---

[2]Well, not quite *any* number. We'll want $|b| > 1$, as we will explain later.

or in fancier notation,

$$x = \sum_{k=0}^{\infty} d_k 10^{-k}.$$

Now, let's work in a generic base b, where b is any number (real or complex). Suppose $d_1, d_2, d_3, \ldots$ is a sequence of 0's and 1's, then by $x = 0.d_1 d_2 d_3 \cdots_b$ we mean

$$x = \sum_{k=0}^{\infty} \frac{d_k}{b^k}.$$

For example, let's take $b = 1 + i$ and let $x = 0.10101010 \cdots_b$. What number is this in conventional notation? We rewrite this as

$$x = \frac{1}{b} + \frac{1}{b^3} + \frac{1}{b^5} + \cdots,$$

which is a geometric series with ratio $1/b^2$. Now

$$|b^2| = |(1 + i)^2| = |2i| = 2 > 1,$$

so $|1/b^2| < 1$; hence the sequence defining x converges. Finally, we can sum this series and we have

$$x = \frac{1/b}{1 - 1/b^2} = \frac{1}{5} - \frac{3}{5}i.$$

Recall that $C(b)$ denotes the set of all numbers we can express in the form $0.d_1 d_2 d_3 \cdots_b$ where the d's are 0's and 1's. Thus we have shown that $\frac{1}{5} - \frac{3}{5}i \in C(1 + i)$.

Now, for some values of b the set $C(b)$ is undefined; let's illustrate this with an example. Take $b = \frac{1}{2}$. Then

$$0.11111 \cdots_{1/2} = 2 + 4 + 8 + 16 + 32 + \cdots,$$

which is ∞ (or undefined).

The question becomes, Does $0.d_1 d_2 d_3 \cdots_b$ always converge for any sequence of 0's and 1's? We can take the absolute values of the terms and we get

$$\left| \frac{d_1}{b} \right| + \left| \frac{d_2}{b^2} \right| + \left| \frac{d_3}{b^3} \right| + \left| \frac{d_3}{b^4} \right| + \cdots.$$

This is bounded by taking all the d's equal to 1, i.e.,

$$\left| \frac{1}{b} \right| + \left| \frac{1}{b^2} \right| + \left| \frac{1}{b^3} \right| + \left| \frac{1}{b^4} \right| + \cdots,$$

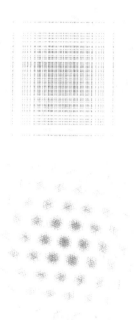

Figure 6.21. Images of the sets $C(1.1i)$ and $C(-0.5 + 0.9i)$.

When $|b| > 1$, then $C(b)$ is defined.

and this is a geometric series. The latter series converges provided $|1/b| < 1$, or equivalently $|b| > 1$. Thus if $|b| > 1$, then we know that $C(b)$ is defined.

What do the sets $C(b)$ look like? We can plot all the points

$$0.\underbrace{0 \cdots 0}_{12}\,_b \qquad \text{through} \qquad 0.\underbrace{1 \cdots 1}_{12}\,_b$$

This is a total of $2^{12} \approx 4000$ points (you can do more). Each of these values is a complex number, and so we can plot them in the plane. Two examples are shown in Figure 6.21. The figure shows $C(1.1i)$ and $C(-0.5 + 0.9i)$.

6.4.2   IFSs revisited

The pretty pictures in Figure 6.21 are fractals; let's understand them as attractors of iterated function systems.

Let b be a complex number with $|b| > 1$. Now, let's consider the following pair of functions

$$g_1(z) = z/b \quad \text{and} \quad g_2(z) = (z+1)/b.$$

Since $|b| > 1$ one can prove (in this case *one* is *you*—see problem 1 on page 340) that g_1 and g_2 are contraction maps. We can then ask, What compact set is the attractor of $g_1 \cup g_2$? It should be no surprise that the answer is exactly $C(b)$. Here's why.

First, it helps to understand how g_1 and g_2 act on numbers in $C(b)$. Let $x = 0.d_1d_2d_3 \cdots_b$, where the d's are 0's and 1's. First we compute $g_1(x)$. Since $g_1(x) = x/d$, we simply slide the digits of x one place to the right and fill in a 0 in the first place. The action of $g_2(x)$ is similar; we can write $g_2(x)$ as $x/d + 1/d$, which means we slide the digits of x one place to the right, and then fill in a 1 as the new first digit. Symbolically, we have

Understanding how g_1 and g_2 act symbolically.

$$g_1(0.d_1d_2d_3 \cdots_b) = 0.0d_1d_2d_3 \cdots_b, \quad \text{and}$$

$$g_2(0.d_1d_2d_3 \cdots_b) = 0.1d_1d_2d_3 \cdots_b.$$

Now we ask, What is $(g_1 \cup g_2)(C(b))$? (What do you *want* the answer to be?) Since $C(b)$ is the set $\{0.d_1d_2d_3 \cdots_b : d_i = 0, 1\}$, we see that

$$g_1(C(b)) = \{0.0d_1d_2d_3 \cdots_b : d_i = 0, 1\}, \quad \text{and}$$

$$g_2(C(b)) = \{0.1d_1d_2d_3 \cdots_b : d_i = 0, 1\}$$

and therefore $(g_1 \cup g_2)(C(b)) = g_1(C(b)) \cup g_2(C(b)) = C(b)$. Thus $C(b)$ is (by the contraction mapping theorem) the unique stable fixed point of $g_1 \cup g_2$.

Now, g_1 and g_2 are not—at first glance—functions of the form

$$\begin{bmatrix} x \\ y \end{bmatrix} \mapsto \begin{bmatrix} a & b \\ c & d \end{bmatrix} \begin{bmatrix} x \\ y \end{bmatrix} + \begin{bmatrix} e \\ f \end{bmatrix}.$$

However, we can think of a complex number $z = x + yi$ as a point in the plane written as a vector $\begin{bmatrix} x \\ y \end{bmatrix}$. Suppose that $b = c + di$. Then

$$\frac{1}{b} = \frac{c}{c^2 + d^2} - \frac{d}{c^2 + d^2}i = c' + d'i.$$

Thus

$$g_1(z) = g_1(x + yi) = (c' + d'i)(x + yi) = (c'x - d'y) + (d'x + c'y)i,$$

so we can rewrite g_1 as

$$g_1 \begin{bmatrix} x \\ y \end{bmatrix} = \begin{bmatrix} c' & -d' \\ d' & c' \end{bmatrix} \begin{bmatrix} x \\ y \end{bmatrix}.$$

In a similar way, $g_2(z) = (z+1)/b$ can be written as

$$g_2 \begin{bmatrix} x \\ y \end{bmatrix} = \begin{bmatrix} c' & -d' \\ d' & c' \end{bmatrix} \begin{bmatrix} x \\ y \end{bmatrix} + \begin{bmatrix} c' \\ d' \end{bmatrix}.$$

The conclusion is that sets $C(b)$ are special cases of the fractals we examined in the previous chapter: attractors of iterated function systems of contractive affine transformations of the plane.

Problems for §6.4

◆1. Suppose b is a complex number with $|b| > 1$. Prove that the functions

$$g_1(z) = z/b \quad \text{and} \quad g_2(z) = (z+1)/b$$

are contraction maps with contractivity $|1/b|$.

Are the functions g_1 and g_2 similitudes? Prove or give a counterexample.

◆2.* Suppose b is a complex number with $|b| > 1$. Prove that $C(b)$ is a compact set.

◆3. Let b be a real number with $b > 2$. What is the dimension of $C(b)$?

◆4. Let b be a real number with $1 < b \leq 2$. Describe (exactly) the set $C(b)$.

◆5. Let b be a real number with $b > 1$. For some values of b, the members of $C(b)$ can be expressed in more than one way (as a sum of powers of b) and for some values of b the members of $C(b)$ have a unique 0,1 base b representation. Determine which values of b have each of these behaviors. Examine your results in light of the previous two problems.

◆6. What is the dimension of $C(3i)$?

Appendix A

Background Material

In this book we use ideas from linear algebra, complex numbers, and calculus. In this section we briefly recall some of the ideas from those subjects for your easy reference. We also discuss some of the rudiments of differential equations.

We are deliberately terse; see other texts for a more complete treatment.

A.1 Linear algebra

A.1.1 Much ado about 0

Be careful when you see the numeral 0; it has at least three different meanings in this book. It can stand for

- the ordinary number 0, or

- a vector of all zeros, in which case we write it in boldface

$$\mathbf{0} = \begin{bmatrix} 0 \\ 0 \\ \vdots \\ 0 \end{bmatrix}, \quad \text{or}$$

- a matrix for all zeros, in which case we again write it in boldface

$$\mathbf{0} = \begin{bmatrix} 0 & 0 & \cdots & 0 \\ 0 & 0 & \cdots & 0 \\ \vdots & \vdots & \ddots & \vdots \\ 0 & 0 & \cdots & 0 \end{bmatrix}.$$

So when you see 0 or **0**, don't think nothing of it. Consider which kind of zero it is!

A.1.2 Linear independence

Let X be a finite set of vectors, i.e.,

$$X = \{\mathbf{x}_1, \mathbf{x}_2, \ldots, \mathbf{x}_m\} \subset \mathbf{R}^n.$$

A *linear combination* of vectors in X is any vector of the form

$$c_1\mathbf{x}_1 + c_2\mathbf{x}_2 + \cdots + c_m\mathbf{x}_m, \tag{A.1}$$

where the c's are scalars (i.e., numbers). The set of all linear combinations of the vectors in X is called their *span*. The vectors in X are said to *span* $\mathbf{R}^n$ (or, simply, to *span*) if every vector in $\mathbf{R}^n$ is a linear combination of vectors in X.

Linear (in)dependence.   The vectors in X are called *linearly dependent* if the zero vector, **0**, is a *nontrivial* linear combination of the **x**'s; this means that there is a linear combination in which the coefficients (the c's in equation (A.1)) are *not all zero*. [Note: "not all zero" is very different from "all not zero".]

The vectors in X are called *linearly independent* provided they are not linearly dependent. Less cryptically, they are called linearly independent if the *only* way to make **0** as a linear combination of the **x**'s is to take all coefficients equal to 0.

A.1.3 Eigenvalues/vectors

Let A be a square matrix. A nonzero vector **x** is called an *eigenvector* of A provided

$$A\mathbf{x} = \lambda\mathbf{x} \tag{A.2}$$

for some scalar λ. In other words, $A\mathbf{x}$ must be a scalar multiple of **x**. The number λ is called an *eigenvalue* of A, and **x** is its associated eigenvector.

The *characteristic polynomial* of A is $\det(xI - A)$, where x is a variable. The roots of the characteristic polynomial are exactly the eigenvalues of A.

Some roots of the characteristic equation may be complex. Such complex roots are eigenvalues, but their associated eigenvectors have complex entries. Further, if A is a real matrix with a complex eigenvalue $z = a + bi$, then the *conjugate* of z, $\bar{z} = a - bi$, is also an eigenvalue with its own distinct associated eigenvector.

A.1.4   Diagonalization

Let A be an $n \times n$ matrix. Suppose that A has n *linearly independent* eigenvectors $\mathbf{s}_1, \mathbf{s}_2, \ldots, \mathbf{s}_n$ with associated eigenvalues $\lambda_1, \lambda_2, \ldots, \lambda_n$. (Thus $A\mathbf{s}_i = \lambda_i \mathbf{s}_i$.)

Diagonalizable if and only if the eigenvectors are linearly independent.

Collect the n eigenvectors of A as the columns of a matrix S (i.e., the ith column of S is $\mathbf{s}_i$). Collect the n eigenvalues as a diagonal matrix Λ (i.e., the ii entry of Λ is λ_i). Then observe that

$$AS = S\Lambda. \qquad (A.3)$$

Since the columns of S are linearly independent, we know that S is invertible. Thus we can rewrite equation (A.3) as

$$\Lambda = S^{-1}AS \qquad \text{or} \qquad A = S\Lambda S^{-1}.$$

This says that A is similar to a diagonal matrix.[1]

It is necessary and sufficient that there be n linearly independent eigenvectors for a matrix to be diagonalizable. If the n eigenvalues of an $n \times n$ matrix are all distinct (no repeated roots of the characteristic equation), then the associated eigenvectors must be linearly independent, and hence the matrix diagonalizes. This condition (distinct eigenvalues) is sufficient for diagonalization but not necessary; some matrices with repeated eigenvalues are diagonalizable.

No repeated eigenvalues implies diagonalizable.

Not all matrices are diagonalizable; for example $\begin{bmatrix} 2 & 1 \\ 0 & 2 \end{bmatrix}$ is not similar to a diagonal matrix.

If A is a real symmetric (i.e., $A = A^T$) matrix, then the eigenvalues of A are all real, and A has linearly independent eigenvectors. Further, we can choose the n eigenvectors to be all of length 1 and pairwise orthogonal.

Symmetric matrices diagonalize.

A.1.5   Jordan canonical form*

Not all square matrices can be diagonalized, that is, not all are similar to a diagonal matrix.

If a square matrix is not diagonalizable, it is similar (by Schur's theorem) to an upper triangular matrix. Further, it is similar to an upper triangular matrix of a rather particular form.

A square matrix is called *upper triangular* if all entries below the diagonal are 0.

---

[1] Square matrices A and B are called *similar* provided $A = MBM^{-1}$ for some invertible matrix M.

A *Jordan block* is a square matrix of the form

$$
J_{n,\lambda} = \begin{bmatrix}
\lambda & 1 & 0 & 0 & \cdots & 0 & 0 \\
0 & \lambda & 1 & 0 & \cdots & 0 & 0 \\
0 & 0 & \lambda & 1 & \cdots & 0 & 0 \\
0 & 0 & 0 & \lambda & \cdots & 0 & 0 \\
\vdots & \vdots & \vdots & \vdots & \ddots & \vdots & \vdots \\
0 & 0 & 0 & 0 & \cdots & \lambda & 1 \\
0 & 0 & 0 & 0 & \cdots & 0 & \lambda
\end{bmatrix}.
$$

In words, $J_{n,\lambda}$ is an $n \times n$ matrix whose diagonal entries are all λ, whose entries just above the diagonal are 1, and all of whose other entries are 0.

A matrix is said to be in *Jordan canonical form* if it can be decomposed into diagonal blocks, each of which is a Jordan block. For example, the following matrix is in Jordan canonical form:

$$
M = \left[\begin{array}{cccc|ccc}
2 & 1 & 0 & 0 & 0 & 0 & 0 \\
0 & 2 & 1 & 0 & 0 & 0 & 0 \\
0 & 0 & 2 & 1 & 0 & 0 & 0 \\
0 & 0 & 0 & 2 & 0 & 0 & 0 \\
\hline
0 & 0 & 0 & 0 & 4 & 1 & 0 \\
0 & 0 & 0 & 0 & 0 & 4 & 1 \\
0 & 0 & 0 & 0 & 0 & 0 & 4
\end{array}\right].
$$

The diagonal blocks are $J_{4,2}$ and $J_{3,4}$.

Every square matrix is similar to a matrix in Jordan canonical form; this similarity is unique up to rearrangement of the blocks.

A.1.6   Basic linear transformations of the plane

A *linear transformation* from $\mathbf{R}^2$ to itself is a function $f: \mathbf{R}^2 \to \mathbf{R}^2$ with the following properties:

$$f(\mathbf{x} + \mathbf{y}) = f(\mathbf{x}) + f(\mathbf{y}), \quad \text{and}$$

$$f(a\mathbf{x}) = af(\mathbf{x}),$$

where $\mathbf{x}$, $\mathbf{y}$ are vectors (elements of $\mathbf{R}^2$), and a is a scalar (number). Equivalently, we can express f as $f(\mathbf{x}) = A\mathbf{x}$, where A is a 2×2 matrix, $\begin{bmatrix} a & b \\ c & d \end{bmatrix}$.

We can think of a linear transformation as a motion which takes a point $\mathbf{x}$ and moves it to the point $f(\mathbf{x}) = A\mathbf{x}$. There are some basic linear

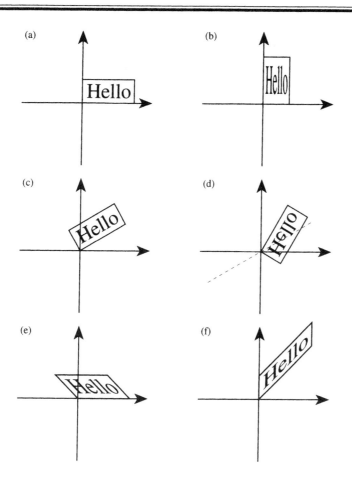

Figure A.1. Basic linear transformations of the plane: (a) identity, (b) axis rescal-ing, (c) rotation, (d) reflection, (e) and (f) shearing.

transformations which we can combine to make any linear transformation we desire. Here we describe these basic transformations. They are also illustrated in Figure A.1.

- *Identity.* The identity transformation is simply $f(\mathbf{x}) = I\mathbf{x}$, where $I = \begin{bmatrix} 1 & 0 \\ 0 & 1 \end{bmatrix}$ is the identity matrix. It leaves all points unmoved. See part (a) in Figure A.1.

- *Axis rescaling.* This is the transformation $f(\mathbf{x}) = A\mathbf{x}$, where $A =$

$\begin{bmatrix} a & 0 \\ 0 & b \end{bmatrix}$. The effect is to rescale the x-axis by a factor of a and the

y-axis by a factor of b. In Figure A.1 part (b) we use $A = \begin{bmatrix} \frac{1}{2} & 0 \\ 0 & 2 \end{bmatrix}$.

The identity transformation is a special case of axis rescaling where we take both a and b equal to 1.

If just one of a or b is zero, we collapse the entire plane into the y- or x-axis, respectively. If both a and b are zero, everything collapses into the origin, **0**.

- *Rotation.* In this transformation, points are moved through an angle θ about the origin. We have $A = \begin{bmatrix} \cos\theta & -\sin\theta \\ \sin\theta & \cos\theta \end{bmatrix}$. In part (c) of Figure A.1 we illustrate a rotation with $\theta = \pi/6$ (30°).

- *Reflection.* Let ℓ be a line through the origin. Let **x** and **y** be points on opposite sides of ℓ so that the line ℓ is the perpendicular bisector of the line segment joining **x** and **y**. A reflection though ℓ is the linear transformation which exchanges these two points (and leaves points on the line ℓ unmoved). See Figure A.1 part (d), where we illustrate a reflection through a line ℓ which makes a 30° angle with the horizontal.

  An axis rescaling with $a = -1$ and $b = 1$ (i.e., $A = \begin{bmatrix} -1 & 0 \\ 0 & 1 \end{bmatrix}$) is a reflection through the y-axis. (Likewise, when $a = 1$ and $b = -1$, we have a reflection through the x-axis.) All reflections can be created by combining a reflection through the x-axis with a rotation. Every rotation can be created by combining two reflections.

- *Shearing* (or *skewing*). Imagine a rectangle constructed from rigid poles and flexible joints at the corners. This rectangle is anchored to the x-axis, i.e., the bottom side is held fixed. How can this device move? We can slide the opposite horizontal side of this rectangle, transforming the rectangle into a parallelogram. This type of motion is known as a *shear* and is illustrated in Figure A.1 part (e). A shear is a linear transformation of the form $f(\mathbf{x}) = A\mathbf{x}$, where $A = \begin{bmatrix} 1 & a \\ 0 & 1 \end{bmatrix}$; in (e) we take $a = -1$. Likewise, $A = \begin{bmatrix} 1 & 0 \\ b & 1 \end{bmatrix}$ is also a shear; see part (f) of Figure A.1, where we take $b = 1$.

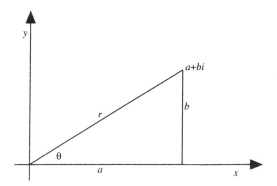

Figure A.2. A geometric view of the complex number $a + bi$.

Any linear transformation of the plane can be constructed using a combination of these basic linear transformations.

A.2 Complex numbers

A complex number is a number z of the form $a + bi$, where a and b are real numbers and $i = \sqrt{-1}$.

Although $z = a + bi$ is the traditional way to write a complex number, it is also very useful to write complex numbers in *polar notation*, which we now describe.

Each complex number $a + bi$ can be associated with a point in the plane (a, b). The *absolute value* of $a + bi$ is the distance of the point (a, b) from the origin. Thus $|a + bi| = \sqrt{a^2 + b^2}$.

We can also specify the *angle* (also called the *argument*) of the complex number $a + bi$ as follows. Draw a line segment from the point (a, b) to the origin. The angle this segment makes (measured in the usual counter-clockwise fashion) with the positive x-axis is the angle θ of the complex number z. See Figure A.2. Observe that

$$a = r \cos \theta, \quad \text{and} \quad b = r \sin \theta,$$

and therefore $b/a = \tan \theta$, or $\theta = \arctan(b/a)$.

If we wish to express $z = a + bi$ succinctly in terms of r and θ, we use   Euler's formula.

Euler's wonderful formula:

$$e^{i\theta} = \cos\theta + i\sin\theta. \tag{A.4}$$

Here is a justification of Euler's formula. We use the power series representations for sin, cos, and e^x:

$$\cos x = 1 - \frac{x^2}{2!} + \frac{x^4}{4!} - \frac{x^6}{6!} + \cdots,$$

$$\sin x = x - \frac{x^3}{3!} + \frac{x^5}{5!} - \frac{x^7}{7!} + \cdots,$$

$$e^x = 1 + x + \frac{x^2}{2!} + \frac{x^3}{3!} + \frac{x^4}{4!} + \frac{x^5}{5!} + \cdots.$$

Now we substitute $i\theta$ for x in the series for e^x to get

$$e^{i\theta} = 1 + (i\theta) + \frac{(i\theta)^2}{2!} + \frac{(i\theta)^3}{3!} + \frac{(i\theta)^4}{4!} + \frac{(i\theta)^5}{5!} + \frac{(i\theta)^6}{6!} + \frac{(i\theta)^7}{7!} + \cdots$$

$$= 1 + i\theta - \frac{\theta^2}{2!} - i\frac{\theta^3}{3!} + \frac{\theta^4}{4!} + i\frac{\theta^5}{5!} - \frac{\theta^6}{6!} - i\frac{\theta^7}{7!} + \cdots$$

$$= \left(1 - \frac{\theta^2}{2!} + \frac{\theta^4}{4!} - \frac{\theta^6}{6!} + \cdots\right) + i\left(\theta - \frac{\theta^3}{3!} + \frac{\theta^5}{5!} - \frac{\theta^7}{7!} + \cdots\right)$$

$$= \cos\theta + i\sin\theta.$$

Thus if a complex number $a + bi$ has absolute value r and argument θ, then

$$a + bi = re^{i\theta}.$$

Sin and cos in terms of exp.   Euler's formula [equation (A.4)] is useful in writing the sine and cosine functions in terms of the e^x function. Since $e^{i\theta} = \cos\theta + i\sin\theta$, it follows that $e^{-i\theta} = \cos(-\theta) + i\sin(-\theta) = \cos\theta - i\sin\theta$. In short

$$e^{i\theta} = \cos\theta + i\sin\theta, \quad \text{and}$$

$$e^{-i\theta} = \cos\theta - i\sin\theta.$$

If we add these two equations, we can solve for $\cos\theta$; if we subtract them, we can solve for $\sin\theta$. Doing this, we get

$$\cos\theta = \frac{e^{i\theta} + e^{-i\theta}}{2}, \quad \text{and}$$

$$\sin\theta = \frac{e^{i\theta} - e^{-i\theta}}{2i}.$$

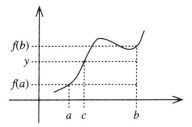

Figure A.3. Understanding the intermediate value theorem.

Writing $z = a + bi$ in the form $re^{i\theta}$ is useful for computing z^k. Observe then that

$$z^k = \left(re^{i\theta}\right)^k = r^k e^{ik\theta}. \qquad (A.5)$$

Thus $|z^k| = |z|^k$.

A.3 Calculus

A.3.1 Intermediate and mean value theorems

Let $f: \mathbf{R} \to \mathbf{R}$ be a continuous function. Let a and b be two real numbers. Suppose y is a real number between $f(a)$ and $f(b)$. Then the *intermediate value theorem* guarantees us that there is a number c between a and b for which $f(c) = y$. See Figure A.3.

Now, let $f: \mathbf{R} \to \mathbf{R}$ be a differentiable function, and let a and b be two numbers. The slope of the line segment from the point $(a, f(a))$ to the point $(b, f(b))$ is

$$\frac{\Delta y}{\Delta x} = \frac{f(b) - f(a)}{b - a}.$$

The *mean value theorem* states that there is a number c between a and b for which

$$f'(c) = \frac{f(b) - f(a)}{b - a}.$$

See Figure A.4.

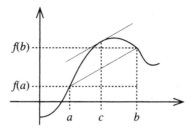

Figure A.4. Understanding the mean value theorem.

A.3.2 Partial derivatives

Let $f: \mathbf{R}^2 \to \mathbf{R}$, i.e., f is a real-valued function of two variables, say, $f(x, y)$. The *partial derivative* of f with respect to x, denoted by $\partial f/\partial x$, is computed by imagining y is a constant and computing the derivative with respect to x. We define $\partial f/\partial y$ in a similar fashion. For example, if $f(x, y) = x^2 \cos y$, then

$$\frac{\partial f}{\partial x} = 2x \cos y \quad \text{and} \quad \frac{\partial f}{\partial y} = -x^2 \sin y. \qquad (A.6)$$

If x and y are, in turn, functions of some variable t, then f can be considered as a function of just t. We find the derivative of f with respect to t via the formula

$$\frac{df}{dt} = \frac{\partial f}{\partial x} \frac{dx}{dt} + \frac{\partial f}{\partial y} \frac{dy}{dt}.$$

Continuing our previous example, if $x(t) = e^t$ and $y(t) = t^2$, then

$$\frac{df}{dt} = \frac{\partial f}{\partial x} \frac{dx}{dt} + \frac{\partial f}{\partial y} \frac{dy}{dt}$$

$$= (2x \cos y)(e^t) + (-x^2 \sin y)(2t)$$

$$= 2e^t (\cos t^2) e^t - e^{2t} (\sin t^2) 2t$$

$$= 2e^{2t} \cos t^2 - 2t e^{2t} \sin t^2,$$

which you can check is the derivative of $e^{2t} \cos t^2$.

A.4   Differential equations

A.4.1   Equations

An *equation* is a mathematical expression with an equal sign somewhere in the middle. For example,

$$3x + 1 = 10$$

is an equation. It tells us the number on the left equals the number on the right.

Typically the equation (such as the preceding one) contains a letter representing an unknown quantity. In this case x stands for a number. To *solve* this equation means to find the number x which makes the equation true. In this easy example, the answer is $x = 3$.

Here are two more examples of equations:

$$x^2 = 4 \quad \text{and} \quad x + 2 = x.$$

The first has two different solutions (2 and -2), while the second doesn't have any solutions.

You are undoubtedly quite familiar with these kinds of algebraic equations. They relate two *numbers* (the expression on the left and the expression on the right) and have a variable (which stands for a number) whose value(s) you want to determine. Sometimes there is just one answer. Other times, there may be several answers or no answer at all.

A.4.2   What is a differential equation?

A *differential equation* is a bit different. The equation does not assert that two numbers are the same; rather, it asserts that two *functions* are the same. And, in addition to the usual mathematical operations (such as $+$ and $\times$), the derivative operation is used. Let's look at an example:

$$f'(t) = g(t) \quad \text{where} \quad g(t) = 2t + 1. \tag{A.7}$$

In this equation we have two functions (the one on the left and the one on the right). The one on the left is $f'(x)$; it is the derivative of some function f. We don't know what the function f is. It is the "variable" whose "value" we want to find. The function on the right is g and we are told what g is; it is the function $g(t) = 2t + 1$. The letter t is a place holder. It is useful for writing down the functions, but it isn't of interest to us.

A solution to equation (A.7) is a function. Here are two possible solutions to this differential equation:

$$f_1(t) = t^2 + t, \quad \text{and}$$

$$f_2(t) = \sin t.$$

Which (if either) is a solution to equation (A.7)?

Let's consider f_1 first. The derivative of f_1 is $f_1'(t) = 2t + 1$. Notice that $f_1'(t)$ is exactly $g(t)$. So f_1 is a solution to differential equation (A.7).

Now, consider f_2. The derivative of f_2 is $f_2'(t) = \cos t$. Is the function $\cos t$ the same as the function $g(t)$? The answer is, of course, no. These two functions are different. Now, it is true that when $t = 0$ we have $\cos 0 = 1 = g(0)$, but that does *not* mean that f_2' and g are the *same* function. Function f_2 is *not* a solution to differential equation (A.7).

We know that f_1 is a solution to differential equation (A.7). Notice that we said *a* solution and not *the* solution. There are others. For example, let $f_3(t) = t^2 + t - 5$. To see that f_3 is also a solution, we take its derivative, $f_3'(t) = 2t + 1$, and notice that f_3' is the same function as g. So f_3 is another solution to equation (A.7).

Are there other solutions? Yes. Consider $f_4(t) = t^2 + t + 17$ and $f_5(t) = t^2 + t - \pi$. These are also solutions to equation (A.7). Indeed, I am sure you realize that there are infinitely many different solutions to differential equation (A.7). A convenient way to write this whole family of solutions is

$$f(t) = t^2 + t + C,$$

where C is a constant.

You might be worried and ask, Are there any *more* solutions to equation (A.7)? The answer is no; let's see why. We are told that $f'(t) = g(t)$; i.e., that $f'(t)$ and $g(t)$ are the same function. So if we integrate $f'(t)$ or if we integrate $g(t)$ we get the same answer:

$$\int f'(t)\, dt = \int g(t)\, dt.$$

Now, we know that $g(t) = 2t + 1$, so $\int g(t)\, dt = t^2 + t + C$, where C is a constant. Also, we know that $\int f'(t)\, dt = f(t) + C$, where C is a (perhaps different) constant. It follows then that $f(t) = t^2 + t + C$.

A.4.3   Standard notation

Differential equation (A.7) is *not* written in standard form. There is nothing wrong about how we wrote it; it's just a bit verbose. Here is the same differential equation but written in another form:

$$y' = 2t + 1. \tag{A.8}$$

There are two letters in equation (A.8): y and t. The letter y stands for the *function* we want to find, and the letter t is a place holder. The equation says, "The derivative of the unknown function y equals the function $2t + 1$."

The letter y is called the *dependent* variable, and the letter t is called the *independent* variable. The dependent variable is the function we are trying to find; in equation (A.8) we know that the dependent variable is y because it has a prime symbol (') on it. The independent variable, t, is a place holder which is used to express functions.

Let's look at another example. Consider the following differential equation:

$$y'' = y. \tag{A.9}$$

What this equation is asking us to find is a function f that is exactly the same as its own second derivative! The dependent variable is y, but what is the *independent* variable? The answer is, it doesn't matter. You may use any letter you wish (except y, of course!). In this book, we like to use the letter t for the independent variable (other books like to use x for the independent variable). The symbol t is just a place holder—a "dummy" variable—used to specify functions.

What are the solutions to equation (A.9)? Here's one solution: $y = e^t$ (understand that y stands for a *function*, and it would also be correct to say that $y(t) = e^t$ is a solution to (A.9)). To verify that $y = e^t$ is a solution, we notice that $y' = e^t$, and $y'' = e^t$, so $y'' = y$ as required.

What about $y = e^t + 4$? Is that a solution? No. Notice that $y' = e^t$ and $y'' = e^t$, which is *not* the same as $e^t + 4$.

Here is another solution to equation (A.8): $y = e^{-t}$. To see why, note that $y' = -e^{-t}$, and $y'' = e^{-t}$, so $y'' = y$.

Are there other solutions? Yes, here's another: $y = 2e^t + 3e^{-t}$. And another: $y = \frac{1}{2}e^t - e^{-t}$. Indeed, the entire family of solutions is

$$y = c_1 e^t + c_2 e^{-t},$$

where c_1 and c_2 are constants.

Sometimes we want to know a *specific* solution to a differential equation. In order to do this—in order to select a specific function from an infinite family of choices—we need more information. For example, for differential equation (A.9), $y'' = y$, we might also be told that

$$y(0) = 1 \quad \text{and} \quad y'(0) = 2.$$

Nailing down the constants.

We can use this additional information to select which function of the form $y(t) = c_1 e^t + c_2 e^{-t}$ we want.

We are told that $y(0) = 1$. Now, $y(0) = c_1 e^0 + c_2 e^0$, so we know that

$$c_1 + c_2 = 1. \tag{A.10}$$

We are also told that $y'(0) = 2$. Since $y(t) = c_1 e^t + c_2 e^{-t}$, it follows that $y'(t) = c_1 e^t - c_2 e^{-t}$. Thus $y'(0) = c_1 e^0 - c_2 e^0$, so we have

$$c_1 - c_2 = 2. \tag{A.11}$$

We can solve equations (A.10) and (A.11) to find

$$c_1 = \frac{3}{2} \quad \text{and} \quad c_2 = -\frac{1}{2}.$$

Thus the function we want is

$$y(t) = \frac{3}{2} e^t - \frac{1}{2} e^{-t}.$$

How does one solve a differential equation? There are a variety of techniques, but you do not need to know how to solve differential equations to use this book. Indeed, it is easy to write a differential equation that is so complicated that no one will know how to solve it. Indeed, one of the concerns of this book is understanding how to handle differential equations that are too difficult to solve.

In Appendix B (see §B.1.1 and §B.1.2) we discuss how to use a computer to get either exact (analytical) and/or approximate (numerical) solutions to differential equations.

Appendix B

Computing

Mathematics advances through discovery and proof. Without proof, our observations are merely conjectural. We do not stress rigorous proofs in this book because our aim is to *invite* readers into the world of dynamical systems.

Proofs, however, are the second step in the process.[1] We first need to discover the truths we hope to prove. And there is no better exploration tool for dynamical systems than the computer.

In this appendix we first consider (§B.1) how to use the computer to solve differential equations. We show how to find both analytic and numerical solutions. In §B.2 we give programs to perform the triangle dance of §5.5.2. Finally, in §B.3 we discuss the software which accompanies this book.

Discovery and proof.

B.1 Differential equations

A computer can be used to solve differential equations. There are two types of solutions we can hope for. Computer algebra packages (such as *Maple* and *Mathematica*) can find *analytic* solutions, i.e., formulas, in nearly standard mathematical notation, for the functions which solve the differential equation.

Sometimes analytic solutions are beyond the grasp of computer or human differential equation solvers. In this case, numerical methods can be useful. We show (§B.1.2) how to use *Maple*, MATLAB, and *Mathematica* to find numerical solutions to differential equations.

---

[1]Please take the sequence "discovery first, proof second" with the proverbial grain of salt. In research, I have had the experience of figuring out a proof and *then* spending a few hours trying to figure out just what the theorem was that was proved.

B.1.1 Analytic solutions

Suppose we are given a differential equation, such as equation (A.8) (from page 353) or equation (A.9). We repeat these equations here for your convenience:

$$y' = 2t + 1 \tag{A.8}$$

$$y'' = y. \tag{A.9}$$

In this section we show how to use *Maple* and *Mathematica* to solve these equations.

Maple

The *Maple* command
dsolve.

To solve these equations using *Maple* type,

```
dsolve( diff(y(t),t) = 2*t+1, y(t) );
```

and the computer responds:

```
         2
y(t) = t   + t + _C1
```

The `dsolve` command tells the computer you want to solve a differential equation. For more detail on how to use it, give the command

```
help(dsolve);
```

Here is how to solve equation (A.9) in *Maple*. Give the command

```
dsolve( diff(y(t),t$2) = y, y(t) );
```

and the computer responds:

```
y(t) = _C1 exp(t) + _C2 exp(- t)
```

The `diff(y(t),t$2)` stands for the second derivative of *y* with respect to *t*.

Mathematica

The *Mathematica* command
DSolve.

Mathematica can also solve differential equations (A.8) and (A.9). To solve equation (A.8), type

```
DSolve[ y'[t] == 2t+1, y[t], t]
```

and the computer responds:

```
            2
{{y[t] -> t + t  + C[1]}}
```

To solve equation (A.9), type

```
DSolve[ y''[t] == y[t], y[t], t]
```

and the computer replies:

```
          C[1]     t
{{y[t] -> ---- + E   C[2]}}
            t
          E
```

In *Mathematica*, to get more information on how to solve differential equations, give the command ?DSolve and press ENTER.

B.1.2 Numerical solutions

Consider the following differential equation:

A differential equation that's too hard to solve.

$$y' + y^2 = t. \tag{B.1}$$

This is a difficult equation to solve. Indeed, it stumps *Mathematica*. If we type

```
DSolve[y'[t] + y[t]^2 == t, y[t], t]
```

the computer responds (rather unhelpfully):

```
          2
DSolve[y[t]  + y'[t] == t, y[t], t]
```

Mathematica is telling us that it doesn't know how to solve this differential equation analytically and simply returns it to us unsolved.

This is very disappointing. However, let us try for a more modest goal. Suppose we are given that $y(0) = 1$, and all we really want to know is $y(1.2)$. It would seem that without a formula for $y(t)$ we are stuck. However, using numerical methods we can still answer our question.

Here we show how to use numerical methods to compute $y(t)$ and to plot its graph.

Mathematica

The *Mathematica* command
NDSolve.

In *Mathematica* we use the NDSolve command as follows:

```
ans = NDSolve[{y'[t] + y[t]^2 == t, y[0]==1}, y, {t,0,2}]
```

The NDSolve command takes three arguments. The first is a list of equations (in braces); in this case the equations $y' + y^2 = t$ and $y(0) = 1$ are entered as y'[t] + y[t]^2 == t and y[0]==1. The second argument is the name of the independent variable, y in this case. The third argument is a three-element list giving the name of the dependent variable and the first and last values the variable takes. We want to deal with $y(t)$ over the range $0 \le t \le 2$, so we give {t,0,2} as the third argument to NDSolve. We begin the command with "ans =" so the computer will save the result in a variable called ans.

After we give the NDSolve command, *Mathematica* responds:

```
{{y -> InterpolatingFunction[{0., 2.}, <>]}}
```

This is rather cryptic. Suffice it to say that a function approximating $y(t)$ is saved in ans. Now, how do we compute (say) $y(1.2)$? We give the command

```
y[1.2] /. ans
```

This means (roughly): Please compute $y(1.2)$ subject to the transformation rules in ans. The computer responds to this command with the answer we want:

```
{0.903008}
```

We also might like to plot a graph of the function $y(t)$. To do so, we give the command

```
Plot[ y[t] /. ans, {t,0,2} ]
```

and the computer plots the graph of $y(t)$ over the range $0 \le t \le 2$. The output of this command is shown in Figure B.1.

A more complicated example
for *Mathematica*.

Mathematica can also give numerical solutions to systems of differential equations. Recall equation (1.16) on page 14:

$$\begin{bmatrix} \theta \\ \omega \end{bmatrix}' = \begin{bmatrix} \omega \\ -\sin\theta \end{bmatrix}.$$

Suppose that this system starts with $\theta(0) = 3$ and $\omega(0) = 0$. We can compute the trajectory of this system via the *Mathematica* commands

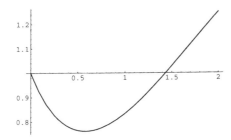

Figure B.1. The graph of $y(t)$ for the function y defined by equation (B.1). This graph was produced by *Mathematica*.

```
ans2 = NDSolve[
                { theta'[t]  ==  omega[t],
                  omega'[t]  ==  -Sin[theta[t]],
                  omega[0]   ==  0,
                  theta[0]   ==  3},
     {omega,theta},
       {t,0,30}
];

Plot[{theta[t] /. ans2, omega[t] /. ans2} , {t,0,30}];
```

The first command (which begins `ans2 = NDSolve[` and ends a few lines later at the close bracket–semicolon) uses `NDSolve` to find approximating functions for $\omega(t)$ and $\theta(t)$. The result is saved in a variable called `ans2`.

The second command `Plot[...` tells the computer to plot the graphs of $\theta(t)$ and $\omega(t)$ over the range $0 \le t \le 30$. The result is very similar to Figure 1.5 on page 15.

MATLAB

Now we show how to use MATLAB to solve equations (B.1) and (1.16).

Recall that equation (B.1) is $y' + y^2 = t$. In order for MATLAB to work with this equation, we first need to solve for y'. This is easy: $y' = t - y^2$. Next, we need to tell MATLAB how to compute y' given t and y. We do this by creating a file such as the following:

The MATLAB command `ode45`.

```
function yprime = example(t,y)
%
% We use this function, together with ode45,
% to numerically solve differential equation (B.1):
%
%   y' + y^2 = t
%
yprime = t - y^2;
```

We save this file as `example.m` (the lines beginning with a percent sign %
are comments and may be omitted).

Next we have to tell MATLAB what to do with this file. We use the
ode45 command to compute the numerical values of $y(t)$. In MATLAB
we give the command

```
[t,y] = ode45('example', 0, 2, 1);
```

Let's examine this piece by piece. The ode45 command gives two results.
The "`[t,y]` =" saves the two outputs in the variables t and y, respectively.
The first argument to ode45 is the name of the file which computes the
right-hand side of the differential equation $y' = \cdots$. In this case the file is
`example.m`; we don't type the `.m`, and we enclose the name of the file in
single quotes.

The second and third arguments (0 and 2) are the initial and final values
of t we wish to consider.

The fourth argument, 1, is the initial value of y, i.e., $y(0) = 1$.

Finally, the command ends with a semicolon—this tells MATLAB not
to print the values of t and y on the screen.

Now we would like to compute $y(1.2)$ and to plot the graph of $y(t)$ on
the screen. Here is how we do each of these tasks.

Using the MATLAB command `interp1`, but poorly. To interpolate the value of $y(1.2)$ we use the MATLAB function `interp1`.
We type

```
interp1(t,y,1.2)
```

and the computer responds:

```
0.9037
```

You might notice that this answer disagrees with the $y(1.2)$ we computed
using *Mathematica* (where we found $y(1.2) \approx 0.903008$). The inaccuracy
is in how we used `interp1`. When ode45 computes y, it does so only for
a modest number of values of t. To find $y(t)$ for another value, we use

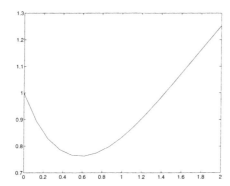

Figure B.2. The graph of $y(t)$ for the function y defined by equation (B.1). This graph was produced by MATLAB.

interp1 to interpolate. The default action of interp1 is to approximate the curve $y(t)$ by a straight line between the two values of t nearest 1.2. A better method of interpolation is to use *cubic splines*. The interp1 command can use this method in place of linear interpolation. To do so, we type

Using the MATLAB command interp1 for greater accuracy.

```
interp1(t,y,1.2,'spline')
```

and the computer responds: 0.9030. This agrees with *Mathematica's* answer to the number of digits shown. (Internally, MATLAB has interpolated $y(1.2)$ to be 0.90300610924786, which is extremely close to the value found by *Mathematica*.)

Next, to plot a graph of $y(t)$ we give the following command to MATLAB:

```
plot(t,y)
```

and the result is shown in Figure B.2.

To compute the motion of the rigid pendulum of §1.2.3 we again use the MATLAB command ode45.

The first thing MATLAB needs is a version of equation (1.16) it can understand. As before, we do this by creating a file we call pendulum.m.

```
function xprime = pendulum(t,x)
%
% This function computes the right side
% of equation (1.16) for use with the ode45
% differential equation solver.
% the first component of the state vector x is theta
% and the second component is omega
theta = x(1);
omega = x(2);
thetaprime = omega;
omegaprime = -sin(theta);
% the result is placed in xprime with the first
% component of xprime being omega and the second
% component being sin(theta)
xprime = [thetaprime; omegaprime];
```

Next, we need to give MATLAB the ode45 command according to the following syntax:

$$[\texttt{t,x}] = \texttt{ode45('pendulum'}, t_{\text{start}}, t_{\text{finish}}, \mathbf{x_0});$$

where t_{start} is the initial time (we use 0), t_{finish} is the ending time, and $\mathbf{x_0}$ is the initial state vector. We can then call plot(t,x) to obtain a picture of how the state variables (θ and ω) vary with time.

Let's try this starting with $\mathbf{x_0} = \begin{bmatrix} \theta(0) \\ \omega(0) \end{bmatrix} = \begin{bmatrix} 0.1 \\ 0 \end{bmatrix}$. The command we give is

```
[t,x] = ode45('pendulum',0,20,[0.1;0]);
```

We then give the command plot(t,x), and the result is shown in Figure 1.4 on page 14.

For more information about the ode45 command, give MATLAB the command help ode45.

Maple

The *Maple* command dsolve with the numeric option.

The dsolve in *Maple*, which produces analytic solutions to differential equations, can also be used to compute numerical solutions. To solve equation (B.1) we give the command

```
yy :=
dsolve( {D(y)(t) + (y(t))^2 = t, y(0)=1}, y(t), numeric):
```

This defines yy to be a function which approximates the solution to equation (B.1). To find $y(1.2)$, we give *Maple* the command

`yy(1.2);`

and the computer responds:

`1.200000000, .9030065770`

Do it yourself

If you do not have access to MATLAB, *Mathematica*, or the like, it is not difficult to write your own program (in any common computer language such as Pascal or Basic) to find numerical solutions to differential equations.

   While the Euler method (§1.2.10 on page 28) is the simplest to understand and to code, it is slow and inaccurate. Here we present the Runge-Kutta method, a sample program (written in C), and the output.

The Runge-Kutta method.

   The Runge-Kutta method is a simple-to-program, accurate method for computing numerical solutions to differential equations. As in MATLAB, we need to put the differential equation into standard form, namely

$$y' = f(t, y).$$

For our example, equation (B.1), we write

$$y' = t - y^2.$$

We also need to know the value of y at a particular value of t. In our example, we know that at $t = 0$ the value of y is 1.

   The method computes y at increasing values of t. At each step, we increase t by a small amount—called the *step size*—and we denote this quantity by h. The smaller h is, the more accurate the results are, but more computations are required.

   Given specific numbers for t and $y(t)$, the method computes the values $t + h$ and $y(t + h)$ as follows:

$$a \leftarrow f(t, y),$$
$$b \leftarrow f(t + h/2, y + ha/2),$$
$$c \leftarrow f(t + h/2, y + hb/2),$$
$$d \leftarrow f(t + h, y + hc),$$
$$t \leftarrow t + h,$$
$$y \leftarrow y + h(a + 2b + 2c + d)/6.$$

At the end of one pass through these calculations, t and $y = y(t)$ have their new values. The steps are repeated until the desired range of t has been covered.

Here we give a C language program which uses the Runge-Kutta method to compute values of $y(t)$ for differential equation (B.1).

```
#include <stdio.h>
#define  STEPSIZE  0.05     /* "h" in the method */
#define  TSTART    0.       /* initial value for t */
#define  TLAST     2.       /* last value for t */
#define  YSTART    1.       /* initial value for y */

double f(t,y)    /* a subroutine to compute the RHS */
double t,y; {
    return(t - y*y);
}

printout(t,y)    /* a subroutine to print out t and y */
double t,y; {
    printf("%g\t%g\n",t,y);
}

main() {         /* the main program starts here */
    /* declare variables */
    double t,y;         /* t and y from algorithm */
    double a,b,c,d;     /* intermediate steps in RK */
    double h;           /* h = STEPSIZE */

    /* initialize */
    y = YSTART;
    t = TSTART;
    h = STEPSIZE;

    printout(t,y);

    while (t <= TLAST) {        /* the main RK loop */
        a = f(t,y);
        b = f(t+h/2,y+h*a/2);
        c = f(t+h/2,y+h*b/2);
        d = f(t+h,y+h*c);
        t = t + h;
        y = y + h*(a+2*b+2*c+d)/6;
        printout(t,y);
    }
}
```

When compiled and run, this program produces the following output:

```
0       1
0.05    0.953592
0.1     0.913794
0.15    0.879866
0.2     0.851191
. . . . . . . . . . . . . . . . .
1.15    0.884177
1.2     0.903007     <-- the value we want
1.25    0.9226
1.3     0.942854
. . . . . . . . . . . . . . . . .
1.9     1.20749
1.95    1.2295
2       1.25132
```

B.2   Triangle Dance

The following MATLAB program performs the triangle dance from §5.5.2 on page 279. Save the following instructions in a file called dance.m, and then give the command dance to MATLAB.

```
% Dancing on a triangle

% set up the graphics screen
axis('square');
axis([0 1 0 1]);
hold on;

% number of points to plot
% (increase for a better picture)
npoints = 500;

corners = [ 0 1 .5; 0 0 1];    % corners of the triangle
points = zeros(2,npoints);     % points to plot
choice = fix(3*rand(1,npoints)+1); % random choices recorded

for k=2:npoints
  points(:,k) = (points(:,k-1) + corners(:,choice(k-1)))/2;
end;

plot(points(1,:),points(2,:),'.') % plot the figure
```

If you run the program several times (without clearing or closing the graphics window), the diagram will be more filled in.

You can also use *Mathematica* to do the triangle dance. Give the following commands to *Mathematica* and it will draw the result:

```
mat = {{.5,0},{0,.5}};
f1[vec_] := mat . vec
f2[vec_] := (mat . vec) + {.5,0}
f3[vec_] := (mat . vec) + {.25,.5}
flist = {f1,f2,f3};
dancer[npts_] := Block[{outlist,k},
   outlist = Table[{0.,0.},{k,npts}];
   For[k=1, k<npts, k++,
     pick = Random[Integer,{1,3}];
     outlist[[k+1]] = outlist[[k]] // (flist[[pick]])
   ];
   outlist
]
pts = dancer[1000];
ListPlot[pts,AspectRatio->Automatic];
```

Alternatively, you can save this file as, say, dancer and then load it into *Mathematica* by typing <<dancer and pressing ENTER.

B.3   About the accompanying software

The diskette available for this book includes a variety of MATLAB programs which enhance the utility of this book. Please read the file README for more specific information.

Why MATLAB?

There is a proliferation of computer platforms: Macintosh, MS-Dos, Unix, etc. To write programs which run in all environments is a headache. We chose to use MATLAB for the programs with this book for the following reasons. First, MATLAB is portable. The same ".m" file (which is a plain text file) should run the same on any machine with MATLAB. Second, MATLAB is easy to learn. Students can start using MATLAB in much less time than other, more sophisticated packages. Likewise, MATLAB programs are fairly readable, and students should be able to modify them to suit their needs. Finally, MATLAB is useful for many other mathematics and engineering courses (from linear algebra to signal processing), so it seems sensible to stick with a tool that students know.

This said, other popular environments (such as *Mathematica* or *Maple*) can be used for computer experimentation. In particular, spreadsheet software—such as *Excel*—is nicely suited for working with discrete time dynamical systems. These business packages produce nice graphs, too.

Obtaining the software

There are three ways to obtain the software which accompanies this book:

1. Fill out the postcard that comes with this book and mail it.

2. Via FTP from The MathWorks (the manufacturers of MATLAB). Use your FTP software to connect to `ftp.mathworks.com`. Log in as anonymous giving your e-mail address as your password. The software is in the `/pub/books/scheinerman` directory. The files are contained in a set of subdirectories. You should replicate the directory structure of the stored files on your own machine.

   This software is also archived on `brutus.mts.jhu.edu` in the directory `/pub/scheinerman/invitation/software`.

3. Another way to access the software over the Internet is via the World Wide Web. Connect your web browser (e.g., Mosaic) to

   `http://www.mts.jhu.edu/~ers/invite.html`

   and read the information there.

Bibliography

[1] Abraham, Ralph H., and Christopher D. Shaw, *Dynamics: The Geometry of Behavior*, Addison-Wesley (1992).

[2] Barnsley, Michael, *Fractals Everywhere*, Academic (1988).

[3] Beltrami, Edward, *Mathematics for Dynamic Modeling*, Academic (1987).

[4] Blanchard, Olivier Jean, and Stanley Fischer, *Lectures on Macroeconomics*, MIT Press (1989).

[5] Boyce, William E., and Richard C. DiPrima, *Elementary Differential Equations and Boundary Value Problems*, Wiley (1977).

[6] Çambel, A.B., *Applied Chaos Theory: A Paradigm for Complexity*, Academic Press (1993).

[7] Diaconis, Persi, *Group Representations in Probability and Statistics*, Institute of Mathematical Sciences Lecture Notes Monograph Series, vol. 11 (1988).

[8] Devaney, Robert L., *An Introduction to Chaotic Dynamical Systems*, 2d ed., Addison-Wesley (1989).

[9] Devaney, Robert L., *Chaos, Fractals and Dynamics: Computer Experiments in Mathematics*, Addison-Wesley (1990).

[10] Dornbusch, Rudiger, and Stanley Fischer, *Macroeconomics*, McGraw-Hill (1981).

[11] Field, Michael, and Martin Golubitsky, *Symmetry in Chaos*, Oxford (1992).

[12] Gleick, J., *Chaos: Making a New Science*, Viking (1987).

[13] Hofbauer, Josef, and Karl Sigmund, *The Theory of Evolution and Dynamical Systems*, Cambridge University Press (1984).

[14] Horton, W., et. al., eds., *Recent Trends in Physics: Chaotic Dynamics and Transport in Fluids and Plasmas*, American Institute of Physics (1992).

[15] Jackson, E. Atlee, *Perspectives of Nonlinear Dynamics*, Cambridge University Press (1989).

[16] Kaye, Brian, *Chaos & Complexity: Discovering the Surprising Patterns of Science and Technology*, VCH Publishers (1993).

[17] Kapitaniak, Tomasz, *Chaotic Oscillations in Mechanical Systems*, Manchester University Press (1991).

[18] Luenberger, David G., *Introduction to Dynamic Systems: Theory, Models & Applications*, Wiley (1979).

[19] Moon, Francis C., *Chaotic and Fractal Dynamics: An Introduction for Applied Scientists and Engineers*, Wiley (1992).

[20] Parker, Thomas S., and Leon O. Chua, *Practical Algorithms for Chaotic Systems*, Springer-Verlag (1989).

[21] Priebe, C.E., et. al., The application of fractal analysis to mammorgraphic tissue classifications, *Cancer Letters* **77** (1994), 183-189.

[22] Ruelle, David, *Chaotic Evolution and Strange Attractors*, Cambridge University Press (1989).

[23] Rugh, Wilson J., *Linear System Theory*, 2d ed., Prentice-Hall (1996).

[24] Sandefur, James T., *Discrete Dynamical Systems: Theory and Applications*, Clarendon Press (1990).

[25] Shaw, Robert, *The Dripping Faucet as a Model Chaotic System*, Aerial Press (1984).

[26] Strogatz, Steven, *Nonlinear Dynamics and Chaos*, Addison-Wesley (1994).

[27] Tsonis, Anastasios A., *Chaos: From Theory to Applications*, Plenum Press (1992).

Index

Mathematics–Bestsellers

HANDBOOK OF MATHEMATICAL FUNCTIONS: with Formulas, Graphs, and Mathematical Tables, Edited by Milton Abramowitz and Irene A. Stegun. A classic resource for working with special functions, standard trig, and exponential logarithmic definitions and extensions, it features 29 sets of tables, some to as high as 20 places. 1046pp. 8 x 10 1/2. 0-486-61272-4

ABSTRACT AND CONCRETE CATEGORIES: The Joy of Cats, Jiri Adamek, Horst Herrlich, and George E. Strecker. This up-to-date introductory treatment employs category theory to explore the theory of structures. Its unique approach stresses concrete categories and presents a systematic view of factorization structures. Numerous examples. 1990 edition, updated 2004. 528pp. 6 1/8 x 9 1/4. 0-486-46934-4

MATHEMATICS: Its Content, Methods and Meaning, A. D. Aleksandrov, A. N. Kolmogorov, and M. A. Lavrent'ev. Major survey offers comprehensive, coherent discussions of analytic geometry, algebra, differential equations, calculus of variations, functions of a complex variable, prime numbers, linear and non-Euclidean geometry, topology, functional analysis, more. 1963 edition. 1120pp. 5 3/8 x 8 1/2. 0-486-40916-3

INTRODUCTION TO VECTORS AND TENSORS: Second Edition–Two Volumes Bound as One, Ray M. Bowen and C.-C. Wang. Convenient single-volume compilation of two texts offers both introduction and in-depth survey. Geared toward engineering and science students rather than mathematicians, it focuses on physics and engineering applications. 1976 edition. 560pp. 6 1/2 x 9 1/4. 0-486-46914-X

AN INTRODUCTION TO ORTHOGONAL POLYNOMIALS, Theodore S. Chihara. Concise introduction covers general elementary theory, including the representation theorem and distribution functions, continued fractions and chain sequences, the recurrence formula, special functions, and some specific systems. 1978 edition. 272pp. 5 3/8 x 8 1/2. 0-486-47929-3

ADVANCED MATHEMATICS FOR ENGINEERS AND SCIENTISTS, Paul DuChateau. This primary text and supplemental reference focuses on linear algebra, calculus, and ordinary differential equations. Additional topics include partial differential equations and approximation methods. Includes solved problems. 1992 edition. 400pp. 7 1/2 x 9 1/4. 0-486-47930-7

PARTIAL DIFFERENTIAL EQUATIONS FOR SCIENTISTS AND ENGINEERS, Stanley J. Farlow. Practical text shows how to formulate and solve partial differential equations. Coverage of diffusion-type problems, hyperbolic-type problems, elliptic-type problems, numerical and approximate methods. Solution guide available upon request. 1982 edition. 414pp. 6 1/8 x 9 1/4. 0-486-67620-X

VARIATIONAL PRINCIPLES AND FREE-BOUNDARY PROBLEMS, Avner Friedman. Advanced graduate-level text examines variational methods in partial differential equations and illustrates their applications to free-boundary problems. Features detailed statements of standard theory of elliptic and parabolic operators. 1982 edition. 720pp. 6 1/8 x 9 1/4. 0-486-47853-X

LINEAR ANALYSIS AND REPRESENTATION THEORY, Steven A. Gaal. Unified treatment covers topics from the theory of operators and operator algebras on Hilbert spaces; integration and representation theory for topological groups; and the theory of Lie algebras, Lie groups, and transform groups. 1973 edition. 704pp. 6 1/8 x 9 1/4. 0-486-47851-3

Browse over 9,000 books at www.doverpublications.com

A SURVEY OF INDUSTRIAL MATHEMATICS, Charles R. MacCluer. Students learn how to solve problems they'll encounter in their professional lives with this concise single-volume treatment. It employs MATLAB and other strategies to explore typical industrial problems. 2000 edition. 384pp. 5 3/8 x 8 1/2. 0-486-47702-9

NUMBER SYSTEMS AND THE FOUNDATIONS OF ANALYSIS, Elliott Mendelson. Geared toward undergraduate and beginning graduate students, this study explores natural numbers, integers, rational numbers, real numbers, and complex numbers. Numerous exercises and appendixes supplement the text. 1973 edition. 368pp. 5 3/8 x 8 1/2. 0-486-45792-3

A FIRST LOOK AT NUMERICAL FUNCTIONAL ANALYSIS, W. W. Sawyer. Text by renowned educator shows how problems in numerical analysis lead to concepts of functional analysis. Topics include Banach and Hilbert spaces, contraction mappings, convergence, differentiation and integration, and Euclidean space. 1978 edition. 208pp. 5 3/8 x 8 1/2. 0-486-47882-3

FRACTALS, CHAOS, POWER LAWS: Minutes from an Infinite Paradise, Manfred Schroeder. A fascinating exploration of the connections between chaos theory, physics, biology, and mathematics, this book abounds in award-winning computer graphics, optical illusions, and games that clarify memorable insights into self-similarity. 1992 edition. 448pp. 6 1/8 x 9 1/4. 0-486-47204-3

SET THEORY AND THE CONTINUUM PROBLEM, Raymond M. Smullyan and Melvin Fitting. A lucid, elegant, and complete survey of set theory, this three-part treatment explores axiomatic set theory, the consistency of the continuum hypothesis, and forcing and independence results. 1996 edition. 336pp. 6 x 9. 0-486-47484-4

DYNAMICAL SYSTEMS, Shlomo Sternberg. A pioneer in the field of dynamical systems discusses one-dimensional dynamics, differential equations, random walks, iterated function systems, symbolic dynamics, and Markov chains. Supplementary materials include PowerPoint slides and MATLAB exercises. 2010 edition. 272pp. 6 1/8 x 9 1/4. 0-486-47705-3

ORDINARY DIFFERENTIAL EQUATIONS, Morris Tenenbaum and Harry Pollard. Skillfully organized introductory text examines origin of differential equations, then defines basic terms and outlines general solution of a differential equation. Explores integrating factors; dilution and accretion problems; Laplace Transforms; Newton's Interpolation Formulas, more. 818pp. 5 3/8 x 8 1/2. 0-486-64940-7

MATROID THEORY, D. J. A. Welsh. Text by a noted expert describes standard examples and investigation results, using elementary proofs to develop basic matroid properties before advancing to a more sophisticated treatment. Includes numerous exercises. 1976 edition. 448pp. 5 3/8 x 8 1/2. 0-486-47439-9

THE CONCEPT OF A RIEMANN SURFACE, Hermann Weyl. This classic on the general history of functions combines function theory and geometry, forming the basis of the modern approach to analysis, geometry, and topology. 1955 edition. 208pp. 5 3/8 x 8 1/2. 0-486-47004-0

THE LAPLACE TRANSFORM, David Vernon Widder. This volume focuses on the Laplace and Stieltjes transforms, offering a highly theoretical treatment. Topics include fundamental formulas, the moment problem, monotonic functions, and Tauberian theorems. 1941 edition. 416pp. 5 3/8 x 8 1/2. 0-486-47755-X

Browse over 9,000 books at www.doverpublications.com

Mathematics–Logic and Problem Solving

PERPLEXING PUZZLES AND TANTALIZING TEASERS, Martin Gardner. Ninety-three riddles, mazes, illusions, tricky questions, word and picture puzzles, and other challenges offer hours of entertainment for youngsters. Filled with rib-tickling drawings. Solutions. 224pp. 5 3/8 x 8 1/2. 0-486-25637-5

MY BEST MATHEMATICAL AND LOGIC PUZZLES, Martin Gardner. The noted expert selects 70 of his favorite "short" puzzles. Includes The Returning Explorer, The Mutilated Chessboard, Scrambled Box Tops, and dozens more. Complete solutions included. 96pp. 5 3/8 x 8 1/2. 0-486-28152-3

THE LADY OR THE TIGER?: and Other Logic Puzzles, Raymond M. Smullyan. Created by a renowned puzzle master, these whimsically themed challenges involve paradoxes about probability, time, and change; metapuzzles; and self-referentiality. Nineteen chapters advance in difficulty from relatively simple to highly complex. 1982 edition. 240pp. 5 3/8 x 8 1/2. 0-486-47027-X

SATAN, CANTOR AND INFINITY: Mind-Boggling Puzzles, Raymond M. Smullyan. A renowned mathematician tells stories of knights and knaves in an entertaining look at the logical precepts behind infinity, probability, time, and change. Requires a strong background in mathematics. Complete solutions. 288pp. 5 3/8 x 8 1/2.

0-486-47036-9

THE RED BOOK OF MATHEMATICAL PROBLEMS, Kenneth S. Williams and Kenneth Hardy. Handy compilation of 100 practice problems, hints and solutions indispensable for students preparing for the William Lowell Putnam and other mathematical competitions. Preface to the First Edition. Sources. 1988 edition. 192pp. 5 3/8 x 8 1/2. 0-486-69415-1

KING ARTHUR IN SEARCH OF HIS DOG AND OTHER CURIOUS PUZZLES, Raymond M. Smullyan. This fanciful, original collection for readers of all ages features arithmetic puzzles, logic problems related to crime detection, and logic and arithmetic puzzles involving King Arthur and his Dogs of the Round Table. 160pp. 5 3/8 x 8 1/2. 0-486-47435-6

UNDECIDABLE THEORIES: Studies in Logic and the Foundation of Mathematics, Alfred Tarski in collaboration with Andrzej Mostowski and Raphael M. Robinson. This well-known book by the famed logician consists of three treatises: "A General Method in Proofs of Undecidability," "Undecidability and Essential Undecidability in Mathematics," and "Undecidability of the Elementary Theory of Groups." 1953 edition. 112pp. 5 3/8 x 8 1/2. 0-486-47703-7

LOGIC FOR MATHEMATICIANS, J. Barkley Rosser. Examination of essential topics and theorems assumes no background in logic. "Undoubtedly a major addition to the literature of mathematical logic." — *Bulletin of the American Mathematical Society.* 1978 edition. 592pp. 6 1/8 x 9 1/4. 0-486-46898-4

INTRODUCTION TO PROOF IN ABSTRACT MATHEMATICS, Andrew Wohlgemuth. This undergraduate text teaches students what constitutes an acceptable proof, and it develops their ability to do proofs of routine problems as well as those requiring creative insights. 1990 edition. 384pp. 6 1/2 x 9 1/4. 0-486-47854-8

FIRST COURSE IN MATHEMATICAL LOGIC, Patrick Suppes and Shirley Hill. Rigorous introduction is simple enough in presentation and context for wide range of students. Symbolizing sentences; logical inference; truth and validity; truth tables; terms, predicates, universal quantifiers; universal specification and laws of identity; more. 288pp. 5 3/8 x 8 1/2. 0-486-42259-3

Mathematics–Algebra and Calculus

VECTOR CALCULUS, Peter Baxandall and Hans Liebeck. This introductory text offers a rigorous, comprehensive treatment. Classical theorems of vector calculus are amply illustrated with figures, worked examples, physical applications, and exercises with hints and answers. 1986 edition. 560pp. 5 3/8 x 8 1/2. 0-486-46620-5

ADVANCED CALCULUS: An Introduction to Classical Analysis, Louis Brand. A course in analysis that focuses on the functions of a real variable, this text introduces the basic concepts in their simplest setting and illustrates its teachings with numerous examples, theorems, and proofs. 1955 edition. 592pp. 5 3/8 x 8 1/2. 0-486-44548-8

ADVANCED CALCULUS, Avner Friedman. Intended for students who have already completed a one-year course in elementary calculus, this two-part treatment advances from functions of one variable to those of several variables. Solutions. 1971 edition. 432pp. 5 3/8 x 8 1/2. 0-486-45795-8

METHODS OF MATHEMATICS APPLIED TO CALCULUS, PROBABILITY, AND STATISTICS, Richard W. Hamming. This 4-part treatment begins with algebra and analytic geometry and proceeds to an exploration of the calculus of algebraic functions and transcendental functions and applications. 1985 edition. Includes 310 figures and 18 tables. 880pp. 6 1/2 x 9 1/4. 0-486-43945-3

BASIC ALGEBRA I: Second Edition, Nathan Jacobson. A classic text and standard reference for a generation, this volume covers all undergraduate algebra topics, including groups, rings, modules, Galois theory, polynomials, linear algebra, and associative algebra. 1985 edition. 528pp. 6 1/8 x 9 1/4. 0-486-47189-6

BASIC ALGEBRA II: Second Edition, Nathan Jacobson. This classic text and standard reference comprises all subjects of a first-year graduate-level course, including in-depth coverage of groups and polynomials and extensive use of categories and functors. 1989 edition. 704pp. 6 1/8 x 9 1/4. 0-486-47187-X

CALCULUS: An Intuitive and Physical Approach (Second Edition), Morris Kline. Application-oriented introduction relates the subject as closely as possible to science with explorations of the derivative; differentiation and integration of the powers of x; theorems on differentiation, antidifferentiation; the chain rule; trigonometric functions; more. Examples. 1967 edition. 960pp. 6 1/2 x 9 1/4. 0-486-40453-6

ABSTRACT ALGEBRA AND SOLUTION BY RADICALS, John E. Maxfield and Margaret W. Maxfield. Accessible advanced undergraduate-level text starts with groups, rings, fields, and polynomials and advances to Galois theory, radicals and roots of unity, and solution by radicals. Numerous examples, illustrations, exercises, appendixes. 1971 edition. 224pp. 6 1/8 x 9 1/4. 0-486-47723-1

AN INTRODUCTION TO THE THEORY OF LINEAR SPACES, Georgi E. Shilov. Translated by Richard A. Silverman. Introductory treatment offers a clear exposition of algebra, geometry, and analysis as parts of an integrated whole rather than separate subjects. Numerous examples illustrate many different fields, and problems include hints or answers. 1961 edition. 320pp. 5 3/8 x 8 1/2. 0-486-63070-6

LINEAR ALGEBRA, Georgi E. Shilov. Covers determinants, linear spaces, systems of linear equations, linear functions of a vector argument, coordinate transformations, the canonical form of the matrix of a linear operator, bilinear and quadratic forms, and more. 387pp. 5 3/8 x 8 1/2. 0-486-63518-X

Browse over 9,000 books at www.doverpublications.com

Mathematics–Probability and Statistics

BASIC PROBABILITY THEORY, Robert B. Ash. This text emphasizes the probabilistic way of thinking, rather than measure-theoretic concepts. Geared toward advanced undergraduates and graduate students, it features solutions to some of the problems. 1970 edition. 352pp. 5 3/8 x 8 1/2. 0-486-46628-0

PRINCIPLES OF STATISTICS, M. G. Bulmer. Concise description of classical statistics, from basic dice probabilities to modern regression analysis. Equal stress on theory and applications. Moderate difficulty; only basic calculus required. Includes problems with answers. 252pp. 5 5/8 x 8 1/4. 0-486-63760-3

OUTLINE OF BASIC STATISTICS: Dictionary and Formulas, John E. Freund and Frank J. Williams. Handy guide includes a 70-page outline of essential statistical formulas covering grouped and ungrouped data, finite populations, probability, and more, plus over 1,000 clear, concise definitions of statistical terms. 1966 edition. 208pp. 5 3/8 x 8 1/2. 0-486-47769-X

GOOD THINKING: The Foundations of Probability and Its Applications, Irving J. Good. This in-depth treatment of probability theory by a famous British statistician explores Keynesian principles and surveys such topics as Bayesian rationality, corroboration, hypothesis testing, and mathematical tools for induction and simplicity. 1983 edition. 352pp. 5 3/8 x 8 1/2. 0-486-47438-0

INTRODUCTION TO PROBABILITY THEORY WITH CONTEMPORARY APPLICATIONS, Lester L. Helms. Extensive discussions and clear examples, written in plain language, expose students to the rules and methods of probability. Exercises foster problem-solving skills, and all problems feature step-by-step solutions. 1997 edition. 368pp. 6 1/2 x 9 1/4. 0-486-47418-6

CHANCE, LUCK, AND STATISTICS, Horace C. Levinson. In simple, non-technical language, this volume explores the fundamentals governing chance and applies them to sports, government, and business. "Clear and lively ... remarkably accurate." — *Scientific Monthly.* 384pp. 5 3/8 x 8 1/2. 0-486-41997-5

FIFTY CHALLENGING PROBLEMS IN PROBABILITY WITH SOLUTIONS, Frederick Mosteller. Remarkable puzzlers, graded in difficulty, illustrate elementary and advanced aspects of probability. These problems were selected for originality, general interest, or because they demonstrate valuable techniques. Also includes detailed solutions. 88pp. 5 3/8 x 8 1/2. 0-486-65355-2

EXPERIMENTAL STATISTICS, Mary Gibbons Natrella. A handbook for those seeking engineering information and quantitative data for designing, developing, constructing, and testing equipment. Covers the planning of experiments, the analyzing of extreme-value data; and more. 1966 edition. Index. Includes 52 figures and 76 tables. 560pp. 8 3/8 x 11. 0-486-43937-2

STOCHASTIC MODELING: Analysis and Simulation, Barry L. Nelson. Coherent introduction to techniques also offers a guide to the mathematical, numerical, and simulation tools of systems analysis. Includes formulation of models, analysis, and interpretation of results. 1995 edition. 336pp. 6 1/8 x 9 1/4. 0-486-47770-3

INTRODUCTION TO BIOSTATISTICS: Second Edition, Robert R. Sokal and F. James Rohlf. Suitable for undergraduates with a minimal background in mathematics, this introduction ranges from descriptive statistics to fundamental distributions and the testing of hypotheses. Includes numerous worked-out problems and examples. 1987 edition. 384pp. 6 1/8 x 9 1/4. 0-486-46961-1

Mathematics–Geometry and Topology

PROBLEMS AND SOLUTIONS IN EUCLIDEAN GEOMETRY, M. N. Aref and William Wernick. Based on classical principles, this book is intended for a second course in Euclidean geometry and can be used as a refresher. More than 200 problems include hints and solutions. 1968 edition. 272pp. 5 3/8 x 8 1/2.   0-486-47720-7

TOPOLOGY OF 3-MANIFOLDS AND RELATED TOPICS, Edited by M. K. Fort, Jr. With a New Introduction by Daniel Silver. Summaries and full reports from a 1961 conference discuss decompositions and subsets of 3-space; n-manifolds; knot theory; the Poincaré conjecture; and periodic maps and isotopies. Familiarity with algebraic topology required. 1962 edition. 272pp. 6 1/8 x 9 1/4.   0-486-47753-3

POINT SET TOPOLOGY, Steven A. Gaal. Suitable for a complete course in topology, this text also functions as a self-contained treatment for independent study. Additional enrichment materials make it equally valuable as a reference. 1964 edition. 336pp. 5 3/8 x 8 1/2.   0-486-47222-1

INVITATION TO GEOMETRY, Z. A. Melzak. Intended for students of many different backgrounds with only a modest knowledge of mathematics, this text features self-contained chapters that can be adapted to several types of geometry courses. 1983 edition. 240pp. 5 3/8 x 8 1/2.   0-486-46626-4

TOPOLOGY AND GEOMETRY FOR PHYSICISTS, Charles Nash and Siddhartha Sen. Written by physicists for physics students, this text assumes no detailed background in topology or geometry. Topics include differential forms, homotopy, homology, cohomology, fiber bundles, connection and covariant derivatives, and Morse theory. 1983 edition. 320pp. 5 3/8 x 8 1/2.   0-486-47852-1

BEYOND GEOMETRY: Classic Papers from Riemann to Einstein, Edited with an Introduction and Notes by Peter Pesic. This is the only English-language collection of these 8 accessible essays. They trace seminal ideas about the foundations of geometry that led to Einstein's general theory of relativity. 224pp. 6 1/8 x 9 1/4. 0-486-45350-2

GEOMETRY FROM EUCLID TO KNOTS, Saul Stahl. This text provides a historical perspective on plane geometry and covers non-neutral Euclidean geometry, circles and regular polygons, projective geometry, symmetries, inversions, informal topology, and more. Includes 1,000 practice problems. Solutions available. 2003 edition. 480pp. 6 1/8 x 9 1/4.   0-486-47459-3

TOPOLOGICAL VECTOR SPACES, DISTRIBUTIONS AND KERNELS, François Trèves. Extending beyond the boundaries of Hilbert and Banach space theory, this text focuses on key aspects of functional analysis, particularly in regard to solving partial differential equations. 1967 edition. 592pp. 5 3/8 x 8 1/2.

0-486-45352-9

INTRODUCTION TO PROJECTIVE GEOMETRY, C. R. Wylie, Jr. This introductory volume offers strong reinforcement for its teachings, with detailed examples and numerous theorems, proofs, and exercises, plus complete answers to all odd-numbered end-of-chapter problems. 1970 edition. 576pp. 6 1/8 x 9 1/4.   0-486-46895-X

FOUNDATIONS OF GEOMETRY, C. R. Wylie, Jr. Geared toward students preparing to teach high school mathematics, this text explores the principles of Euclidean and non-Euclidean geometry and covers both generalities and specifics of the axiomatic method. 1964 edition. 352pp. 6 x 9.   0-486-47214-0

Browse over 9,000 books at www.doverpublications.com